Statistics for Biology and Health

Statistics for Biology and Health

Borchers/Buckland/Zucchini: Estimating Animal Abundance: Closed Populations.

Burzykowski/Molenberghs/Buyse: The Evaluation of Surrogate Endpoints.

Everitt/Rabe-Hesketh: Analyzing Medical Data Using S-PLUS.

Ewens/Grant: Statistical Methods in Bioinformatics: An Introduction. 2nd ed.

Gentleman/Carey/Huber/Irizarry/Dudoit: Bioinformatics and Computational Biology Solutions using R and Bioconductor

Hougaard: Analysis of Multivariate Survival Data.

Keyfitz/Caswell: Applied Mathematical Demography, 3rd ed.

Klein/Moeschberger: Survival Analysis: Techniques for Censored and Truncated Data, 2nd ed.

Kleinbaum/Klein: Survival Analysis: A Self-Learning Text. 2nd ed.

Kleinbaum/Klein: Logistic Regression: A Self-Learning Text, 2nd ed.

Lange: Mathematical and Statistical Methods for Genetic Analysis, 2nd ed.

Manton/Singer/Suzman: Forecasting the Health of Elderly Populations.

Nielsen: Statistical Methods in Molecular Evolution.

Moyé: Multiple Analyses in Clinical Trials: Fundamentals for Investigators.

Parmigiani/Garrett/Irizarry/Zeger: The Analysis of Gene Expression Data: Methods and Software.

Salsburg: The Use of Restricted Significance Tests in Clinical Trials.

Simon/Korn/McShane/Radmacher/Wright/Zhao: Design and Analysis of DNA Microarray Investigations.

Sorensen/Gianola: Likelihood, Bayesian, and MCMC Methods in Quantitative Genetics.

Stallard/Manton/Cohen: Forecasting Product Liability Claims: Epidemiology and Modeling in the Manville Asbestos Case.

Therneau/Grambsch: Modeling Survival Data: Extending the Cox Model.

Vittinghoff/Glidden/Shiboski/McCulloch: Regression Methods in Biostatistics: Linear, Logistic, Survival, and Repeated Measures Models

Zhang/Singer: Recursive Partitioning in the Health Sciences.

Robert Gentleman   Rafael A. Irizarry
Vincent J. Carey   Sandrine Dudoit
Wolfgang Huber

Editors

Bioinformatics and Computational Biology Solutions Using R and Bioconductor

With 128 Illustrations

Springer

Editors

Robert Gentleman
Program in Computational Biology
Division of Public Health Sciences
Fred Hutchinson Cancer Research Center
1100 Fairview Ave. N, M2-B876
PO Box 19024
Seattle, Washington 98109-1024 USA
rgentlem@fhcrc.org

Vincent J. Carey
Channing Laboratory
Brigham and Women's Hospital
Harvard Medical School
181 Longwood Ave Boston MA 02115 USA
stvjc@channing.harvard.edu

Wolfgang Huber
European Bioinformatics Institute
European Molecular Biology
    Laboratory
Cambridge, CB10 1SD UK
huber@ebi.ac.uk

Rafael A. Irizarry
Department of Biostatistics
Johns Hopkins Bloomberg
    School of Public Health
615 North Wolfe Street
Baltimore, MD 21205 USA
rafa@jhu.edu

Sandrine Dudoit
Division of Biostatistics
School of Public Health
University of California,
    Berkeley
140 Earl Warren Hall, #7360
Berkeley, CA 94720-7360
    USA
sandrine@stat.berkeley.edu

Series Editors

Wing Wong
Department of Statistics
Stanford University
Stanford, CA 94305
USA

M. Gail
National Cancer Institute
Rockville, MD 20892
USA

K. Krickeberg
Le Châtelet
F-63270 Manglieu
France

A. Tsiatis
Department of Statistics
North Carolina State University
Raleigh, NC 27695
USA

J. Samet
Department of Epidemiology
School of Public Health
Johns Hopkins University
615 Wolfe Street
Baltimore, MD 21205
USA

Library of Congress Control Number: 2005923843

ISBN-10: 0-387-25146-4   Printed on acid-free paper.
ISBN-13: 978-0387-25146-2

Printed in China.   (EVB)

9 8 7 6 5 4 3 2

springer.com

Preface

During the past few years, there have been enormous advances in genomics and molecular biology, which carry the promise of understanding the functioning of whole genomes in a systematic manner. The challenge of interpreting the vast amounts of data from microarrays and other high throughput technologies has led to the development of new tools in the fields of computational biology and bioinformatics, and opened exciting new connections to areas such as chemometrics, exploratory data analysis, statistics, machine learning, and graph theory.

The Bioconductor project is an open source and open development software project for the analysis and comprehension of genomic data. It is rooted in the open source statistical computing environment R. This book's coverage is broad and ranges across most of the key capabilities of the Bioconductor project. Thanks to the hard work and dedication of many developers, a responsive and enthusiastic user community has formed. Although this book is self-contained with respect to the data processing and data analytic tasks covered, readers of this book are advised to acquaint themselves with other aspects of the project by touring the project web site `www.bioconductor.org`.

This book represents an innovative approach to publishing about scientific software. We made a commitment at the outset to have a fully *computable book*. Tables, figures, and other outputs are dynamically generated directly from the experimental data. Through the companion web site, `www.bioconductor.org/mogr`, readers have full access to the source code and necessary supporting libraries and hence will be able to see how every plot and statistic was computed. They will be able to reproduce those calculations on their own computers and should be able to extend most of those computations to address their own needs.

Acknowledgments

This book, like so many projects in bioinformatics and computational biology, is a large collaborative effort. The editors would like to thank the chapter authors for their dedication and their efforts in producing widely used software, and also in producing well-written descriptions of how to use that software.

We would like to thank the developers of R, without whom there would be no Bioconductor project. Many of these developers have provided additional help and engaged in discussions about software development and design. We would like to thank the many Bioconductor developers and users who have helped us to find bugs, think differently about problems, and whose enthusiasm has made the long hours somewhat more bearable.

We would also like to thank Dorit Arlt, Michael Boutros, Sabina Chiaretti, James MacDonald, Meher Majety, Annemarie Poustka, Jerome

Ritz, Mamatha Sauermann, Holger Sültmann, Stefan Wiemann, and Seth Falcon, who have contributed in many different ways to the production of this monograph. Much of the preliminary work on the **MLInterfaces** package, described in Chapter 16, was carried out by Jess Mar, Department of Biostatistics, Harvard School of Public Health. Ms Mar's efforts were supported in part by a grant from Insightful Corporation.

The Bioconductor project is supported by grant 1R33 HG002708 from the NIH as well as by institutional funds at both the Dana Farber Cancer Institute and the Fred Hutchinson Cancer Research Center. W.H. received project-related funding from the German Ministry for Education and Research through National Genome Research Network (NGFN) grant FKZ 01GR0450.

Seattle   *Robert Gentleman*
Boston   *Vincent Carey*
Cambridge (UK)   *Wolfgang Huber*
Baltimore   *Rafael Irizarry*
Berkeley   *Sandrine Dudoit*
                                                  February 2005

Contents

6 SELDI-TOF Mass Spectrometry Protein Data

X. Li, R. Gentleman, X. Lu, Q. Shi, J. D. Iglehart, L. Harris, and
A. Miron

II Meta-data: biological annotation and visualization   111

7 Meta-data Resources and Tools in Bioconductor

R. Gentleman, V. J. Carey, and J. Zhang

8 Querying On-line Resources   135

V. J. Carey, D. Temple Lang, J. Gentry, J. Zhang, and R. Gentleman

9 Interactive Outputs   147

C. A. Smith, W. Huber, and R. Gentleman

10 Visualizing Data   161

W. Huber, X. Li, and R. Gentleman

III   Statistical analysis for genomic experiments   181

11   Analysis Overview   183
V. J. Carey and R. Gentleman

12   Distance Measures in DNA Microarray Data Analysis.   189
R. Gentleman, B. Ding, S. Dudoit, and J. Ibrahim

13   Cluster Analysis of Genomic Data   209
K. S. Pollard and M. J. van der Laan

24 Classification with Gene Expression Data   421
M. Dettling

25 From CEL Files to Annotated Lists of Interesting Genes   431
R. A. Irizarry

A Details on selected resources   443

List of Contributors

B. M. Bolstad, Department of Statistics, University of California, Berkeley, CA, USA

J. Brettschneider, Department of Statistics, University of California, Berkeley, CA, USA

P. Buhlmann, Swiss Federal Institute of Technology, Zürich, CH

V. J. Carey, Channing Laboratory, Brigham and Women's Hospital, Harvard Medical School, Boston, MA, USA

F. Collin, Department of Statistics, University of California, Berkeley, CA, USA

L. Cope, Division of Oncology Biostatistics, The Sidney Kimmel Comprehensive Cancer Center, Johns Hopkins Medical School, Baltimore, MD, USA

M. Dettling, Division of Oncology and Biostatistics, The Sidney Kimmel Comprehensive Cancer Center, Johns Hopkins Medical School, Baltimore, MD, USA

B. Ding, Medical Affairs Biostatistics, Amgen Inc., Thousand Oaks, CA, USA

S. Dudoit, Department of Biostatistics, University of California, Berkeley, CA, USA

L. Gautier, Independent investigator, Copenhagen, DK

R. Gentleman, Program in Computational Biology, Fred Hutchinson Cancer Research Center, Seattle, WA, USA

J. Gentry, Center for Cancer Research, Massachusetts General Hospital, Boston, MA, USA

F. Hahne, Division of Molecular Genome Analysis, German Cancer Research Center, Heidelberg, FRG

L. Harris, Department of Cancer Biology, Dana Farber Cancer Institute, Boston, MA, USA

T. Hothorn, Institut für Medizininformatik, Biometrie und Epidemiologie, Friedrich-Alexander-Universität Erlangen-Nürnberg, FRG

W. Huber, European Molecular Biology Laboratory, European Bioinformatics Institute, Cambridge, UK

J. Ibrahim, Department of Biostatistics, University of North Carolina, Chapel Hill, NC, USA

J. D. Iglehart, Department of Cancer Biology, Dana Farber Cancer Institute, Boston, MA, USA

R. A. Irizarry, Department of Biostatistics, Johns Hopkins Bloomberg School of Public Health, Baltimore, MD, USA

X. Li, Department of Biostatistics and Computational Biology, Dana Farber Cancer Institute, Boston, MA, USA

X. Lu, Department of Biostatistis, Harvard School of Public Health, Boston, MA, USA

A. Miron, Department of Cancer Biology, Dana Farber Cancer Institute, Boston, MA, USA

A. C. Paquet, Department of Biostatistics, University of California, San Francisco, CA, USA

K. S. Pollard, Center for Biomolecular Science and Engineering, University of California, Santa Cruz, USA

D. Scholtens, Department of Preventive Medicine, Northwestern University, Chicago, IL, USA

Q. Shi, Department of Cancer Biology, Dana Farber Cancer Institute, Boston, MA, USA

K. Simpson, The Walter and Eliza Hall Institute of Medical Research, Melbourne, Australia

C. A. Smith, Department of Molecular Biology, The Scripps Research Institute, La Jolla, CA, USA

G. K. Smyth, The Walter and Eliza Hall Institute of Medical Research, Melbourne, Australia

T. P. Speed, Department of Statistics, University of California, Berkeley, CA, USA

D. Temple Lang, Department of Statistics, University of California, Davis, CA, USA

M. J. van der Laan, Department of Biostatistics, University of California, Berkeley, CA, USA

A. von Heydebreck, Global Technologies, Merck KGaA, Darmstadt, FRG

Z. Wu, Department of Biostatistics, Johns Hopkins Bloomberg School of Public Health, Baltimore, MD, USA

Y. H. Yang, Department of Biostatistics, University of California, San Francisco, CA, USA

J. Zhang, Department of Medical Oncology, Dana-Farber Cancer Institute, Boston, MA, USA

Part I

Preprocessing data from genomic experiments

1

Preprocessing Overview

W. Huber, R. A. Irizarry, and R. Gentleman

Abstract

In this chapter, we give a brief overview of the tasks of microarray data preprocessing. There are a variety of microarray technology platforms in use, and each of them requires specific considerations. These will be described in detail by other chapters in this part of the book. This overview chapter describes relevant data structures, and provides with some broadly applicable theoretical background.

1.1 Introduction

Microarray technology takes advantage of hybridization properties of nucleic acid and uses complementary molecules attached to a solid surface, referred to as *probes*, to measure the quantity of specific nucleic acid transcripts of interest that are present in a sample, referred to as the *target*. The molecules in the target are labeled, and a specialized scanner is used to measure the amount of hybridized target at each probe, which is reported as an intensity. Various manufacturers provide a large assortment of different platforms. Most manufacturers, realizing the effects of optical noise and non-specific binding, include features in their arrays to directly measure these effects. The raw or *probe-level* data are the intensities read for each of these components. In practice, various sources of variation need to be accounted for, and these data are heavily manipulated before one obtains the genomic-level measurements that most biologists and clinicians use in their research. This procedure is commonly referred to as *preprocessing*.

The different platforms can be divided into two main classes that are differentiated by the type of data they produce. The *high-density oligonucleotide array* platforms produce one set of probe-level data per microarray with some probes designed to measure specific binding and others to measure non-specific binding. The two-color spotted platforms produce two sets of probe-level data per microarray (the red and green channels), and local background noise levels are measured from areas in the glass slide not containing probe.

Despite the differences among the different platforms, there are some tasks that are common to all microarray technology. These tasks are described in Section 1.2. The data structures needed to effectively preprocess microarray data are described in Section 1.3. In Section 1.4 we present statistical background that serves as a mathematical framework for developing preprocessing methodology. Detailed description of the preprocessing tasks for this platforms are described in Chapters 2 and 3. The specifics for the two-color spotted platforms are described in Chapter 4. Chapters 5 and 6 describe preprocessing methodology for related technologies where similar principles apply.

1.2   Tasks

Preprocessing can be divided into 6 tasks: image analysis, data import, background adjustment, normalization, summarization, and quality assessment. *Image analysis* permits us to convert the pixel intensities in the scanned images into probe-level data. Flexible data import methods are needed because data come in different formats and are often scattered across a number of files or database tables from which they need to be extracted and organized. *Background adjustment* is essential because part of the measured probe intensities are due to non-specific hybridization and the noise in the optical detection system. Observed intensities need to be adjusted to give accurate measurements of specific hybridization. Without proper *normalization*, it is impossible to compare measurements from different array hybridizations due to many obscuring sources of variation. These include different efficiencies of reverse transcription, labeling, or hybridization reactions, physical problems with the arrays, reagent batch effects, and laboratory conditions. In some platforms, *summarization* is needed because transcripts are represented by multiple probes. For each gene, the background adjusted and normalized intensities need to be summarized into one quantity that estimates an amount proportional to the amount of RNA transcript. *Quality assessment* is an important procedure that detects divergent measurements beyond the acceptable level of random fluctuations. These data are usually flagged and not used, or down weighted, in subsequent statistical analyses.

The complex nature of microarray data and data formats makes it necessary to have flexible and efficient statistical methodology and software. This part of the book describes what Bioconductor has to offer in this capacity. In the rest of this section, we describe prerequisites necessary to perform these tasks and two general approaches to preprocessing.

1.2.1   Prerequisites

A number of important steps are involved in the generation of the raw data. The experimental design includes the choice and collection of samples (tissue biopsies or cell lines exposed to different treatments); the choice of probes and array platform; the choice of controls, RNA extraction, amplification, labeling, and hybridization procedures; the allocation of replicates; and the scheduling of the experiments. The experimental design must take into account technical, logistic, and financial boundary conditions. Its quality determines to a large extent the utility of the data. A fundamental guideline is the avoidance of *confounding* between different biological factors of interest or between a biological factor of interest and a technical factor that is anticipated to affect the measurements. The experiment then has to be carried out, which requires great skill and expertise.

In the *image analysis* step, we extract probe intensities out of the scanned images containing pixel-level data. The arrays are scanned by the detector at a high spatial resolution to produce a digitized image in which each probe is represented by dozens of pixels. To obtain a single overall intensity value for each probe, the associated pixels need to be identified (segmentation) and their intensities summarized (quantification). In addition to the overall probe intensity, further auxiliary quantities may be calculated, such as an estimate of apparent unspecific "local background" intensity, or spot quality measures. Various software packages offer different segmentation and quantification methods. They differ in their robustness against irregularities and in the amount of human interaction that they require. The different platforms present different problems which implies that the types of image analysis algorithms used are quite different. Currently, Bioconductor does not offer image processing software. Thus, the user will need alternative software to process the image pixel-level data. However, import functions that are compatible with most of the existing image analysis products are available. For an evaluation of image analysis methods for two-color spotted arrays see, for example, the study of Yang et al. (2002a). Details on image analysis methodology for high-density oligonucleotide arrays were described by Schadt et al. (2001).

1.2.2   Stepwise and integrated approaches

The *stepwise* approach to microarray data preprocessing starts with probe-level data as input, performs the tasks sequentially and produces an *expression matrix* as output. In this matrix, rows correspond to gene transcripts, columns to conditions, and each element represents the abundance or relative abundance of a transcript. Subsequent biological analyses work off the expression matrix and generally do not consider the statistical manipulations performed on the probe-level data. The preprocessing task are divided into a set of sequential instructions: for example, subtract the background, then normalize the intensities, then summarize replicate probes, then summarize replicate arrays. The modularity of this approach allows us to structure the analysis work-flow. Software, data structures, and methodology can be easily re-used. For example, the same machine learning algorithm can be applied to an expression matrix irrespective of whether the data were obtained on high-density oligonucleotide chips or two-color spotted arrays. A potential disadvantage of the stepwise approach is that each step is independently optimized without considering the effect of previous or subsequent steps. This could lead to sub-optimal bottom-line results.

In contrast, *integrated* approaches solve specific problems by carrying out the analysis in one unified estimation procedure. This approach has the potential of using the available data more efficiently. For example, rather than calculating an expression matrix, one might fit an ANOVA-type linear model to the probe-level data, which includes both technical covariates, such as dye and sample effects, and biological covariates, such as treatment effects (Kerr et al., 2000). In the affyPLM package, the weighting and summarization of the multiple probes per transcript on Affymetrix chips is integrated with the detection of differential expression. Another example is the vsn method (Huber et al., 2002), which integrates background subtraction and normalization in a non-linear model.

Stepwise approaches are often presented as modular data processing pipelines; integrated approaches are motivated by statistical models with parameters representing quantities of interest. In practice, data analysts will often choose to use a combination of both approaches. For example, a researcher may start with the stepwise approach and do a first round of high-level analyses that motivates an integrated approach that is applied to obtain final results. Bioconductor software allows users to explore, adapt, and combine stepwise and integrated methods.

1.3 Data structures

1.3.1 Data sources

The basic data types that we deal with in microarray data preprocessing are probe and background intensities, probe annotations, array layout, and sample annotations. Typically, they come in the form of rectangular tables, stored either in flat files or in a database server. The probe intensities are the result of image processing. The format in which they are reported varies between different vendors of image processing software. Examples are discussed in Sections 2 and 4.

The probe annotations are usually provided by the organization that selected the probes for the array. This may be a commercial vendor, another laboratory, or the experimenters themselves. For high-density oligonucleotide arrays, the primary annotation is the sequence. In addition, there may be a database identifier of the gene transcript that the probe is intended to match and possibly the exact location. Often, the probe sequences are derived from cDNA sequence clusterings such as Unigene (Pontius et al., 2003). For spotted cDNA arrays, the primary probe identifier is often a clone ID in a nucleotide sequence database. The largest public nucleotide sequence databases are EMBL in Europe, DDBJ in Japan, and Genbank in the United States. Through a system of cross-mirroring, their contents are essentially equivalent. These databases contain full or partial sequences of a large number of expressed sequences. Their clone identifiers can be mapped to genomic databases such as Entrez Gene, H-inv, or Ensembl. Further annotations of the genes that are represented by the probes are provided by various genomic database, for example genomic locus, disease associations, participation in biological processes, molecular function, cellular localization. This will be discussed in Part II of the book.

The array layout is provided by the organization that produced the array. As a minimum, the layout specifies the physical position of each probe on the array. In principle, this can be done through its x- and y-coordinates. For spotted arrays, it is customary to specify probe coordinates through three coordinates: *block*, *row*, and *column*, where the *block* coordinate addresses a particular sub-sector of the array, and the *row* and *column* coordinates address the probe within that sub-sector. Details are discussed in Sections 2 and 4.

The sample annotations describe the labeled cDNA that has been hybridized to the array. This includes technical information on processing protocols (e.g., isolation, amplification, labeling, hybridization) as well as the biologically more interesting covariates such as treatment conditions and harvesting time points for cell lines or histopathological and clinical data for tissue biopsies and the individuals that the biopsies originated from. A table containing this information can sometimes be obtained from the laboratory information management system (LIMS) of the lab that performed the experiments. Sometimes, it is produced *ad hoc* with office spreadsheet software.

1.3.2 Facilities in R and Bioconductor

Specific data structures and functions for the import and processing of data from different experimental platforms are provided in specialized packages. We will see a number of examples in the subsequent sections. A more general-purpose data structure to represents the data from a microarray experiment is provided by the class *exprSet* in the package **Biobase**.

The design of the *exprSet* class supports the stepwise approach to microarray preprocessing, as discussed in Section 1.2. This class represents a self-documenting data structure, with data separated into logically distinct but substantively interdependent components. Our primary motivation was to link together the large expression arrays with the phenotypic data in such a way that it would be easy to further process the data. Ensuring correct alignment of data when subsets are taken or when resampling schemes are used should be left to well-designed computer code and generally should not be done by hand.

The general premise is that there is an array, or a set of arrays, that are of interest. The *exprSet* structure imposes an order on the sample-specific expression measures in the set, provides convenient access to probe and sample identifier codes, allows coordinated management of standard errors of expression, and couples to this expression information sample- and experiment-level information, following the MIAME standard (Brazma et al., 2001). This data structure is straightforwardly employed with data from single-channel experiments, for ratio quantities derived from double-channel experiments, and for protein mass-spectrometry data. It can be extended, using formal inheritance infrastructure, to accommodate other output formats. One advantage to the use of `exprSets` is demonstrated in Chapter 16 where we describe the use of a uniform calling sequence for many machine learning algorithms (package `MLInterfaces`). This greatly simplifies individual users' interactions and will simplify the design and construction of graphical user interfaces. Establishment of a standardized calling paradigm is most simply accomplished when there are structural standards for the inputs. Both users and developers will profit from closer acquaintance with the `exprSet` structure, especially those who are contemplating complex downstream workflows.

1.4   Statistical background

The purpose of this section is to provide a general statistical framework for the following components of preprocessing: background adjustment, normalization, summarization, and quality assessment. More specific issues relating to the individual technological platforms will be discussed in Chapters 2–4.

With a microarray experiment, we aim to make statements about the abundances of specific molecules in a set of biological samples. However, the quantities that we measure are the fluorescence intensities of the different elements of the array. The measurement process consists of a cascade of biochemical reactions and an optical detection system with a laser scanner or a CCD camera. Biochemical reactions and detection are performed in parallel, allowing up to a million measurements on one array. Subtle variations between arrays, the reagents used, and the environmental conditions lead to slightly different measurements even for the same sample.

The effects of these variations may be grouped in two classes: *systematic effects*, which affect a large number of measurements (for example, the measurements for all probes on one array; or the measurements from one probe across several arrays) simultaneously. Such effects can be estimated and approximately removed. Other kinds of effects are completely random, with no well-understood pattern. These effects are commonly called *stochastic components* or *noise*.

Stochastic models are useful for preprocessing because they permit us to find *optimal* estimates of the systematic effects. We are interested in estimates that are precise and accurate. However, given the noise structure of the data, we sometimes have to sacrifice accuracy for better precision and *vice versa*. An appropriate stochastic model will aid in understanding the accuracy-precision, or bias-variance, trade-off.

Stochastic models are also useful for construction of inferential statements about experimental results. Consider an experiment in which we want to compare gene expression in the colons of mice that were treated with a substance and mice that were not. If we have many measurements from two populations being compared, we can, for example, perform a Wilcoxon test to obtain a p-value for each transcript of interest. But often it is not possible, too expensive, or unethical, to obtain so many replicate measurements for all genes and for all conditions of interest. Often, it is also not necessary. Models that provide good approximations of reality can add power to our statistical results.

Quality assessment is yet another example of the usefulness of stochastic models: if the distribution of a new set of data greatly deviates from the model, this may direct our attention to quality issues with these data. Chapter 3 demonstrates an example of the use of models for quality assessment.

1.4.1 An error model

A generic model for the value of the intensity y of a single probe on a microarray is given by

$$Y = B + \alpha S \tag{1.1}$$

where B is a random quantity due to *background noise*, usually composed of optical effects and non-specific binding, α is a gain factor, and S is the amount of measured specific binding. The signal S is considered a random variable as well and accounts for measurement error and probe effects. The measurement error is typically assumed to be multiplicative so we write:

$$\log(S) = \theta + \phi + \varepsilon. \tag{1.2}$$

Here θ represents the logarithm of the true abundance, ϕ is a probe-specific effect, and ε accounts for measurement error. This is the *additive-multiplicative error model* for microarray data, which was first proposed by Rocke and Durbin (2001) and in a closely related form by Ideker et al. (2000).

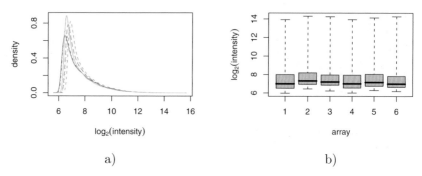

Figure 1.1. a) Density estimates of data from six replicate Affymetrix arrays. The x-axis is on a logarithmic scale (base 2). b) Box-plots.

Different arrays will have different distributions of B and different values of α, resulting in quite different distributions of the values of Y even if S is the same. To see this, let us look at the empirical distribution of six replicate Affymetrix arrays.

```
> library("affy")
> library("SpikeInSubset")
> data("spikein95")
> hist(spikein95)
> boxplot(spikein95)
```

The resulting plots are shown in Figure 1.1.

Part of the task of preprocessing is to eliminate the effect of background noise. Notice in Figure 1.1 that the smallest values attained are around 64, with slight differences between the arrays. We know that many of the probes are not supposed to be hybridizing to anything (as not all genes are expressed), so many measurements should indeed be 0. A bottom line effect of not appropriately removing background noise is that estimates of differential expression are biased. Specifically, the ratios are attenuated toward 1. This can be seen using the Affymetrix spike-in experiment, where genes were spiked in at known concentrations. Figure 1.2a shows the observed concentrations versus nominal concentrations of the spiked-in genes. Measurements with smaller nominal concentrations appear to be affected by attenuation bias. To see this, notice that the curve has a slope of about 1 for high nominal concentrations but gets flat as the nominal concentration gets closer to 0. This is consistent with the additive background noise model (1.1). Mathematically, it is easy to see that if s_1/s_2 is the true ratio and b_1 and b_2 are approximately equal positive numbers, then $(s_1 + b_1)/(s_2 + b_2)$ is closer to 1 than the true ratio, and the more so the smaller the absolute values of the s_i are compared to the b_i.

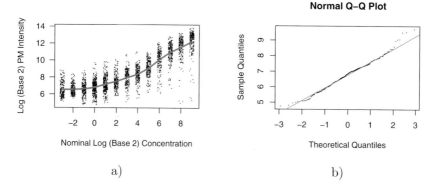

Figure 1.2. a) Plot of observed against nominal concentrations. Both axes are on the logarithmic scale (base 2). The curve represents the average value of all probes at each nominal concentration. Nominal concentrations are measured in picomol. b) Normal quantile-quantile plot of the logarithmic (base 2) intensities for all probes with the same nominal concentration of 1 picomol.

Figure 1.2b shows a normal quantile-quantile plot of logarithmic intensities of probes for genes with the same nominal concentration. Note that these appear to roughly follow a normal distribution. Figure 1.2 supports the multiplicative error assumption of model 1.1.

1.4.2   The variance-bias trade-off

A typical problem with many preprocessing algorithms is that much precision is sacrificed to remove background effects and improve accuracy. Model (1.1) can be used to show that subtracting unbiased estimates of background effects leads to exaggerated variance for genes with small values of a. In fact, background estimates that are often used in practice, such as the "local background values" from many image analysis programs for two-color spotted arrays and the mismatch (MM) value from Affymetrix arrays, tend to be *over*-estimates, which makes the problem even worse.

Various researchers have used models similar to Equation (1.1) to develop preprocessing algorithms that improve both accuracy and precision in a balanced way. Some of these methods propose variance stabilizing transformations (Durbin et al., 2002; Huber et al., 2002, 2004), others use estimation procedures that improve mean squared error (Irizarry et al., 2003b). Some examples will be provided in Chapters 2 and 4.

1.4.3   Sensitivity and specificity of probes

The probes on a microarray are intended to measure the abundance of the particular transcript that they are assigned to. However, probes may differ in terms of their sensitivity and specificity. This fact is represented by the existence of ϕ in model (1.2). Here, sensitivity means that a probe's fluorescence signal indeed responds to changes in the transcript abundance; specificity, that it does not respond to other transcripts or other types of perturbations.

Probes may lack sensitivity. Some probes initially identified with a gene do not actually hybridize to any of its products. Some probes will have been developed from information that has been superseded. In some cases, the probe may correspond to a different gene or it may in fact not represent any gene. There is also the possibility of human error (Halgren et al., 2001; Knight, 2001).

A potential problem especially with short oligonucleotide technology is that the probes may not be specific, that is, in addition to matching the intended transcript, they may also match some other gene(s). In this case, we expect the observed intensity to be a composite from all matching transcripts. Note that here we are limited by the current state of knowledge of the human transcriptome. As our knowledge improves, the information about sensitivity of probes should also improve.

1.5   Conclusion

Various academic groups have demonstrated that the use of modern statistical methodology can substantially improve accuracy and precision of bottom-line results, relative to *ad hoc* procedures introduced by designers and manufacturers of the technology. In the following chapters, we provide some details of how Bioconductor tools can be used to do this, not only in microarray platforms, but also in other related technologies.

2

Preprocessing High-density Oligonucleotide Arrays

B. M. Bolstad, R. A. Irizarry, L. Gautier, and Z. Wu

Abstract

High-density oligonucleotide expression arrays are a widely used microarray platform. Affymetrix GeneChip arrays dominate this market. An important distinction between the GeneChip and other technologies is that on GeneChips, multiple short probes are used to measure gene expression levels. This makes preprocessing particularly important when using this platform. This chapter begins by describing how to import probe-level data into the system and how these data can be examined using the facilities of the *AffyBatch* class. Then we will describe background adjustment, normalization, and summarization methods. Functionality for GeneChip probe-level data is provided by the affy, affyPLM, affycomp, gcrma, and affypdnn packages. All these tools are useful for preprocessing probe-level data stored in an *AffyBatch* object into expression-level data stored in an *exprSet* object. Because there are many competing methods for this preprocessing step, it is useful to have a way to assess the differences. In Bioconductor, this can be carried out using the affycomp package, which we discuss briefly.

2.1 Introduction

The most popular microarray application is measuring genome-wide expression levels. High-density oligonucleotide expression arrays are a commonly used technology for this purpose. Affymetrix GeneChip arrays dominate this market. In this platform, the choice of preprocessing method can have enormous influence on the quality of the ultimate results. Many preprocessing methods have been proposed for high-density oligonucleotide array data. In this chapter, we discuss methodology and Bioconductor tools that are useful for preprocessing Affymetrix GeneChip probe-level data. However, many of the procedures described here apply to high-density oligonucleotide platforms in general.

A number of mass-produced arrays or array sets are commercially available from Affymetrix. These include arrays for the organisms human, mouse, rat, arabidopsis, *Drosophila*, yeast, zebrafish, canine, and *E. coli*. It is also possible to purchase custom arrays with user-specified sequences on the array. However, there are no substantive differences in the processing of the data from the chips, so the discussion here is applicable regardless of the organism being studied.

For more details about the GeneChip platform, we refer the reader to the Affymetrix Web site and general overviews provided by Lipshutz et al. (1999) and Warrington et al. (2000). For detailed information about sample preparation, hybridization, scanning, and basic default analysis, see the Affymetrix Microarray Suite Users Guide (Affymetrix, b) and the GeneChip Expression Analysis Technical Manual (Affymetrix, c). To learn about an alternative high-density oligonucleotide technology, we refer the reader to an overview of *Nimblegen* arrays (Singh-Gasson et al., 1999).

Affymetrix GeneChip arrays use short oligonucleotides to probe for genes in an RNA sample. Genes are represented by a set of oligonucleotide probes each with a length of 25 bases. Because of their short length, multiple probes are used to improve specificity. Affymetrix arrays typically use between 11 and 20 probe pairs, referred to as a *probeset*, for each gene. One component of these pairs is referred to as a *perfect match probe* (*PM*) and is designed to hybridize only with transcripts from the intended gene (specific hybridization). However, hybridization to the PM probes by other mRNA species (non-specific hybridization) is unavoidable. Therefore, the observed intensities need to be adjusted to be accurately quantified. The other component of a probe pair, the *mismatch probe* (*MM*), is constructed with the intention of measuring only the non-specific component of the corresponding PM probe. Affymetrix's strategy is to make MM probes identical to their PM counterpart except that the 13-th base is exchanged with its complement.

The outline for this chapter is as follows. First, in Section 2.2, we describe methods for importing and accessing the probe-level data. In Section 2.3, we describe background adjustment and normalization procedures. Next, in Section 2.4 we describe different methods available for summarization. In Section 2.5 we describe tools for assessment and comparison of preprocessing methodology. Finally, in Section 2.6, we give some concluding remarks. Chapter 3 then addresses the important subject of quality assessment.

2.2   Importing and accessing probe-level data

In the Affymetrix system, the raw image data are stored in so-called *DAT files*. Currently, Bioconductor software does not handle the image data, and the starting point for most analyses are the *CEL files*. These are the result of processing the DAT files using Affymetrix software to produce estimated probe intensity values. Each CEL file contains data on the intensity at each probe on the GeneChip, as well as some other quantities.

2.2.1   *Importing*

To import CEL file data into the Bioconductor system, we use the `ReadAffy` function. A typical invocation is:

```
> library("affy")
> Data <- ReadAffy()
```

which will attempt to read all the CEL files in the current working directory and return the probe-level data in an object of class *AffyBatch*. The *AffyBatch* class is described in Section 2.2.2. The function `list.celfiles()` can be used to show the CEL files that are located in your current working directory. Note that the functions `getwd` and `setwd` can be used for getting and setting the working directory. If a specific set of CEL files are to be read, the `filenames` argument of `ReadAffy` should be supplied. If the `widget` argument is set `TRUE`, a graphical user interface can be used to choose files and enter sample information.

Affymetrix provides probe information in what are referred to as *CDF files*. These files denote which probe belongs to which probeset and whether the probe is a PM or MM. Bioconductor offers the information via ready made *CDF packages*, which work in conjunction with the *AffyBatch* class and are available for most mass produced chips. The affy package is designed to automatically load a CDF package when it is needed; when the package does not exist on the users system, it will attempt to download and install it. For custom arrays, the package makecdfenv can be used to transform the Affymetrix provided CDF files into a suitable R package. The information in the CDF file can be imported into R using the `read.cdffile` function from the makecdfenv package.

2.2.2   *Examining probe-level data*

The *AffyBatch* object is a container for storing probe-level data and related phenotypic information for a particular experiment. It extends the *exprSet* class described in Chapter 1. The main difference is that instead of containing an expression-level matrix, the *AffyBatch* class stores a probe-level intensity matrix. The `cdfName` component of the *AffyBatch* class contains a character string denoting which CDF package should be used to map the probe to probesets.

The affy package provides a number of methods for the *AffyBatch* class that permit the close examination of probe-level data. We provide various examples based on the Dilution data available in the affydata package. This data set represents a small part of a large Dilution experiment. More information about the experiment can be found by accessing its help page through typing

```
> ? Dilution
```

To load the data, we can use the following code.

```
> library("affydata")
> data(Dilution)
```

The pm and mm accessor functions return the PM and MM probe intensities. A useful feature is the ability to access the probe intensities for one or several probesets by providing the name (or a vector of names) as a second argument in the call to the accessor function. To see the intensities of the first three probes in the probeset with ID 1000_at, we can use:

```
> pm(Dilution, "1001_at")[1:3, ]
           20A   20B 10A   10B
1001_at1 129   93.8 130  73.8
1001_at2 223 129.0 174 112.8
1001_at3 194 146.8 155  93.0
```

If no probeset name is specified, then a matrix containing all the PM probe intensities is returned. To associate rows of the matrix with probesets, the function probeNames returns a vector of probeset IDs where the *i*th entry corresponds to the *i*th row of the PM matrix. The same conventions apply to the mm methods. The functions sampleNames and geneNames access the names of the samples and Affymetrix probeset IDs stored in the *AffyBatch* object, respectively.

The probe-response pattern for a particular probesets can be examined visually by plotting the probe intensities across probes or across arrays. Such plots are shown in Figure 2.1. The code follows:

```
> matplot(pm(Dilution, "1001_at"), type = "l", xlab = "Probe No.",
+     ylab = "PM Probe intensity")
> matplot(t(pm(Dilution, "1001_at")), type = "l",
+     xlab = "Array No.", ylab = "PM Probe intensity")
```

The phenoData slot of the *AffyBatch* class is where phenotypic data is stored. The function pData can be used to access this information.

```
> pData(Dilution)
     liver sn19 scanner
20A     20    0       1
```

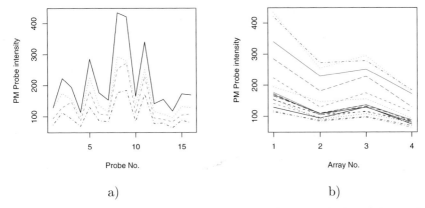

Figure 2.1. Examining the probe response pattern for a particular probeset a) across probe or b) across arrays.

20B	20	0	2
10A	10	0	1
10B	10	0	2

For the `Dilution`, object we see that the phenotypic data consists of the concentrations of RNA from two different samples, obtained from liver and central nervous system total RNA, along with the ID of the scanner used. For the four arrays (`20A`, `20B`, `10A` and `10B`) in the `Dilution` data set, all entries in the `sn19` columns are zero, indicating that no central nervous system RNA was used. The A samples were scanned using scanner 1, and the B samples were scanned using scanner 2. The 10 and 20 in the sample names represent the concentration of RNA used.

To examine probe intensity behavior for a number of different arrays we can use the `hist` and `boxplot` methods. Applying these methods to an instance of *AffyBatch* produces plots such as those in Figure 1.1 of Chapter 1. In general, these boxplots are useful for identifying differences in the level of raw probe-intensities between arrays. Differences between arrays in the shape or center of the distribution often highlight the need for normalization.

The *MA-plot* of two vectors Y_1 and Y_2 is a 45-degree rotation and axis scaling of their scatter plot. Instead of $Y_{2,j}$ versus $Y_{1,j}$ for $j = 1, \ldots, J$, we plot $M_j = Y_{2,j} - Y_{1,j}$ versus $A_j = (Y_{2,j} + Y_{1,j})/2$. If Y_1 and Y_2 are logarithmic expression values, then this plot is particularly useful for microarray data, because M_j will represent the log fold change for gene j, and A_j will represent average log intensity for that gene, and both quantities are of interest. The affy package provides two methods to create MA-plots: `mva.pairs` and `MAplot`. The `mva.pairs` function plots all pairwise MA-plots comparing all arrays in the *AffyBatch*. When comparing many arrays, it becomes prohibitive to look at all pairwise comparisons. A more sensible approach is to examine MA-plots against a reference array, which we can do using the function `MAplot`. A reference array is created by taking probe-wise medians across all the arrays in the *AffyBatch*, then MA-plots are created for each array compared to this reference array.

The inter-quartile range (IQR) and median of the M values are displayed in the plot. Loess curves which differ from the $M = 0$ axis demonstrate overall differences in the intensity level between each array and the reference array. If the assumption that most genes are not differentially expressed holds, then these curves would ideally be close to a horizontal line at $M = 0$. An example of these plots can be seen in Figure 3.3 of Chapter 3. Various arguments make the `MAplot` quite flexible. Details are available from the documentation.

2.3   Background adjustment and normalization

Preprocessing Affymetrix expression arrays usually involves three steps: background adjustment, normalization, and summarization. Bioconductor software implements a wide variety of methods for each of these steps. Self-contained routines for background correction and normalization usually take an *AffyBatch* as input and return a processed *AffyBatch*. Routines for summarization produce *exprSet* objects containing expression summary values. These will be discussed in more details in Section 2.4.

2.3.1   Background adjustment

Many background adjustment methods have been proposed in the microarray literature. In this section we describe the three background adjustment method that have been implemented in the affy package. The methods implemented in the gcrma package are described in Section 2.4.4.

RMA convolution. This is an implementation of the background adjustment carried out as part of the *RMA* method developed by Irizarry et al. (2003b). These authors found various problems with using the MM probes in the preprocessing steps and proposed a procedure that uses only the PM intensities. In this procedure, the PM values are corrected, array by array, using a global model for the distribution of probe intensities. The model was motivated by the empirical distribution of probe intensities. In particular the observed PM probes are modeled as the sum of a Gaussian noise component, B, with mean μ and variance σ^2 and an exponential signal component, S, with mean α. To avoid the possibility of negatives expression values, the normal distribution is truncated at zero. Given that we have the observed intensity, Y, then this leads to the following adjustment.

$$E\left(S|Y=y\right)=a+b\frac{\phi\left(\frac{a}{b}\right)-\phi\left(\frac{y-a}{b}\right)}{\Phi\left(\frac{a}{b}\right)+\Phi\left(\frac{y-a}{b}\right)-1},\tag{2.1}$$

where $a=s-\mu-\sigma^2\alpha$ and $b=\sigma$. Note that ϕ and Φ are the standard normal density and distribution functions, respectively. Although in principle MM probe intensities could be adjusted by this procedure, this implementation of RMA's background adjustment only transforms the PM intensities.

To produce a background adjusted *AffyBatch* for the `Dilution` data set, the following code can be used.

```
> Dilution.bg.rma <- bg.correct(Dilution, method = "rma")
```

MAS 5.0 background. This background adjustment method is outlined in the Statistical Algorithms Description Document (Affymetrix, 2002) and used in the MAS 5.0 software (Affymetrix, b). The chip is divided into a grid of k (default $k = 16$) rectangular regions. For each region, the lowest 2% of probe intensities are used to compute a background value for that grid. Then each probe intensity is adjusted based upon a weighted average of each of the background values. The weights are dependent on the distance between the probe and the centroid of the grid. In particular, the weights are:

$$w_k\left(x,y\right)=\frac{1}{d_k^2\left(x,y\right)+s_0}$$

where $d_k\left(x,y\right)$ is the Euclidean distance from location (x,y) to the centroid of region k. The default value for the smoothing coefficient s_0 is 100. Special care is taken to avoid negative values or other numerical problems for low-intensity regions. Note this method corrects both PM and MM probes. A background adjusted *AffyBatch* could be produced using the following code:

```
> Dilution.bg.mas <- bg.correct(Dilution, method = "mas")
```

Ideal mismatch. Originally, the suggested purpose of the MM probes was that they could be used to adjust the PM probes (Affymetrix, a) for probe-specific non-specific binding by subtracting the intensity of the MM probe from the intensity of the corresponding PM probes. However, this becomes problematic because, for data from a typical array, as many as 30% of MM probes have intensities higher than their corresponding PM probe (Naef et al., 2001). Thus, when raw MM intensities are subtracted from the PM intensities many negative expression values result, which makes little sense, because an expression value should not be below zero. Another drawback is that the negative values preclude the use of logarithms.

To remedy the negative impact of using raw MM values, Affymetrix introduced the concept of an *Ideal Mismatch* (IM) (Affymetrix, 2001), which was guaranteed, by design, to be smaller than the corresponding

PM intensity. The goal is to use MM when it is physically possible and a quantity smaller than the PM in other cases. This is done by computing the *specific background*, (SB), for each probeset. This is a robust average of the log ratios of PM to MM for each probe pair in the probeset. If i is the probe and k is the probeset, then for the probe pair indexed by i and k the ideal mismatch IM is given by

$$IM_i^{(k)} = \begin{cases} MM_i^{(k)} & \text{when } MM_i^{(k)} < PM_i^{(k)}, \\ \frac{PM_i^{(k)}}{2^{SB_k}} & \text{when } MM_i^{(k)} \geq PM_i^{(k)} \text{ and } SB_k > \tau_c, \\ \frac{PM_i^{(k)}}{2^{\tau_c/(1+(\tau_c-SB_k)/\tau_s)}} & \text{when } MM_i^{(k)}ij \geq PM_i^{(k)} \text{ and } SB_i \leq \tau_c, \end{cases}$$

where τ_c and τ_s are tuning constants, referred to as the *contrast* τ (with a default value of 0.03) and the *scaling* τ (with a default value of 10), respectively. The adjusted PM intensity is obtained by subtracting the corresponding IM from the observed PM intensity.

2.3.2   *Normalization*

As described in Chapter 1 normalization refers to the task of manipulating data to make measurements from different arrays comparable. A comparison and analysis of a number of normalization methods, as applied to high-density oligonucleotide data, was carried out by Bolstand et al. (2003). To perform a normalization procedure on an *AffyBatch* the generic function `normalize` may be used. Many methods have been proposed. In this Section we describe the methods implemented in the affy package.

   Scaling. For this normalization method, which is used by Affymetrix in versions 4.0 and 5.0 of their software (Affymetrix, a,b), a baseline array is chosen and all the other arrays are scaled to have the same mean intensity as this array. This is equivalent to selecting a baseline array and then fitting a linear regression, without an intercept term, between each array and the chosen array. Then, the fitted regression line is used as the normalizing relationship. This method is outlined in Table 2.1. One modification is to remove the highest and lowest intensities when computing the mean, that is, a trimmed mean is used. Affymetrix removes the highest and lowest 2% of the data. Affymetrix has proposed using scaling normalization after the computation of expression values, but it may also be used on probe-level data. Another modification is to use a target mean value in place of the mean intensity on the baseline array.

   An *AffyBatch* can be scale normalized using the following code:

```
> Dilution.norm.scale <- normalize(Dilution, method = "constant")
```

   Non-linear methods. Methods that perform non-linear adjustments between arrays have been proposed and tend to out-perform linear adjustments such as the scaling method. Numerous non-linear relationships have

---

Pick a column of X to serve as baseline array, say column j.
Compute the (trimmed) mean of column j. Call this $\tilde{X}_j$.
for $i = 1$ to n, $i \neq j$ **do**
    Compute the (trimmed) mean of column i. Call this $\tilde{X}_i$.
    Compute $\beta_i = \tilde{X}_j / \tilde{X}_i$.
    Multiply elements of column i by β_i.
end for

Table 2.1. Scaling normalization algorithm.

---

been proposed including cross-validated splines (Schadt et al., 2001), running median lines (Li and Wong, 2001b), and loess smoothers (Bolstand et al., 2003). For a typical implementation, the normalizing relationship is fitted using a rank-invariant set of points, that is, a set of points that has same rank ordering on each array.

An outline of the procedure is given in Table 2.2.

Pick a column of X to serve as the baseline array, say column j.
for $i = 1$ to n, $i \neq j$ **do**
    Fit a smooth non-linear relationship mapping column i to the baseline j. Call this $\hat{f}_i$
    Normalized values for column j are given by $\hat{f}_i(X_j)$
end for

Table 2.2. Non-linear normalization algorithm.

Non-linear normalization can be performed using the code below.

```
> Dilution.norm.nl <- normalize(Dilution, method = "invariantset")
```

Quantile normalization. The goal of quantile normalization, as discussed by Bolstand et al. (2003), is to impose the same empirical distribution of intensities to each array. A quantile-quantile plot will have a straight diagonal line, with slope 1 and intercept 0, if two data vectors have the same distribution. Thus, plotting the quantiles of two data vectors against each other and then projecting each point onto the 45-degree diagonal line yields a transformation that gives the same distribution to both data vectors.

In n dimensions, a quantile-quantile plot where all data vectors have the same distribution would have the points lying on the line described by the vector $\left(\frac{1}{\sqrt{n}}, \ldots, \frac{1}{\sqrt{n}}\right)$. This extension to n dimensions motivates the quantile normalization algorithm described in Table 2.3.

The quantile normalization method is a specific case of the transformation $x_i' = F^{-1}[G(x_i)]$, where G is estimated by the empirical distribution of each array and F is the empirical distribution of the averaged sample quantiles. This transformation is illustrated in Figure 2.2. Extensions of

Given n vectors of length p, form X, of dimension $p \times n$, where each array is a column.

Sort each column of X separately to give X_s.

Take the mean, across rows, of X_s and create X'_s, an array of the same dimension as X, but where all values in each row are equal to the row means of X_s.

Get X_n by rearranging each column of X'_s to have the same ordering as the corresponding input vector.

Table 2.3. Quantile normalization algorithm.

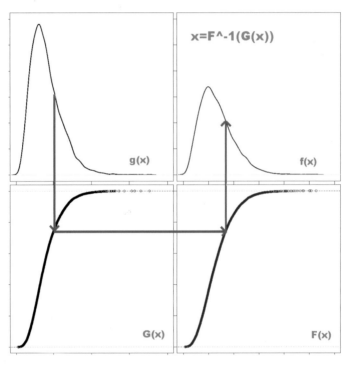

Figure 2.2. The quantile normalization method transforms the distribution of intensities from one distribution to another.

the method can be implemented where F^{-1} and G are more smoothly estimated. However, for high-density oligonucleotide data, the current method has been found to perform satisfactorily in practice.

The quantile normalization method discussed here is not the only normalization method based upon quantiles. Workman et al. (2002) fit splines to subsets of quantiles to estimate the normalizing relation, Sidorov et al. (2002) implemented a non-parametric method of giving each array the same distribution, and Amaratunga and Cabrera (2001) describe a procedure very similar to quantile normalization. To apply the procedure described by Bolstand et al. (2003) use the code below.

```
> Dilution.norm.quantile <- normalize(Dilution,
+     method = "quantiles")
```

Cyclic loess. The cyclic loess method is a generalization of the global loess method, which is described in Yang et al. (2002b), where Cy5 and Cy3 channel intensities are normalized on cDNA microarrays by using MA-plots. When dealing with single-channel array data, it is pairs of arrays that are normalized to each other. The cyclic loess method normalizes intensities for a set of arrays by working in a pairwise manner. With only two arrays, the algorithm is identical to that in Yang et al. (2002b). With more than two arrays, only part of the adjustment is made. In this case, the procedure cycles through all pairwise combinations of arrays, repeating the entire process until convergence. One drawback is that this procedure requires $O\left(n^2\right)$ loess normalizations although usually only one or two complete cycles through the data are required.

The cyclic loess algorithm is outlined in Table 2.4. The indices i and j in this algorithm index arrays, and k is used to represent probes or probesets. Convergence is measured by how much additional adjustment has occurred on that iteration through the data set. Using a subset of the data to fit the loess normalization curves can dramatically reduce the computational burden.

Let X be a $p \times n$ matrix with columns representing arrays and rows probes or probesets.

log transform the data: $X \leftarrow \log X$

repeat
    for $i = 1$ to $n - 1$ **do**
        for $j = i + 1$ to n **do**
            for $k = 1$ to p **do**
                Compute $M_k = x_{ki} - x_{kj}$ and $A_k = \frac{1}{2}\left(x_{ki} + x_{kj}\right)$
            end for
            fit a loess curve for M on A. Call this $\hat{f}$.
            for $k = 1$ to p **do**
                $\hat{M}_k = \hat{f}(A_K)$
                set $a_k = (M_k - \hat{M}_k)/n$
                $x_{ki} = x_{ki} + a_k$ and $x_{kj} = x_{ki} - a_k$
            end for
        end for
    end for
until convergence or the maximum number of iterations is reached
Revert to the original scale $X \leftarrow \exp(X)$

Table 2.4. Cyclic Loess Algorithm.

This procedure can be carried out using the **affy** package by specify the method to be `loess` in the call to the `normalize` function.

```
> Dilution.norm.loess <- normalize(Dilution, method = "loess")
```

Contrast normalization. The contrast normalization method, described by Astrand (2003), is another generalization of the methods described by Yang et al. (2002b). In brief, the data are transformed to a set of contrasts, a non-linear MA-plot normalization is performed, and then a reverse transformation is applied. It requires only $O(n)$ loess normalizations, which is considerably fewer than with the cyclic loess method. As with the cyclic loess method, a subset of the data can be used to fit the loess curves, leading to considerably reduced running times. One way that the subset may be chosen is to use a rank-invariant set of probes; see Schadt et al. (2001) for a method to select such a set. Further details about the algorithm are given by Astrand (2003). The calling sequence is given below.

```
> Dilution.norm.contrast <- normalize(Dilution,
+      method = "contrast")
```

2.3.3   vsn

The **vsn** method combines background correction and normalization into one single procedure (Huber et al., 2002, 2003). This is in contrast to the other methods, which consider background correction and normalization as separate tasks. A possible advantage of the combined approach is that information across arrays can be shared to estimate the background correction parameters, which are otherwise estimated separately for each array.

For a data matrix x_{ki}, with k counting over the probes and i over the arrays, it fits a normalization transformation

$$x_{ki} \mapsto h_i(x_{ki}) = \mathrm{glog}\left(\frac{x_{ki} - a_i}{b_i}\right), \tag{2.2}$$

where b_i is the scale parameter for array i, a_i is a background offset, and glog is the so-called generalized logarithm or attenuated logarithm (Rocke and Durbin, 2003). One of the nice properties of the glog function is that with appropriate values of a_i and b_i, the data from the different arrays are not just adjusted to each other, but also the variances across replicates are approximately independent of the mean.

Software for fitting the model (2.2) and applying the transformation is provided in the **vsn** package. We can use the following code to normalize an *AffyBatch*.

```
> library("vsn")
> Dil.vsn <- normalize(Dilution, method = "vsn")
```

The transformation parameters are returned in the `preprocessing` slot of the `description` slot of the returned *AffyBatch* object, and users are referred to the manual page of the helper function `vsnh` and the package vignette for details.

2.4   Summarization

Summarization is the final stage in preprocessing Affymetrix GeneChip data. It is the process of combining the multiple probe intensities for each probeset to produce an expression value. Bioconductor packages provide a number of functions that carry out summarization to produce gene expression values. Some of these functions also perform background correction and normalization. In particular, these functions accept raw probe intensity data stored in *AffyBatch* objects and return expression value stored in an *exprSet*. Two general functions, `expresso` and `threestep`, provide the ability to produce expression measures using a wide variety of user specified preprocessing methods. Functions optimized for computing specific expression measures such as `rma`, `gcrma`, and `expressopdnn` are also available.

2.4.1   expresso

The `expresso` function provides quite general facilities for computing expression summary values. In particular it allows most background adjustment, normalization, and summarization methods to be combined. The trade-off is that `expresso` is often considerably slower than the functions that have been optimized for producing specific expression measures. The names of background correction, PM correction, and summarization methods available to `expresso` can be found by typing `bgcorrect.methods` `pmcorrect.methods` and `express.summary.stat.methods`, respectively. For example, calling the function `normalize.methods` on an *AffyBatch* will list the available normalization methods:

```
> normalize.methods(Dilution)
```

```
[1] "constant"          "contrasts"    "invariantset"
[4] "loess"             "qspline"      "quantiles"
[7] "quantiles.robust" "vsn"
```

To control which background correction, normalization, PM correction and summarization method is used, the `bgcorrect.method`, `normalize.method`, `pmcorrect.method`, and `summary.method` arguments should be used. For instance, suppose we want to calculate an expression summary where we use the RMA convolution background, a scaling normalization, and then summarization by averaging the PM probes. The following call to `expresso` produces such expression values:

```
> eset <- expresso(Dilution, bgcorrect.method = "rma",
+      normalize.method = "constant", pmcorrect.method = "pmonly",
+      summary.method = "avgdiff")
```

An implementation of the PM-only model based expression index developed by Li and Wong (2001a) can be performed using `expresso` in this way:

```
> eset <- expresso(Dilution, normalize.method = "invariantset",
+      bg.correct = FALSE, pmcorrect.method = "pmonly",
+      summary.method = "liwong")
```

The reduced model can be obtained by changing the `pmcorrect.method` to `"subtractMM"`.

MAS 5.0 expression values (Affymetrix, b) can also be reproduced using `expresso`. However the function `mas5`, a wrapper for `expresso`, is available:

```
> eset <- mas5(Dilution)
```

2.4.2   threestep

The affyPLM package provides the `threestep` function that provides the user with a great deal of control and the ability to compute very general expression measures. Because `threestep` is primarily implemented in compiled code, it is typically faster than `expresso`. Although there is some overlap in functionality between the two functions, there are also some expression measures that may only be computed by one of the functions. Another difference is that `threestep` always returns expression measures in $\log_2$ scale, whereas `expresso` imposes no such restriction.

The `background.method`, `normalize.method`, and `summary.method` arguments of `threestep` control which preprocessing methods are used at each stage. For `background.method`, some of the available methods are `RMA.2`, the RMA convolution model; `IdealMM`, the ideal mismatch; `MAS`, MAS 5 background correction; and `MASIM`, which is the MAS 5 background followed by the ideal mismatch correction. Normalization options include `quantile`, `quantile.probeset`, and `scaling`. The summarization methods can be divided into two classes, those that are single array and those that are multi-array. The single array methods include `average.log`, averaging on the $\log_2$ scale; `log.average`, log of the average on the natural scale; `median.log`, median on the log scale; `log.median`, log of the median on the natural scale; `tukey.biweight`, a robust average on the $\log_2$ scale; and `log.2nd.largest`, which gives the log of the second largest probe intensity in the probeset. The multiple array methods include: `lm`, a linear model fit on the $\log_2$ scale; `rlm`, the same model fit using M-estimation; and `median.polish`, which fits a similar model using an alternative robustness procedure.

For example, to compute expression measures where the ideal mismatch is subtracted from PM, then quantile normalization between arrays is carried out, and probesets are summarized by using a robust average, one can use the following code:

```
> library("affyPLM")
> eset <- threestep(Dilution, background.method = "IdealMM",
+      normalize.method = "quantile", summary.method = "tukey.biweight")
```

2.4.3 RMA

RMA (Irizarry et al., 2003a,b) is an expression measure consisting of three particular preprocessing steps: convolution background correction, quantile normalization, and a summarization based on a multi-array model fit robustly using the median polish algorithm (Emerson and Hoaglin, 1983). In Bioconductor, the easiest way to compute RMA expression values is to use the rma function.

```
> eset <- rma(Dilution)
```

The function justRMA can be used instead of rma in cases where there are a large number of CEL files to process and no other low-level analysis is desired. As input, you specify a vector of CEL file names and it will output an *exprSet* containing expression values. Notice that justRMA combines the functionality of ReadAffy and rma. By never creating an *AffyBatch* object, it uses less memory.

2.4.4 GCRMA

The default background noise adjustment, provided as part of the Affymetrix system, is based on the difference $PM - MM$. However, MMs have been observed to detect some specific signal, so the $PM - MM$ transformation is likely to overadjust. In addition, Irizarry et al. (2003b) found that the $PM - MM$ transformation results in expression estimates with exaggerated variance. As an alternative, RMA was introduced with a background adjustment step that ignores the MM intensities. This approach sacrificed some accuracy for large gains in precision. However, because the global background adjustment in RMA ignores the different propensities of probes to undergo non-specific binding (NSB), the background is often underestimated. Therefore, a method that does adequate non-specific binding correction without much sacrifice in precision would be desirable.

The characteristics of each probe are determined by its sequence. Since the probe sequences were released by the manufacturer, their relationship with non-specific hybridization has been partly revealed. A statistical model that uses this information to describe background noise has been proposed by Wu et al. (2005). Using the sequence information an *affinity* measure is computed. A background adjustment method motivated by this model has been implemented and together with quantile normalization and the median polish procedures, used by RMA, define a new expression measure called GCRMA. Details are available from Wu et al. (2005).

To compute the GCRMA expression measure, we need the probe sequences. This information is available, for most mass-produced chip types, from the Bioconductor project via the *probe packages*. The matchprobes package provides utilities to manipulate the information in the probe packages. When using gcrma, the necessary probe package is determined and loaded automatically. If necessary, the package is automatically downloaded from the Internet and installed.

The function gcrma computes GCRMA expression measures from *AffyBatch* objects and returns them in exprSet objects as demonstrated in the following example.

```
> library("gcrma")
> Dil.expr <- gcrma(Dilution)
```

In many cases, researchers will deal with the same type of chip repeatedly. Unless told otherwise, the gcrma computes the sequence dependent affinities every time. The affinity information can be computed once and saved for future analysis in the following way:

```
> ai <- compute.affinities(cdfName(Dilution))
> Dil.expr <- gcrma(Dilution, affinity.info = ai)
```

The default background correction method in GCRMA uses both probe affinity information and the observed MM intensities (type="fullmodel"). Users can choose to use only affinity information by setting type="affinities" or to use only MM intensities with type="mm".

```
> Dil.expr2 <- gcrma(Dilution, affinity.info = ai,
+       type = "affinities")
```

Similar to the justRMA, the function justGCRMA can be used to compute expression measures directly from CEL files.

2.4.5   affypdnn

The package affypdnn implements the position-dependent nearest neighbor (PDNN) approach described by Zhang et al. (2004). This package was not only designed to perform the preprocessing and transform probe intensities to expression values but also to provide easy access to intermediate results and a working environment from which one can develop similar methods.

In the PDNN approach, non-specific-binding and specific-binding are predicted from the sequence of each probe using the PDNN model. In this model, each pair of bases at each pair of locations gets assigned a weight. These weights are then used to estimate expression levels for which one corrects for the non-specific and specific effects of each probe. Please refer to

Zhang et al. (2004) for the details. Software for computing these expression levels is available at the PerfectMatch Web site.

In PDNN, the weights are obtained first. The weights used by PerfectMatch for the HG-U95Av2, HG-U133A, and MG-U74A arrays can be downloaded. We included this information in the affypdnn package, and it can be accessed as follows:

```
> library("affypdnn")
> energy.files <- list.files(system.file("exampleData",
+     package = "affypdnn"), "^pdnn-energy-parameter")
> energyfile <- file.path(system.file("exampleData",
+     package = "affypdnn"), energy.files[1])
> ep <- read.table(energyfile, nrows = 80, header = TRUE)
> Wg <- as.vector(ep[33:56, 2])
> Wn <- as.vector(ep[57:80, 2])
```

Wg and Wn represent the position-dependent weights given to a particular base for the specific and non-specific parts, respectively. By plotting these (figure not shown), we see that the middle base pairs get most of the weight, which is true of the model used by gcrma.

The procedure for obtaining estimates of expression levels is relatively complicated. However, the function expressopdnn is a wrapper that automatically fits the model described by Zhang et al. (2004). We first need to compute the parameters needed for the model. These only depend on the chip type and not on the data at hand:

```
> hgu95av2.pdnn.params <- pdnn.params.chiptype(energyfile,
+     probes.pack = "hgu95av2probe")
> attach(hgu95av2.pdnn.params)
> par.ct <- list(params.chiptype = hgu95av2.pdnn.params)
> eset <- expressopdnn(Dilution[, 1], findparams.param = par.ct)
> detach("hgu95av2.pdnn.params")
```

Notice this function does not rely on expresso because the PerfectMatch algorithm is not modular. However, the function expressopdnn permits us to perform background correction and normalization steps.

2.5 Assessing preprocessing methods

The existing alternatives for background correction, normalization, and summarization yield a great number of different possible methods for preprocessing probe-level data. Identifying which method is best suited to a given inquiry can be an overwhelming task. The affycomp package provides a *graphical tool* for the assessment of preprocessing procedures for Affymetrix probe-level data. Plots and summary statistics offer a picture of how an expression measure performs in several important areas. This picture facilitates the comparison of competing expression measures and the selection of methods suitable for a specific investigation. A benchmark data set consisting of a dilution study and two spike-in studies are used. Because the *truth* is known for these data, it is possible to identify statistical features of the data for which the expected outcome is known in advance. Those features highlighted in a suite of graphs are justified by questions of biological interest and motivated by the presence of appropriate data.

The benchmark data are freely available to the public. URLs for downloading the Affymetrix and GeneLogic data sets are given in Appendix A.2. 1) For the dilution study by GeneLogic, two sources of total RNA, human liver tissue and a central nervous system cell line (CNS), were hybridized to human arrays (HG-U95Av2) in a range of dilutions and proportions. 2) For the spike-in studies, different cRNA fragments were added to the hybridization mixture of the arrays at different picomolar concentrations. The cRNAs were spiked-in at a different concentration on each array (apart from replicates). The experiments were done in a cyclic Latin square design with each concentration appearing once in each row and column. All arrays had a common background RNA. The data can be obtained from the Affymetrix data download Web site. One series was carried out on the HG-U95A chip and another one on the HG-U133Atag chip.

Two `phenoData` objects are included in the affycomp package, and they give specific information on the experimental design.

```
> library("affycomp")
> data(dilution.phenodata)
> data(spikein.phenodata)
> data(hgu133a.spikein.phenodata)
```

2.5.1   Carrying out the assessment

There are three main functions: `assessDilution`, `assessSpikeIn`, and `assessSpikeIn2`. Each one of these performs various assessments; `assessSpikeIn2` was created after feedback from users. These function take as arguments an instance of class *exprSet* and a character string giving the name of the expression measure being assessed.

The argument for `assessDilution` must contain expression measures for the 60 arrays in the dilution experiment, and they must be in the order given by `dilution.phenodata`. If one has the data in a file, the function `read.dilution` can be used to import them. The file must be stored as a comma-delimited text file with the first row containing the CEL file names and the first column containing the Affymetrix probeset identifiers. Example files are available from the affycomp Web site. Similarly, the functions `read.spikein` and `read.newspikein` can be used to read the data from the HG-U95A and HG-U133Atag spike-in experiments, respectively.

Once the data are imported, the assessments can be performed in various ways: The function `assessDilution` works on an `exprSet` containing the data from the dilution experiment. The function `assessSpikeIn` works on an `exprSet` containing the data from either of the spike-in experiments but not both at the same time. The function `assessAll` is a wrapper for the above two and takes two `exprSet` arguments containing the dilution experiment data and the data from either of the spike-in experiments (but not both). The function `affycomp` is a wrapper that works similar to `assessAll` but creates some figures and tables. The function `assessSpikeIn2` works on an `exprSet` containing the data from either of the spike-in but not both.

The package contains some examples of the output from the above functions applied to data from RMA and MAS 5.0. Specifically, examples of the output of `assessAll` using the spike-in data is from HG-U95A.can be loaded using

```
> data(rma.assessment)
> data(mas5.assessment)
```

To obtain results from the HGU133 spike-in data, we can use the data sets rma.assessment.133 and mas5.assessment.133. Finally, the following contain the results from `assessSpikeIn2`: mas5.assessment2, mas5.assessment2.133, rma.assessment2, rma.assessment2.133.

The outputs of these functions are lists. Most the components of the list are themselves list with various necessary information to create the assessment figures and tables. Two exceptions are components containing the type of assessment and the method name. For more details, see the affycomp package documentation.

The package provides 7 different assessment plots. Descriptions of the motivation and interpretation for each of these plots were given by Cope et al. (2004). In this chapter, we demonstrate *assessment figure 1*, an *MA-plot*, and *assessment figure 4a*, a signal detection plot. To produce *assessment figure 1*, shown in Figure 2.3a, we use:

```
> affycompPlot(rma.assessment$MA)
```

The function `affycompPlot` can take more than one assessment as arguments and make comparative figures. To produce *assessment figure 4a*, shown in Figure 2.3b, we use:

```
> affycompPlot(rma.assessment$Signal, mas5.assessment$Signal)
```

One can use the function `affycompTable` to create tables of summary statistics. Various other table-making functions are described in the help file for `affycompTable`.

The affycomp package provides many tools for assessing preprocessing algorithms. These tools benchmark different characteristics of the resulting expression measures. The assessments can be viewed as plots or summary

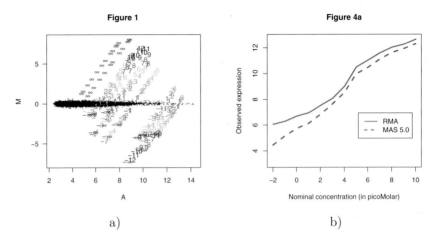

Figure 2.3. affycomp assessment. a) An MA plot with the expected log fold change indicated for spike-in probesets. This plot is called *assessment figure 1* in the **affycomp** package. b) Average observed concentrations versus nominal concentrations (*assessment figure 4a*).

statistics. The details were given by Cope et al. (2004). More information on the implementation is available in the **affycomp** vignette.

2.6   Conclusion

The packages described here provide access to GeneChip probe-level data. This permits users to develop and implement alternative preprocessing algorithms to those provided by the manufacturer. The suite of packages has been designed to balance user control with convenience. Graphical user interfaces, object-oriented programming, and modular function design enhance the convenience of the packages. Nonetheless, the balance is skewed greatly in favor of user control. We believe that this is appropriate. Microarray technology is still quite new and although the ease of use of proprietary analysis platforms for commercial chips is substantial, the cautious user will retain as much hands on control of the process as possible.

3

Quality Assessment of Affymetrix GeneChip Data

B. M. Bolstad, F. Collin, J. Brettschneider, K. Simpson, L. Cope, R. A. Irizarry, and T. P. Speed

Abstract

This chapter covers quality assessment for Affymetrix GeneChip data. The focus is on procedures available from the affy and affy-PLM packages. Initially some exploratory plots provided by the affy package, including images of the raw probe-level data, boxplots, histograms, and M vs A plots are examined. Next methods for assessing RNA degradation are discussed, specifically we compare the standard procedures recommended by Affymetrix and RNA degradation plots. Finally, we investigate how appropriate probe-level models yield good quality assessment tools. Chip pseudo-images of residuals and weights obtained from fitting robust linear models to the probe level data can be used as a visual tool for identifying artifacts on GeneChip microarrays. Other output from the probe-level modeling tools provide summary plots that may be used to identify aberrant chips.

3.1 Introduction

Obtaining gene expression measures for biological samples through the use of Affymetrix GeneChip microarrays is an elaborate process with many potential sources of variation. In addition, the process is both costly and time consuming. Therefore, it is critical to make the best use of the information produced by the arrays, and to ascertain the quality of this information. Unfortunately, data quality assessment is complicated by the sheer volume of data involved and by the many processing steps required to produce the expression measures for each array. Furthermore, quality is a term which has some dependency on context to which it is applied. In this chapter methods for relative quality assessment are discussed. From the viewpoint of a user of GeneChip expression values, lower variability data, with all other things being equal, should be judged to be of higher quality.

Before any use is made of more complex methods, an initial examination of the data can often show evidence of possible quality problems. In some cases arrays are beyond correction, even with normalization, and removing the array from the data set is warranted. Automatic ways to detect problematic arrays are an important, yet difficult, application. In this chapter various graphical tools that can be used to facilitate the decision of whether to remove an array from further analysis are presented.

Throughout this chapter data from a large acute lymphoblastic leukemia (ALL) study (Ross et al., 2004) are used to demonstrate the quality assessment methods. The ALLMLL package contains an instance of *AffyBatch* representing a subset of the original data. For illustrative purposes an even smaller subset of the data is used and the files are renamed. The following code loads the data set, subsets 8 of the arrays from the data set and then names them with letters **a** through **h**. We will use these 8 arrays to demonstrate the different quality assessment procedures.

```
> library("affy")
> library("ALLMLL")
> data(MLL.B)
> Data <- MLL.B[, c(2, 1, 3:5, 14, 6, 13)]
> sampleNames(Data) <- letters[1:8]
```

3.2   Exploratory data analysis

Exploratory data analysis has been the tool of choice for detection of problematic arrays. A typical first step is to look at images of the raw probe-level data. However, for microarray data the largest values are orders of magnitude larger than the bulk of the data and this results in a non-informative image. A simple solution is to examine an image plot of the log intensities. The image function can be used to create chip images of the raw intensities.

```
> palette.gray <- c(rep(gray(0:10/10), times = seq(1,
+      41, by = 4)))
> image(Data[, 1], transfo = function(x) x, col = palette.gray)
> image(Data[, 1], col = palette.gray)
```

Figure 3.1 compares chip images of natural scale and log-scale intensities for chip **a**. The log-scale plot demonstrates that there is a strong spatial artifact on this array not seen in the original scale plot.

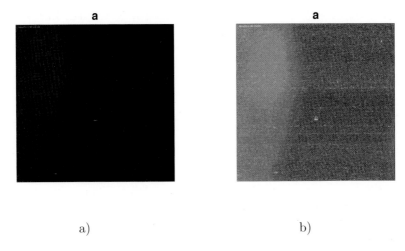

a)   b)

Figure 3.1. a) Image of probe intensities, using a linear mapping from the chips dynamical range to the grayscale. b) As a) but using logarithmically transformed intensities. The letter a above the figures indicates that these plots are for the array with this name.

3.2.1   Multi-array approaches

Looking at the distribution of probe intensities across all arrays at once can sometimes demonstrate that one array is not like the others. The *boxplot* gives a simple summary of the distribution of probes. In the affy package the boxplot method is available for the *AffyBatch* class. The following code creates boxplots of the unprocessed log scale probe-level data for the 8 arrays in the data set.

```
> library("RColorBrewer")
> cols <- brewer.pal(8, "Set1")
> boxplot(Data, col = cols)
```

The RColorBrewer package is used to select a nice color palette. Figure 3.2 shows the results. Notice how array f clearly stands out from the rest. However, a discrepancy in this plot is not conclusive evidence of a quality problem. Many times such differences can be reduced using the normalization procedures discussed in Chapter 2.

Looking at *histograms* of the probe-level data can also reveal problems. The hist function creates smoothed histograms (or density estimators) of the intensities for each array. The following code creates these histograms and adds a legend to aid in the identification of each array.

```
> hist(Data, col = cols, lty = 1, xlab = "Log (base 2) intensities")
> legend(12, 1, letters[1:8], lty = 1, col = cols)
```

Figure 3.2 shows such histograms for this data set. Notice that array a, shown in red, has a bimodal distribution. This is usually indicative of a spatial artifact as that seen in Figure 3.1. The second mode is usually the result of an entire section of the array having abnormally high values.

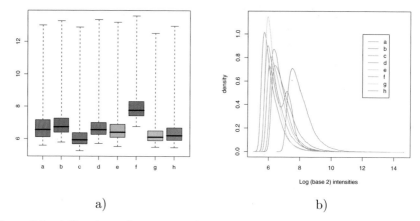

a)   b)

Figure 3.2. a) Boxplots of unprocessed log scale probe intensities for 8 arrays from the **ALLMLL** data set. b) Smoothed histograms of raw log scale intensities.

Another useful exploratory tool for quality assessment purposes is the *MA-plot*, described in Chapter 2. Since these plots show log fold change on the y-axis, the information provided is closely related to the quantities that are eventually used to obtain final results for tasks such as detecting differentially expressed genes. As described in Chapter 2, to avoid making an MA-plot for every pairwise comparison, each array can instead be compared to a *synthetic* array created by taking probe-wise medians. The MA-plot gives us a better idea of how bad quality arrays affect bottom line results. A loess curve is fitted to the scatter-plot to summarize any non-linear relationship. Since there are only 8 arrays in this data set, MA-plots for all 8 can be examined using the following code.

```
> par(mfrow = c(2, 4))
> MAplot(Data, cex = 0.75)
> mtext("M", 2, outer = TRUE)
> mtext("A", 1, outer = TRUE)
```

The results can be seen Figure 3.3. Quality problems are most apparent from an MA-plot where the loess smoother oscillates a great deal or if the variability of the M values seems greater than those of other arrays in the data set. In this case the first MA-plot, corresponding to array **a**, suffers from both these problems.

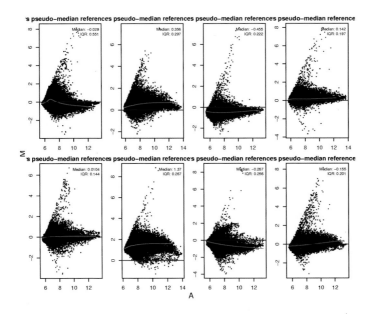

Figure 3.3. MA-plot of eight arrays plotted with common pseudo-array reference.

3.3   Affymetrix quality assessment metrics

Affymetrix software (Affymetrix, b) produces a number of quantities for quality assessment of GeneChip data. These are: Average Background, Scale Factor, Percent Present and $3'/5'$ ratios. Average Background is the average of the 16 background values calculated according to the method discussed in Section 2.3.1. The Scale Factor refers to the constant β_i, from Section 2.3.2 by which every intensity on the chip is multiplied by in the scaling normalization. Percent Present is the percentage of probesets called "present" according to the Affymetrix detection algorithm. GeneChip arrays include several RNA quality control genes, most notably $\beta-$Actin and GAPDH, each represented by 3 probesets, one from the $5'$ end, one from the middle and one from the $3'$ end of the targeted transcript. The ratio of the $3'$ expression to the $5'$ expression for these genes serves as a measure of RNA quality. Using Bioconductor, the simpleaffy package can compute these values. First we load the package and compute the quality assessments as follows:

```
> library("simpleaffy")
> Data.qc <- qc(Data)
```

The average background for each array can then be shown by simply typing:

```
> avbg(Data.qc)
```

a	b	c	d	e	f	g	h
68.2	67.3	42.1	61.3	53.6	128.4	49.4	49.3

According to the guidelines recommended by Affymetrix, the average background values should be comparable to each other. Notice the large background value for array f. This might be indicative of a problem.

The scale factors can be computed using:

```
> sfs(Data.qc)
```

```
[1]  9.77  4.91 10.49  7.05  7.56  2.48 13.53  8.09
```

Affymetrixrecommends that these values be within 3-fold of each other. In this example there appears to be a problem with, for example, arrays f and g.

The percentage of present calls can be obtained with the following code:

```
> percent.present(Data.qc)
```

```
a.present b.present c.present d.present e.present f.present
    21.7      26.5      25.6      23.5      23.4      25.3
g.present h.present
    18.0      24.4
```

These should be similar for replicate samples with extremely low values being a possible indication of poor quality.

Finally, the $3'/5'$ ratios are available via the `ratios` method. In the following piece of code we demonstrate these ratios for the first two quality control probesets.

```
> ratios(Data.qc)[, 1:2]
```

	AFFX-HSAC07/X00351.3'/5'	AFFX-HUMGAPDH/M33197.3'/5'
a	0.970	0.1639
b	0.324	0.0580
c	0.466	-0.1557
d	1.257	0.5755
e	0.604	-0.1402
f	0.672	0.2467
g	0.380	-0.0183
h	0.485	0.2768

Affymetrix suggests 3 as a safe threshold value for the $3'/5'$ ratios, and recommends caution if that threshold is exceeded.

3.4   RNA degradation

As discussed in the previous section, $3'/5'$ ratios for several control genes can be used for quality assessment purposes. However, there are only a few of these genes, and the $3'/5'$ ratios can be quite variable even with high quality data. Therefore, rather than using the expression measures of only a few genes to assess quality, a more global indicator of RNA degradation is desirable.

For every GeneChip probeset, the individual probes are numbered sequentially from the $5'$ end of the targeted transcript. When RNA degradation is sufficiently advanced, PM probe intensities should be systematically elevated at the $3'$ end of a probeset when compared to the $5'$ end. This can in fact be observed in some cases, and has been exploited by the `AffyRNAdeg` function. The approach is very simple. Let Y_{ij} denote the log transformed PM probe data from the jth probe in the ith probeset on a single GeneChip. The log transformation is not crucial to the measurement of degradation itself, and results are quite similar without this transformation. Rather, this step is included out of a conviction that the log scale is approximately the *right* scale for analyzing expression array data.

For any single probeset the probe effects dominate even the most dramatic signs of degradation. Thus, a $3'/5'$ trend only becomes apparent on the average over large numbers of probesets. Accordingly, define $Y_{\cdot j}$ to be the average log intensity taken over all probesets, at probe position j. it has been observed that $Y_{\cdot j}$ is roughly linear in j as shown in Figure 3.4. Typically, the $Y_{\cdot j}$ intensities are slightly lower at both ends relative to the middle, and a distinct $3'/5'$ trend is apparent even with good RNA.

Such plots suggest that a simple linear model may describe the relationship between $Y_{\cdot j}$ and j quite well. A regression estimate of the slope can be used as the starting point for a measure of degradation. The measure should be comparable across chips, so first the data from each chip is rescaled to make $SE(Y_{\cdot j}) \approx 1$. Specifically, suppose that there are N probesets, each having m PM probes. For j in $\{1, 2, ..., m\}$, define $\hat{\sigma}_j$ to be the standard deviation of the N probe intensities at position j. Then $\hat{\sigma}_j/\sqrt{N}$ approximates the standard error for each probe position mean and the scaled mean $Y_{\cdot j}/(\hat{\sigma}_j/\sqrt{N})$ has a standard error of approximately 1. When these quantities are regressed against integer values j, the slope estimate is easily interpreted in terms of standard error. A slope of 1 means that average probe intensity increases by 1 standard error with each *step* toward the $3'$ end of the transcripts.

To illustrate the procedures available for assessing RNA degradation, a data set where three arrays were hybridized with amplified RNA and the other three hybridized with RNA processed using the default procedure suggested by Affymetrix is used. The amplified RNA is likely to have a different $3'/5'$ trend . The affy package includes two functions related to this measure of degradation. The primary function is `AffyRNAdeg`, which calculates probe position means and slopes. Degradation plots can be produced as follows:

```
> library("AmpAffyExample")
> data(AmpData)
```

RNA digestion plot

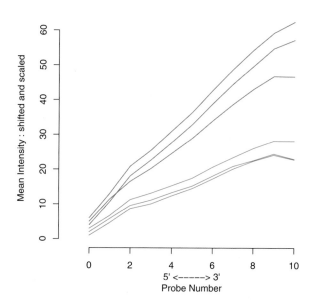

Figure 3.4. Each line represents one of 6 HG-U133A chips. Plotted on the Y axis is mean intensity by probeset position. Intensities have been shifted from original data for a clearer view, but slope is unchanged.

```
> sampleNames(AmpData) <- c("N1", "N2", "N3", "A1",
+       "A2", "A3")
> RNAdeg <- AffyRNAdeg(AmpData)
> plotAffyRNAdeg(RNAdeg, col = c(2, 2, 2, 3, 3,
+       3))
```

Figure 3.4 shows the resultant plot. The lines from amplified RNA data are in green and the standard arrays are in red. Notice that the amplified RNA arrays have much higher slopes than those from the normal procedure. This is likely due to degradation. A summary of the slopes can be viewed in the following way:

```
> summaryAffyRNAdeg(RNAdeg)
```

	N1	N2	N3	A1	A2	A3
slope	2.3e+00	2.21e+00	2.56e+00	5.38e+00	4.32e+00	5.68e+00
pvalue	3.9e-08	8.13e-08	3.12e-09	9.03e-12	4.91e-10	5.35e-12

How large must the slope be to consider the array to have too much degradation? Unfortunately, different chip types have different character-istic slopes. This phenomenon can be attributed to differences in probeset

Chip type	Mean	Q_1	Median	Q_3
HG-U95A	26.1	4	10	32
MG-U74Av2	21.9	3	8	26
HG-U133A	40.7	15	28	53

Table 3.1. The probes in a single probeset are physically near one another on the actual transcript. The distribution of inter-probe distances varies from chip to chip however. All distances are measured in single bases; Q_1 and Q_3 represent the first and third quartiles respectively.

architecture. Some older GeneChips have about 12000 probesets on each chip with probesets containing 16 probe pairs. The newer arrays, on the other hand, have only 11 probe pairs in each probeset, and contain about 22000 probesets on each chip. On any chip, all probes in the probeset are taken from a single continuous sequence that is usually about 600 bases long. The probes in larger probesets are necessarily quite close to one another within the base sequence and sometimes even overlap. The $3'/5'$ trend is of course less pronounced in these probesets. Table 3.1 gives the distribution of inter-probe distances within each probeset on three common array types.

Experience will give the user a sense of what is typical for a given chip type. For high quality RNA, a slope of 0.5 is typical for HG-U95 and MG-U74 chips, while 1.7 is typical for HG-U133A chips. Slopes that exceed these values by a factor of 2 or higher might indicate degradation. However, in general, the actual value is less important than agreement between chips. If all the arrays have similar slopes then comparisons within genes across arrays may still will be valid. You could lose data on some genes altogether - as some mRNA species may have degraded completely, while others only partially. Mixing chips with very different $3'/5'$ trends is very likely to introduce extra bias into the experiment. Therefore, a plot with one or more arrays with very different slope is indicative of a possible problem. Possible causes include poor sample handling or differences in RNA amplification methods.

3.5   Probe level models

For the summarization component, the RMA procedure described in Chapter 2 assumes the following linear model for the the background adjusted normalized probe-level data Y_{gij}:

$$\log(Y_{gij}) = \theta_{gi} + \phi_{gj} + \varepsilon_{gij}, \tag{3.1}$$

with θ_{gi} representing the log-scale expression level for gene g on array i, ϕ_{gj} is the effect of the j-th probe representing gene i, and ε the measurement error. RMA uses median polish to estimate θ robustly. Equation 3.1 is referred to as a probe level model (PLM). Because median polish (Emerson and Hoaglin, 1983) is an ad-hoc procedure, there is no formal way of, for example, obtaining standard errors. However, θ can be robustly estimated with procedures such as those described by Huber (1981), which are implemented in R by the function rlm from the packageMASS (Venables and Ripley, 2002). The affyPLM package provides functions that fit model 3.1 to probe-level data using these robust regression procedures.

The class *PLMset* has been defined to deal with the results of fitting a PLM as described in this section. It is similar to the *AffyBatch* class except it has further components holding the weights, residuals, parameter and standard error estimates, and other related robust model fit information. The fitPLM function permits us to fit these models. The default is to compute an expression measure just like RMA except using M-estimator robust regression instead of the median polish algorithm.

```
> library("affyPLM")
> Pset1 <- fitPLM(AmpData)
> show(Pset1)
```

The are a number of functions for accessing the output of this fitting function including coef and se which return $\hat{\theta}_{gi}$ and SE $\left(\hat{\theta}_{gi}\right)$ respectively. The final weights from the fitting procedure can be accessed using weights. While residuals can be accessed using resid. However, it is more useful to examine these quantities graphically.

3.5.1   Quality diagnostics using PLM

Numerous useful quality assessment tools can be derived from the output of the PLM fitting procedures. The large variability due to the strong probe effect ϕ sometimes makes it difficult to observe artifacts in the data. However, PLMs account for the probe effect and therefore this obscuring variability is not present in, for example, the residuals. Specifically, chip pseudo-images of these residuals, the weights used by robust regression to down-weight outliers, and the sign of the residuals tend to be more visually informative than the image of the raw probe intensities. The following code can be used to draw chip pseudo-images for the third array in the RNA degradation data set:

```
> par(mfrow = c(2, 2))
> image(AmpData[, 3])
> image(Pset1, type = "weights", which = 3)
> image(Pset1, type = "resids", which = 3)
> image(Pset1, type = "sign.resids", which = 3)
```

Figure 3.5 displays the resulting chip pseudo-images. Figure 3.5A is an image of the log intensities with no obvious spatial artifact. Figure 3.5B shows PLM weights and a "ring" artifact is clearly visible. This artifact can been seen in all the PLM images. Weight images uses topographical coloring so that light areas indicate high weights and dark green areas indicate significant down-weighting of mis-performing probes. These weights are obtained from the robust regression procedure for which small weights are associated with outliers. Images based on residuals are colored blue for negative residuals and red for positive residuals. For the raw residuals this means that the most negative residuals are the darkest blue, residuals at, or near, zero are white and the most positive residuals are the most intense red. It is also possible to view the negative or positive residuals in isolation by providing `type="neg.resids"` or `type="pos.resids"` as an argument to `image`. The image of the signs of the residuals can sometimes make visible effects that might not be apparent in the other plots, magnitude is ignored with blue representing negative and red representing positive. These images highlight the power of the PLM procedures at detecting a subtle artifact that might otherwise be missed completely.

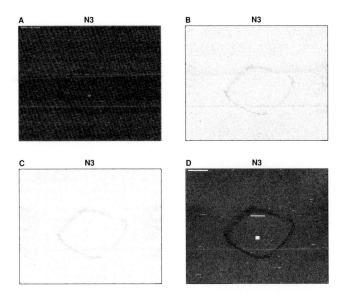

Figure 3.5. Chip pseudo-images based on PLM fit make visible subtle artifacts. Images of A) the raw probe intensities (log transformed) B) weights C) residuals and D) sign of the residuals.

It is not uncommon to have small artifacts on such images. Frequently there are small blemishes, and usually these are inconsequential. This is because in modern array designs, the probes for each probeset are not contiguously next to each other, but rather they are spread throughout the array. Hence, a spatially localized artifact will at most affect a few probes in a probeset. In addition, procedures such as RMA, are robust to outliers.

With larger artifacts, an issue of concern is whether data from a particular array is of poorer quality relative to other arrays in the data set. This question can be addressed using other procedures provided by the affyPLM package. These techniques are demonstrated using the ALLMLL data.

```
> Pset2 <- fitPLM(MLL.B)
```

Recall that array a was observed as having a bimodal histogram in Section 3.2.1. A pseudo chip-image of the residuals for this array, shown in Figure 3.6, makes the bright region along one side of the chip apparent even more clearly than the image of raw intensities. In addition the intense blue regions indicate that the other side of the array is dimmer than expected relative to other arrays in the data set. This array has an obvious large artifact and thus its quality is suspect. Should the array be removed from the data set before beginning further downstream analysis? To help answer this question two graphical procedures based on output from PLM fitting procedure can be used.

JD–ALD051–v5–U133B.CEL

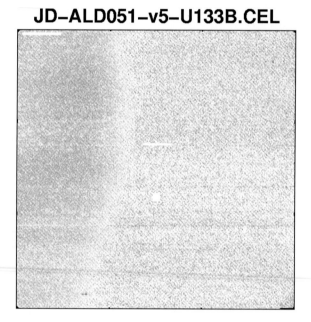

Figure 3.6. An image of the residuals identifies the bright area along one edge of the array.

The first of these procedures is the Relative Log Expression (RLE) plot. This plot is constructed in the following manner. First, start with the log scale estimates of expression $\hat{\theta}_{gi}$ for each gene g on each array i. Next compute the median value across arrays for each gene, m_g, and define the relative expression as $M_{gi} = \hat{\theta}_{gi} - m_g$. These relative expressions are then displayed with a boxplot for each array. In many situations it is not unreasonable to assume that that majority of genes are not changing in expression between experimental conditions. These non-differential genes are reflected on the RLE plot by the boxes. Ideally, these boxes would have small spread and be centered at $M = 0$. An array that has quality problems may result in a box that has greater spread or is not centered near $M = 0$. To create this plot for the ALLMLL data set use:

```
> Mbox(Pset2, ylim = c(-1, 1), col = cols, names = NULL,
+      main = "RLE")
```

The resulting plot, shown in Figure 3.7a, shows that the box for the array with spatial effects deviates considerably from the remaining boxes. Numerical summaries based on these plots, such as $B_i = \mathrm{med}_g\,(M_{gi})$ for a bias measure and $V_i = \mathrm{IQR}_g\,(M_{gi})$ for a variability measure, may be examined as an alternative to the plot.

The other graphical tool is the Normalized Unscaled Standard Error (NUSE) plot. In this case, we start with the standard error estimates obtained for each gene on each array from the PLM fit. To account for the fact that variability differs considerably between genes standardize these standard error estimates so that the median standard error across arrays is 1 for each gene. Specifically, NUSE values are computed using

$$\mathrm{NUSE}\left(\hat{\theta}_{gi}\right) = \frac{\mathrm{SE}\left(\hat{\theta}_{gi}\right)}{\mathrm{med}_i\left(\mathrm{SE}\left(\hat{\theta}_{gi}\right)\right)}.$$

The NUSE values are then shown with a boxplot for each by array. This can be accomplished using the following code:

```
> boxplot(Pset2, ylim = c(0.95, 1.5), col = cols,
+      names = NULL, main = "NUSE", outline = FALSE)
```

Lower quality arrays are indicated on this plot by boxes that are significantly elevated or more spread out relative to the other arrays. For the ALLMLL data set, the NUSE plot in Figure 3.7b further demonstrates the problem with this array. The median NUSE value for each array provide suitable numerical summary values for this method. High values of median NUSE are indicative of a problematic array. However, it should be noted that these values are not comparable across data set since NUSE is relative only within a data set.

Figure 3.8 compares median NUSE with the quality metrics recommended by Affymetrix. On each boxplot the array with the artifact is indicated by a point. The array known to be problematic has an extremely high median NUSE value which is easy to distinguish from the other 19 ar-

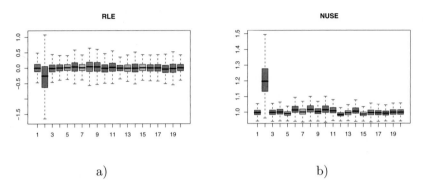

a)   b)

Figure 3.7. a) RLE plot for the 20 HGU-133B arrays in the ALLMLL data set. The array with the spatial effect deviates considerably from the other arrays. b) NUSE plot for the same arrays. The array with the significant spatial effect deviates considerably from the remaining arrays.

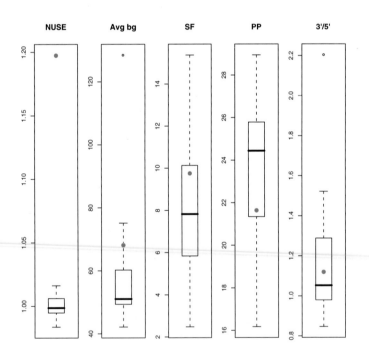

Figure 3.8. Comparing median NUSE with Affymetrix quality metrics for the ALLMLL data set. The aberrant array is indicated on each plot.

rays. In contrast, the Average Background, Scale Factor, Percent Present and the GAPDH $3'/5'$ ratio measures for this array are not even the largest value among the 20 arrays. The advantage that PLM based assessments have over the Affymetrix quality standards is that they are directly related to the quality of the expression measures produced.

3.6   Conclusion

This chapter demonstrated a series of graphical tools and statistical summaries useful for quality assessment of Affymetrix GeneChip data. Thoughtful use of well known summary plots such as histogram and box-plots provides useful first steps in the assessment of array quality. More advanced procedures, such as fitting probe level models and summarizing residuals and weights from these fits, further aid the analyst in the decision of whether an array should be removed from the data set.

4

Preprocessing Two-Color Spotted Arrays

Y. H. Yang and A. C. Paquet

Abstract

Preprocessing of two-color spotted arrays can be broadly divided
in two main categories: quality assessment and normalization. In this
chapter, we will focus on functions from the arrayQuality and marray
packages that perform these tasks. The chapter begins by describing various data structures and tools available in these packages for
reading and storing primary data from two-color spotted arrays. This
is followed by descriptions of various exploratory tools such as MA-
plots, spatial plots, and boxplots to assess data quality of an array.
Finally, algorithms available for performing appropriate normalization to remove sources of systematic variation are discussed. We will
illustrate the above-mentioned functions using a case study.

4.1 Introduction

In this chapter, we discuss the various packages in the Bioconductor project
that focus on preprocessing for two-color spotted arrays. Such packages include marray, arrayQuality, limma, vsn, and arrayMagic. In particular, we
will focus on aspects related to importing data from image analysis software, quality assessment and normalization. In Section 4.2, we give a brief
introduction to the two-color spotted arrays platform and describe a data
set used to illustrate our methodology and software. In Section 4.3, we describe the basic functions of importing and accessing the *probe-level data*.
In Section 4.4, we demonstrate various quality assessment tools available
to the user. Next, in Section 4.5 we describe the main preprocessing steps
of normalization and background adjustment. In Section 4.6, we conclude
by demonstrating the utility of Bioconductor software with a case study.

4.2   Two-color spotted microarrays

Two-color spotted microarrays were first developed in the Brown and Bot-
stein Labs at Stanford. These arrays consist of thousands of different DNA
sequences printed in a high-density array on a glass microscope slide using a robotic arrayer. The probes are obtained from PCR-amplification
of cDNA clones. More recently, synthetic oligonucleotides have been used.
The relative abundance of complementary target molecules in a pair of
samples can be assessed by monitoring the differential hybridization to
the array. For mRNA samples, the two samples or *targets* are reverse-
transcribed into cDNA, labeled using different fluorescent dyes (usually a
red-fluorescent dye, Cyanine 5 or Cy5, and a green-fluorescent dye, Cyanine
3 or Cy3), then mixed in equal proportions and hybridized to the arrayed
DNA probes. After this competitive hybridization, the slides are imaged
using a scanner, and fluorescence measurements are made separately for
each dye at each spot on the array. The ratio of red and green fluorescence
intensities for each spot is intended to be indicative of the relative abundance of the corresponding molecule in the two nucleic acid target samples.
See the Supplement to *Nature Genetics* (1999) for a more detailed introduction to the biology and technology of two-color spotted microarrays and
oligonucleotide chips.

4.2.1   Illustrative data

We will demonstrate our methodology using gene expression data provided
by the Erle Lab from UC San Francisco. Specifically, we will use data
from their integrin $\alpha 4\beta 7$ experiment. This experiment studied the cell adhesion molecule integrin $\alpha 4\beta 7$, which assists in directing the migration of
blood lymphocytes to the intestine and associated lymphoid tissues. The
goal of the study was to identify differentially expressed genes between the
$\alpha 4\beta 7+$ and $\alpha 4\beta 7-$ memory T helper cells. Further details and results of
the experiments can be found in the article by Rodriguez et al. (2004).

For this illustration, we have selected a subset of data from the original
data set consisting of 6 replicated slides from different subjects. Complete
information about the array platform and data from each of the individual
arrays is available from the GEO database at NCBI (series accession number GSE1039). Each hybridization involved $\beta 7+$ cell RNA from a single
subject (labeled with one dye) and $\beta 7-$ cell RNA from the same subject
(labeled with the other dye). Target RNA was hybridized to microarrays
containing 23,184 probes including the Operon Human version 2 set of
70-mer oligonucleotide probes and 1760 control spots. Microarrays were
printed using 12×4 print-tips, and each grid consisted of a 21×23 spots
matrix.

Each array was scanned using an Axon GenePix 4000B scanner, and
images were processed using GenePix 5.0 software. The resulting files

are available from GEO, with sample identifiers GSM16689, GSM16687, GSM16694, GSM16706, GSM16686, and GSM16724, respectively. To retrieve the data from GEO one can use the AnnBuilder package. The code below shows how to retrieve the data for the file `6Hs.166.gpr` (corresponding to sample ID GSM16689) into an R object.

```
> library("AnnBuilder")
> samp.6Hs.166 <- queryGEO(GEO(), "GSM16689")
```

The file contains 23184 rows corresponding to the probes (spots) and its columns correspond to the different statistics from the image analysis output and to basic probe information such as gene names, spot ID, and spot coordinates. Alternatively, the data can be obtained from the Bioconductor experimental data package beta7. The package includes hybridization information, which is stored in a tab-delimited file `Targetbeta7.txt`. Further example data sets and illustrations can also be found in the documentation of the packages marray and limma.

4.3   Importing and accessing probe-level data

Spotted arrays provide flexibility in the choice of the DNA sequences spotted on the slide. Therefore, a specific data structure is needed so that probe annotation information can be easily customized. In Bioconductor there are two main data structures that are used to store two-color spotted array information. These are the *marrayRaw* class from the marray package and the *RGList* class from the limma package. After preprocessing probe-level data, objects are converted to instances of the *exprSet* class from the Biobase package. The convert package allows users to convert between these three data structures. In this section, we focus on the marray package and illustrate how to read in two-color spotted array data. This task can also be performed within the limma package and some details are available in Chapter 23.

4.3.1   Importing

The raw image data are usually stored in pairs of 16-bit *Tagged Image File Format* (TIFF) files, one for each fluorescent dye. Specialized image analysis software is used to determine which pixels in the image correspond to the fluorescence emitted by the labeled samples hybridized to the array (foreground) or to the glass slide (background). Then, intensity measurements and various statistics are extracted for each spot. As mentioned in Section 1, Bioconductor does not provide image processing utilities and relies on other software. Some of the statistical issues and existing solutions were described by Dudoit and Yang (2003). Different image processing programs provide the resulting probe-level data in different file formats. Some of the formats supported by marray include GenePix's .gpr, Spot's .spot, SMD's .xls, and Agilent's .txt files. The import functions can be extended to handle a wide range of other alternatives.

In a two-color spotted array experiment, the information needed to perform an appropriate statistical analysis can be divided into three main components: 1) the sample target information, 2) the probe information, and 3) the probe spot and background intensities. The classes *marrayRaw* and *marrayNorm* allow to store these components and provide methods to link them. These classes permit all the data and metadata related to one microarray experiment to be stored in one object. The class *marrayRaw* is designed to store raw data, whereas an R object of class *marrayNorm* stores normalized data.

4.3.2   Reading target information

We refer to the file that lists the microarray hybridizations and describes which RNA samples were hybridized to each array as the *target file*. A target file is typically a tab-delimited text file that consists of the exact name of each image processing file you would like to include in the data analysis and the corresponding names for the Cy3 and Cy5 labeled target (sample) information. It should also include other variables of interest that are useful for downstream analysis or for quality assessment. Examples include subject identification number, gender, and age; date of hybridization and scanning conditions.

The main function used to read target information in the marray package is read.marrayInfo, which will create an R object of class *marrayInfo*. For example, the target information for the beta7 study can be read with:

```
> datadir <- system.file("beta7", package = "beta7")
> TargetInfo <- read.marrayInfo(file.path(datadir,
+     "TargetBeta7.txt"))

> TargetInfo

An object of class "marrayInfo"
@maLabels
[1] "6Hs.195.1.gpr" "6Hs.168.gpr"   "6Hs.166.gpr"
[4] "6Hs.187.1.gpr" "6Hs.194.gpr"   "6Hs.243.1.gpr"


@maInfo
        FileNames SubjectID  Cy3  Cy5 Date of Blood Draw
1 6Hs.195.1.gpr          1 b7 - b7 +         2002.10.11
2   6Hs.168.gpr          3 b7 + b7 -         2003.01.16
3   6Hs.166.gpr          4 b7 + b7 -         2003.01.16
4 6Hs.187.1.gpr          6 b7 - b7 +         2002.09.16
5   6Hs.194.gpr          8 b7 - b7 +         2002.09.18
6 6Hs.243.1.gpr         11 b7 + b7 -         2003.01.13
  Date of Scan
```

```
1    2003.07.25
2    2003.08.07
3    2003.08.07
4    2003.07.18
5    2003.07.25
6    2003.08.06

@maNotes
[1] "Files were loaded from beta7 package."
```

We see for example that the first array 6Hs.195.1.gpr was done using samples from subject 1, $\beta7-$ cells were labeled with Cy3 and $\beta7+$ cells with Cy5. This information will be used in downstream analysis to account for *dye-swaps* or to check for potential confounding factors.

4.3.3 Reading probe-related information

Probe-related information refers to descriptions of the spotted probe sequences (e. g. gene names and annotations; quality control information on PCR and printing conditions). The array fabrication information includes the dimensions of the spot and grid matrices and for each probe on the array its coordinates. In addition, the plate origin of the probes and information on the spotted control sequences (e. g. negative controls, *housekeeping genes*, spike in controls) are included. This information is stored separately using two objects: an object of class *marrayInfo* on the probe annotation information and an object of class *marrayLayout* to store array fabrication information.

In general, users will find that most image analysis output files include probe related information. If you are using GenePix or Spot image analysis software, array layout, and probe information are stored in what is referred to as a .gal file, which is a tab-delimited text file with rows corresponding to spotted probe sequences and columns containing various coordinates and annotations. The main function to use for reading .gal files is read.Galfile. In most instances, .gpr files will also contain the corresponding gene annotation information stored in the .gal file, and thus the function read.Galfile can read a .gpr file and extract the relevant layout and probe related annotations as well.

```
> galinfo <- read.Galfile("6Hs.166.gpr", path = datadir)
```

Users can modify the arguments of the function to specify which columns represent probe annotations and array fabrication (printer layout) information. The function read.Galfile returns a list of 3 components: an object gnames of class *marrayInfo* that stores probe annotation information; an object layout of class *marrayLayout* that stores array fabrication information; and a numerical vector named neworder, which represents the sequential order of the probes. Probes are assumed to be ordered by blocks, starting from the left to the right and top to bottom, and spots within each block are also ordered from left to right and from top to bottom.

4.3.4   Reading probe and background intensities

Microarray image processing results are usually stored in text files and, by default, they are assumed to be tab-delimited. These can be loaded into R using `read.marrayRaw`. The `marray` package also provides customized functions for specific image analysis outputs: `read.Spot`, `read.Agilent` and `read.GenePix` for Spot, Agilent, and GenePix, respectively. These functions are simply "wrapper" functions around `read.marrayRaw`. The `read` functions will set up the probe annotation and array layout objects. This requires that all arrays in the batch have the same layout and the same spotted probe sequences. The following commands illustrate how to read in the raw expression data for two `.gpr` files, `6Hs.166.gpr` and `6Hs.168.gpr`.

```
> setwd(datadir)
> files <- c("6Hs.166.gpr", "6Hs.187.1.gpr")
> mraw <- read.GenePix(files, name.Gb = NULL, name.Rb = NULL)
```

Here, we have set the arguments `name.Gb` and `name.Rb` to `NULL`, which instructs the function to ignore the "local background" values. If both of its arguments `fnames` or `targets` are not specified, then the function `read.GenePix` will read in all files in the current working directory. Users are encouraged to first read the target file information and use the argument `targets` to read in fluorescence intensities. This will ensure that the foreground and background intensity data are read and stored in the same order as in the target file. The object `mraw` now contains the (unnormalized) intensity data for a batch of arrays. It is an object of class *marrayRaw* and contains slots for the matrices of Cy3 and Cy5 intensities (`maGb`, `maRb`, `maGf`, `maRf`), spot quality weights (`maW`), layout parameters of the arrays (`maLayout`), description of the probes spotted onto the arrays (`maGnames`), and mRNA target information (`maTargets`).

You can verify the integrity of a *marrayRaw* object using:

```
> library("beta7")
> checkTargetInfo(beta7)
```

4.3.5   Data structure: the marrayRaw class

The *marrayRaw* class stores and links information about the probes (genes) and targets (samples), and the measured intensity data for a batch of microarrays. We use the term *batch of microarrays* for a collection of microarrays with the same print layout. Figure 4.1 shows how the *marrayRaw* class structure is related to a two-color spotted array.

The *marrayRaw* class can be divided in three main components:

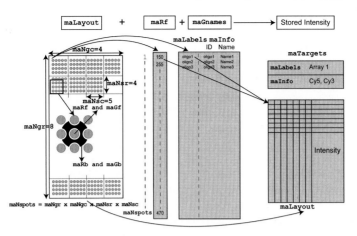

Figure 4.1. Relationship between two-color spotted array information and *marrayRaw* class structure.

1. Target information: An R object of class *marrayInfo* is used to store information on the target samples that were hybridized to each microarray.

2. Probe information: This information is stored using two objects: an object of class *marrayLayout* to store array fabrication information and an object of class *marrayInfo* for the probe annotation information.

    • The *marrayLayout* class is used for the definition of unique coordinates for each spot. It contains the dimensions of the grid matrix representing the print-tip groups (maNgr × maNgc), the dimensions within each print-tip group (maNsr × maNsc), and the total number of spots (maNspots). The slot maSub is a *logical* vector indicating which spots are currently being stored in the slots containing Cy3 and Cy5 background and foreground fluorescence intensities. Its length is equal to the number of spots for the complete array layout, which is defined as maNgr × maNgc × maNsr × maNsc. This feature is often useful to import data from a non-complete array layout.
    • The *marrayInfo* class stores various probe identifiers and annotations related to the probe sequences spotted on the microarray slides.

3. Intensity measures: Foreground and background intensities for both channels are stored as matrices in separate slots (maRf, maGf, maRb, maGb). Another slot maW is provided to store a spot quality weight matrix. For each matrix, rows correspond to spotted probe sequences and columns to arrays in the batch.

4.3.6   Accessing the data

For *marrayRaw* and *marrayNorm* objects, various accessor methods are defined to extract stored information. Methods are also available to extract commonly used statistics such as

$$M = \log_2 \frac{Cy5}{Cy3} = \log_2(Cy5) - \log_2(Cy3) \tag{4.1}$$

$$A = \log_2 \sqrt{Cy5 \cdot Cy3} = \frac{1}{2} \left[\log_2(Cy5) + \log_2(Cy3) \right], \tag{4.2}$$

which are returned by the functions maM and maA. The functions maLG and maLR compute the green and red log intensities respectively. The function maGeneTable creates a *data.frame* of spot coordinates and gene names. It can be used to retrieve gene names or their position on the array. The following command will print the first 4 rows and the first 5 columns of the *data.frame* containing spotted sequences information for the beta7 data.

```
> maGeneTable(beta7)[1:4, 1:5]
```

```
           Grid.R Grid.C Spot.R Spot.C        ID
H200000297      1      1      1      1 H200000297
H200000303      1      1      1      2 H200000303
H200000321      1      1      1      3 H200000321
H200000327      1      1      1      4 H200000327
```

The default *background adjustment* method is to subtract the measured background intensities from the measured probe intensities. Various image processing algorithms produce suboptimal background measurements resulting in background adjustment procedure that increase variance with no gains in accuracy. If one has data created from one of these image processing algorithms, one can avoid subtracting background using the following code:

```
> beta7nbg <- beta7
> beta7nbg@maGb <- beta7nbg@maRb <- 0 * beta7nbg@maRb
```

Users should note that there are other background estimation and adjustment procedures that provide a better trade-off between accuracy and variability. More details on alternate background estimation methods can be found in the references (Dudoit and Yang, 2003; Huber et al., 2002).

4.3.7   Subsetting

In many instances, one is interested in accessing only a subset of arrays in a batch and/or spots in an array. The subsetting method "[" was defined for this purpose. The first index refers to probes and the second to arrays. Thus, to access the first 100 probe sequences in the second and third arrays in the beta7, you can use:

```
> beta7sub <- beta7[1:100, 2:3]
```

Furthermore, subsetting certain information according to the print-run layout can be achieved using the function maCoord2Ind. This can be used to remove some positional controls before further analyses are carried out. For example, to get the index of the first 3 spots on the last row of each grid, you should first calculate the grid and spot coordinates for these spots using maCompCoord, then convert these coordinates back to the index using maCoord2Ind. For more details on these functions, please refer to the on-line help. The following code provides a simple example:

```
> coord <- maCompCoord(1:maNgr(beta7), 1:maNgc(beta7),
+        maNsr(beta7), 1:3)
> ind <- maCoord2Ind(coord, L = maLayout(beta7))
```

4.4 Quality assessment

Before proceeding to normalize the arrays or any higher level analysis, it is important to consider (and ensure) the quality of the data. Quality information is often also a useful feedback to the microarray users. For spotted array experiments, quality assessment can be divided into four components: print-run quality, mRNA quality, general hybridization quality, and spot quality. In this section, we will focus on "general hybridization quality", which refers to the global assessment of the array hybridization performance. Such assessments help to determine if the quality of the experimental data is acceptable or if some hybridizations should be repeated. In this section, we describe the various exploratory data analysis plots and tools provided in the packages marray and arrayQuality which help to identify potential problems with the dyes, uneven hybridizations, or other experimental artifacts.

4.4.1 Diagnostic plots

The package arrayQuality gives users a quick visual way to assess the quality of individual arrays by providing per-slide diagnostic plots. An example is shown in Figure 4.2. It provides a display of various commonly used qualitative array assessment plots. These include *MA-plots*, which can be used to assess intensity biases, spatial plots, which can reveal uneven hybridization artifacts, histograms that assess the signal to noise ratios for each channel, and dot plots to help evaluate the consistency of replicate control elements. If one has already imported the data into an object of class *marrayRaw* or *RGList*, the following command will generate these quality diagnostic plots for each array in the R object.

```
> maQualityPlots(beta7)
```

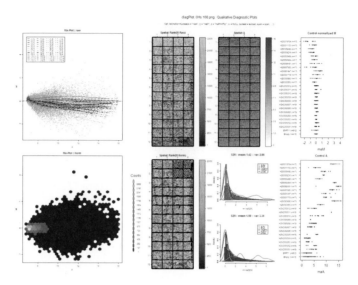

Figure 4.2. Diagnostic plot for qualitative array quality assessment. The display incorporates 8 separate panels that address various aspects of array quality.

In addition, to complement the qualitative aspect of the evaluation process, the package also provides a framework to introduce some quantitative measures. The first step involves generating a database of good-quality arrays from different print-runs and experiments. The next step is to determine a series of quantitative quality measures. New hybridization results are compared to a database of good results based on the series of quantitative measures. The comparison is visualized through a "comparative boxplot" using the function qualBoxplot. All results are compiled in a HTML report, as shown in Figure 4.3.

The main wrapper functions in the arrayQuality package for generating both qualitative and quantitative diagnostic plots for GenePix, Spot and Agilent arrays are respectively: gpQuality, spotQuality and agQuality. These functions will perform data input as well as quality assessment as described in the following two steps.

1. Copy the raw data files from the *same* print-run in to the same directory and start R in this directory.

2. Depending on your file format, use either gpQuality, spotQuality or agQuality. For example, if the data are text files generated from Agilent's image analysis software, the user should use the following command.

    ```
    > agQuality()
    ```

The aim of these functions is to make the quality assessment tools easily accessible to users that are not expert programmers. However, for users who

ArrayQuality General Hybridization report

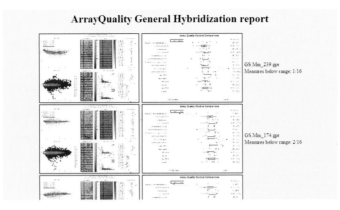

Figure 4.3. HTML report showing quality assessment for a batch of arrays. Each row contains quality assessment results for an individual array, consisting of both the qualitative diagnostic plot and the quantitative comparative boxplot.

like more flexibility, the rest of this section will describe other functions that allow investigators to explore their data more closely.

4.4.2   Spatial plots of spot statistics - image

The generic function image creates an image plot of the array in gray scale or color using statistics such as the intensity log–ratio M, a spot quality measure (e.g., spot size or shape), or a test statistic. This graphical tool can be used effectively to explore spatial defects in the data like print-tip or cover-slip effects. In addition to existing color palette functions, such as rainbow and heat.colors, users can generate more flexible color palettes using the functions brewer.pal from the package RColorBrewer and maPalette from the package marray.

Useful diagnostic plots are images of the Cy3 and Cy5 background intensities; these images may reveal hybridization artifacts such as scratches on the slides, drops, or cover-slip effects. The following command produces an image of the Cy5 background intensities for the fifth array of the beta7 data using a white-to-red color palette

```
> image(beta7[, 5], xvar = "maRb", bar = TRUE)
```

The result is shown in Figure 4.4a. In Figure 4.4b, we see the log–ratio values M for the third array using a blue-to-gray-to-yellow color palette.

```
> RGcol <- maPalette(low = "blue", mid = "gray",
+     high = "yellow", k = 50)
> image(beta7[, 3], xvar = "maM", col = RGcol)
```

Furthermore, we can highlight regions of interest using the argument overlay. Figure 4.4b was generated with the following commands:

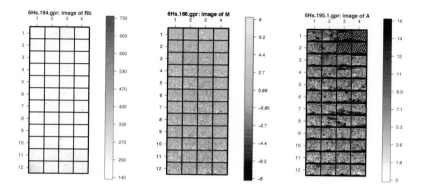

Figure 4.4. a): Spatial plot of the red background intensities for the first beta7 array in the batch using white–to–red color palette. b) M-values for the third array using a blue-to-gray-to-yellow color palette. c) A-values of array 1 using a white–to–blue color palette. White square represent missing spots. Flagged spots (Flags value less than -50) are highlighted in black.

```
> flags <- beta7@maW[, 1] < -50
> image(beta7[, 1], xvar = "maA", overlay = flags)
```

4.4.3   Boxplots of spot statistics - boxplot

Boxplots of spot statistics by plate, print-tip-group, or slide can be useful to identify spots or hybridization artifacts. The function boxplot and associated methods produce boxplots of microarray spot statistics for R objects with classes *marrayRaw* and *marrayNorm*. Figure 4.5a displays boxplots of unnormalized log–intensities A for each of the 61 384-well plates for the third beta7 array. The plot was generated by the following commands:

```
> par(mar = c(5, 3, 3, 3), cex.axis = 0.7)
> boxplot(beta7[, 3], xvar = "maPlate", yvar = "maA",
+      outline = FALSE, las = 2)
```

Notice the use of the function par, which allows us to set the figure margins and the axes font size. The boxplots in Figure 4.5a show that the last few 384-well plates clearly stand out by having lower intensity A values. This is expected as a large collection of negative control spots were placed in these particular plates. This is useful to identify potential printing errors during array fabrication. The function boxplot may also be used to produce boxplots of spot statistics for all arrays in a batch. Such plots are useful for example when assessing the need for between-array normalization to deal with scale differences among different arrays. The following command

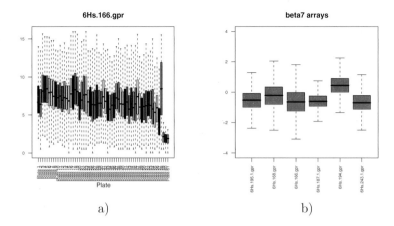

a)   b)

Figure 4.5. a) Boxplots by 384-well plates of the log–intensities A for the third array in the **beta7** data set. b) Boxplots of log–ratios M across all six arrays in **beta7** data set.

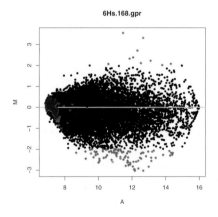

Figure 4.6. Pre–normalization MA-plot for the second **beta7** array with probes corresponding to empty controls highlighted in blue and probes corresponding to large fold change [abs(M) > 2] highlighted in red.

produces a boxplot of the unnormalized intensity log–ratios M for each array, as shown Figure 4.5b.

```
> boxplot(beta7, main = "beta7 arrays", las = 2)
```

4.4.4 Scatter-plots of spot statistics - `plot`

The generic function `plot` produces *scatter-plots* of microarray spot statistics for the classes *marrayRaw* and *marrayNorm*. In addition, the function points, lines, and text allow users to highlight and annotate subsets of points on the plot and display fitted curves from robust local regression or other smoothing procedures. We typically represent microarray data using a *MA-plot*, where we plot the log–ratios (M) on the y-axis and the average log intensities (A) on the x-axis, as described in Section 2.2.2 of Chapter 2. Figure 4.6 displays the MA-plot for the beta7 arrays, with the empty controls highlighted in blue. Probes that have an unnormalized M-value greater than $\log_2(4) = 2$ are highlighted in red. The argument subset is a vector of either logical or numeric values indicating the subset of points to be plotted. Figure 4.6 was generated with the following commands:

```
> plot(beta7nbg[, 2], lines.func = NULL, legend.func = NULL)
> points(beta7nbg[, 2], subset = abs(maM(beta7nbg)[,
+    2]) > 2, col = "red", pch = 18)
> points(beta7nbg[, 2], subset = maControls(beta7nbg) ==
+    "Empty", col = "blue", pch = 18)
```

As the number of probes on the microarray increases, it becomes harder to visualize all spots and print-tip information on the same graph. The function plotPrintTipLoess in limma uses co-plot to display MA-plot information with individual loess curves for separate print-tip groups. This plot allow users to get a better idea of the performance of each print-tip.

4.5   Normalization

The purpose of normalization is to identify and remove systematic technical variation while retaining the biological signal. Among the sources of technical variation are different labeling efficiencies and scanning properties of the Cy3 and Cy5 dyes, different scanning parameters, print-tip, spatial, and plate effects. Normalization procedures aim to ensure that the observed differences in intensity indeed reflect the differential gene expression and not artifactual biases due to such technical factors.

There is a *bias-variance trade-off*: more complex normalization procedures tend to be able to remove more of the technical variation than simple procedures — but they might also remove more of the biological signal. And they might even introduce new sources of variation of their own, for example, due to the uncertainty with which the optimal normalization parameters can be estimated. The choice between different methods and method settings for the optimal trade–off is not simple, and it is currently mostly done manually on an *ad hoc* basis. A more thorough discussion of these issues and a proposal to achieve some automation is given by Xiao et al. (2005).

Normalization is closely related to quality assessment. In an ideal experiment, no normalization would be necessary, as the technical variations would have been avoided in the first place. In a real experiment, a certain amount of technical variations cannot be avoided, and can be accounted for and corrected in the subsequent analysis through a normalization procedure. However, if the technical variations become too large, they turn into a quality problem. The decision of when to discard a set of measurements (e. g., a slide) or when to try to correct it via normalization is not simple.

We group normalization methods for two-color arrays in two classes: *two-channel normalization* methods try to adjust the within–array contrasts M [Equation (4.1)], using the values of A, as well as other factors such as print-tip, PCR-plate, spatial position as covariates. *Separate-channel normalization* methods try to adjust on the original intensities (or suitably transformed versions), using the same factors as above, but not A, as covariates.

4.5.1 Two-channel normalization

The process of *two-channel normalization* can be separated into two main components: *location* and *scale*. In general, methods for location and scale normalization adjust the center and spread of the distribution of log–ratios. The normalized intensity log–ratios M_{norm} are generally given by

$$M_{\mathrm{norm}} = \frac{M - l}{s}, \qquad (4.3)$$

where l and s denote the location and scale normalization values, respectively. Methods differ in how exactly they parameterize Equation (4.3), and in how they estimate the parameters. For example, in global median location normalization (argument m in Table 4.1), the parameter l is assumed to be the same for all spots on an array, whereas in global A-dependent normalization it is assumed to be a smooth function of A, and the function is estimated using the scatter-plot smoother *loess*.

The main functions for two-channel normalization are maNorm in marray and normalizeWithinArrays in limma. Both functions allow the user to choose from a set of four basic location normalization procedures, which are described in Table 4.1. The function maNorm operates on an object of class *marrayRaw* (or possibly *marrayNorm*, if normalization is performed in several steps) and returns an object of class *marrayNorm*. The following command performs print-tip loess normalization:

```
> beta7norm <- maNorm(beta7, norm = "p")
```

Location normalization centers log–ratios around zero by accounting for intensity and spatial dependent bias. However, it does not adjust for differences in scale between multiple arrays. Therefore, *scale normalization* is important when scale differences between multiple arrays can lead to one or more arrays having undue weight in summarizing log–ratios across arrays. The functions for performing scale normalization are maNormScale in marray and normalizeBetweenArrays in limma. Users should note that we

Procedures	Description	Argument
None	No normalization.	n
Median	Global median location normalization.	m
Loess	Global *A*-dependent normalization using the scatter-plot smoother `loess`.	l
Print-tip *loess*	*A*-dependent normalization using the scatter-plot smoother `loess` within print-tip groups.	p
2D loess	2D–spatial normalization using the `loess` function.	twoD

Table 4.1. Different two-channel normalization procedures.

recommend to check the need for performing such normalization manually on a case-to-case basis, as there is a trade–off between the possible gain in bias achieved by scale normalization and the increase in variability introduced by this additional step. In cases where the scale differences are fairly small, it may be preferable to perform only a location normalization. The code below performs a scale normalization across arrays:

```
> beta7norm.scale <- maNormScale(beta7norm)
```

4.5.2   Separate-channel normalization

The normalization methods we have described so far transform the M values directly. The red and green absolute intensities are not normalized separately. This is because spot-to-spot variation introduced by the printing process can be quite large. While in general we do not expect absolute intensities from replicate arrays to be similar, the co-hybridization technique permits the spot effect to be partially or completely canceled out in the M values. Thus, we expect M values from replicate arrays to be closely related. However, as technology improves the spot-to-spot variation is reduced. This has led to the development of normalization strategies that work on the absolute intensities directly (Yang and Thorne, 2003; Huber et al., 2002), referred to as separate-channel normalization methods. These are particularly useful because they improve the quality of across array comparisons.

One approach of a separate-channel normalization method is based on the function `normalizeBetweenArrays` in the limma package. The method proceeds in two stages, a *within-array* step followed by a *between-array* step. The *within-array* step is the same as in the two-channel location described in Section 4.5.1. The *between-array* step addresses the comparability of the distributions of log intensities between arrays. The argument `method` of the function `normalizeBetweenArrays` provides various choices. One possibility is to use the quantile normalization method proposed by Bolstand

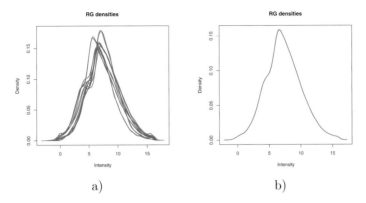

Figure 4.7. a) Single channel densities for red and green single-channel intensity distributions after print-tip group normalization but before separate-channel normalization. b) Single-channel densities after quantile normalization.

et al. (2003). The following commands show how to take the results from within array print-tip normalization and perform a between-array quantile normalization. The resulting density plots are shown in Figure 4.7.

```
> beta7norm@maW <- matrix(0, 0, 0)
> beta7.p <- as(beta7norm, "MAList")
> beta7.pq <- normalizeBetweenArrays(beta7.p, method = "quantile")

> plotDensities(beta7.p)

> plotDensities(beta7.pq)
```

Another approach is the vsn–method proposed by Huber et al. (2002). For a data matrix x_{ki}, it fits the normalization transformation

$$x_{ki} \mapsto h_i(x_{ki}) = \mathrm{glog}\left(\frac{x_{ki} - a_i}{b_i}\right). \tag{4.4}$$

This is the same as Equation (2.2) from Section 2.3.3, but here i runs over arrays *and* color channels; k, as before, indexes the probes. As an extension to Equation (4.4), the function vsn can also accommodate for a more highly parameterized model, in which different subsets of spots, for example different subgrids, require different scale and offset parameters. This can be achieved by setting its argument strata. Please note that vsn expects unnormalized ("raw") data as input.

Users can set the method argument to vsn in normalizeBetweenArrays and the function will return an object of class *MAList* with normalized intensities converted to the log base 2 scale.

```
> library("vsn")
> beta7.vsn <- normalizeBetweenArrays(as(beta7,
+     "RGList"), method = "vsn")
```

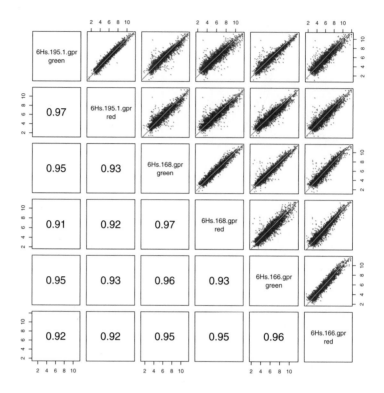

Figure 4.8. The upper panels show pairwise scatter-plots between the red and green intensities from the first three arrays in the beta7 data set. The intensities were normalized with the separate-channel normalization method vsn. The lower panels show the correlation coefficients. Intensities from the same array (and with different colors) tend to be more correlated than those between different arrays.

Alternatively, this can be done by calling the function vsn directly. It will take a *marrayRaw* object as the input and return the transformed intensities in an object of class *exprSet*.

```
> beta7.vsn <- vsn(beta7)
```

The matrix of generalized log transformed intensities can be accessed via

```
> b7 <- exprs(beta7.vsn)
```

b7 is a matrix with 23184 rows and 12 columns. In the following code, we make a *pairs plot* between the first six columns.

```
> upPan <- function(...) {
+      points(..., col = "darkblue")
+      abline(a = 0, b = 1, col = "red")
+ }
```

```
> lowPan <- function(x, y, ...) {
+       text(mean(par("usr")[1:2]), mean(par("usr")[3:4]),
+            signif(cor(x, y), 2), cex = 2)
+ }
> pairs(b7[, 1:6], pch = ".", lower.panel = lowPan,
+       upper.panel = upPan)
```

The result is shown in Figure 4.8.

4.6   Case study

In this section, we describe a complete analysis of the beta7 data set. We start with the probe-level data and end with a list of genes of possible scientific interest.

1. To begin, users should create a directory and move all the relevant image processing output files (e.g., .gpr files) and a file containing target (or samples) descriptions (e.g., Targetbeta7.txt file) to that directory. For this illustration, the data has been gathered in the data directory beta7.

2. Start R in the desired working directory, and load the beta7 package.

   ```
   > library("beta7")
   ```

3. **Load preprocessing packages:** The following command will load arrayQuality, marray and limma.

   ```
   > library("arrayQuality")
   ```

4. **Data input:** Read in the target file containing information about the hybridizations.

   ```
   > TargetInfo <- read.marrayInfo("TargetBeta7.txt")
   ```

5. Read in the raw fluorescent intensities data. By default we assume that the file names are provided in the **first** column of the target file.

   ```
   > mraw <- read.GenePix(targets = TargetInfo)
   ```

6. **Array quality assessment:** The following command generates diagnostic plots for a qualitative assessment of slide quality. An example of diagnostic plot is shown in Figure 4.2.

   ```
   > maQualityPlots(mraw)
   ```

7. **Normalization:** The next step is to perform some normalization of the data. This step aims at removing sources of systematic variation other than differential expression.

   ```
   > normdata <- maNorm(mraw)
   ```

8. **Data output (a):** When the preprocessing of your data is completed, you can export your normalized log–ratios M data to a text file. This can be useful if you would like to perform further analysis using other bioinformatics software or packages.

```
> write.marray(normdata)
```

9. **Data output (b):** An R object of class *exprSet* is the main input used in most Bioconductor packages for downstream analysis. Users will be able to use the package convert to coerce the object normdata from class *marrayNorm* to the object mdata of class *exprSet*.

```
> library("convert")
> mdata <- as(normdata, "exprSet")
```

10. **Identify differentially expressed genes:** The next few steps in this case study illustrate a typical downstream analysis which aims to identify differentially expressed genes between $\beta7+$ and $\beta7-$. To begin, we will estimate the log–ratios between these two samples using the lmFit function in the limma package. Users can specify *dye-swaps* samples using the argument design, which allows for appropriate averaging across multiple arrays. More details on linear models will be discussed in Chapter 23.

```
> LMres <- lmFit(normdata, design = c(1, -1, -1,
+     1, 1, -1), weights = NULL)
```

11. Compute moderated t-statistics and log–odds of differential expression by empirical Bayes methods from limma. Theoretical details can be found in Smyth (2004), and we refer readers to Chapter 23 for more examples.

```
> LMres <- eBayes(LMres)
```

12. **HTML output:** You can show the top 10 differentially expressed genes based on log–odds ratios (default) and export them into an HTML file. Figure 4.9 shows an example. If probe-related annotations such as oligo IDs or Genbank accession numbers are provided, the function table2html will create hyper-links to various external database web sites.

```
> restable <- topTable(LMres, number = 10, resort.by = "M")
> table2html(restable, disp = "file")
```

After we have carefully assessed the quality of individual arrays and performed appropriate normalization procedures, we are now in a position to address the main question for which the beta7 microarray experiment was designed, that is, the identification of genes that are differentially expressed between $\beta7+$ and $\beta7-$ memory T helper cells. Other Bioconductor packages such as EBarrays, genefilter, limma and siggenes may be used to address this question.

BioConductor Gene Listing

ID	Name	M	A	t	P.Value	B
H200012024	ITGA1 - Integrin, alpha 1	1.27	6.98	8.39	1	0.59
H200014446	P2Y5 - Purinergic receptor (family A group 5)	-0.92	12.23	-6.15	1	-0.34
H200001079	EGFL5 - EGF-like-domain, multiple 5	-0.96	7.74	-6.27	1	-0.28
H200001929	EPLIN - Epithelial protein lost in neoplasm beta	-1.1	8.62	-6.29	1	-0.26
H200003977	F5 - Coagulation factor V (proaccelerin, labile factor)	-1.1	7.5	-6.55	1	-0.14
H200007427	CENTG2 - Centaurin, gamma 2	-1.19	6.09	-8.29	1	0.56
H200004937	Homo sapiens cDNA FLJ12815 fis, clone NT2RP2002546	-1.24	6.3	-6.85	1	0.01
H200003784	SEMA5A - Sema domain, seven thrombospondin repeats (type 1 and type 1-like), transmembrane domain (TM) and sh	-1.35	6.81	-6.89	1	0.03
H200018884	Homo sapiens cDNA FLJ11375 fis, clone HEMBA1000411, weakly similar to ANKYRIN	-1.6	6.62	-8.76	0.77	0.71
H200017286	GPR2 - G protein-coupled receptor 2	-2.45	7.79	-10.77	0.18	1.17

Figure 4.9. Example of a HTML report displaying differentially expressed genes for the beta7 experiment.

5

Cell-Based Assays

W. Huber and F. Hahne

Abstract

This chapter describes methods and tools for processing and visualizing data from high-throughput cell-based assays. Such assays are used to examine the contribution of genes to a biological process or phenotype (Carpenter and Sabatini, 2004). In principle, this can be done for any gene or combination of genes and for any biological process of interest. There is a variety of technologies, but all of them rely on the availability of genomic resources such as whole genome sequences, full-length cDNA libraries, siRNA collections; or on libraries of protein-specific ligands (compounds). Typically, all or at least large parts of the experimental procedures and data collection are automated. Cell-based assays offer the potential for clustering of genes based on their functional profiles (Piano et al., 2002) and epistatic analyses to elucidate complex genetic networks (Tong et al., 2004).

5.1 Scope

The special-purpose software used in this chapter is from the package prada. It provides facilities for importing and storing data from cell-based assays, for visualization, and for initial quality control and preprocessing. The focus is on data that was obtained through flow cytometry. For the subsequent statistical inference and modeling, we use general purpose tools such as linear and local regression and hypothesis testing.

5.2 Experimental technologies

We start by describing some high-throughput technologies that can be used to perturb the activity of proteins and to monitor the cellular response.

5.2.1   Expression assays

Expression assays probe the role of a protein in a cellular process or pathway of interest by increasing its abundance. Cells are transfected with a vector that contains the encoding DNA sequence. Usually, a short sequence encoding for the Green Fluorescent Protein (GFP) or one of its variants is attached at one of the ends, resulting in either an N- or C-terminally tagged protein. This way, the protein's overabundance in each individual cell can be monitored through its fluorescence.

This is an elegant and versatile technology. However, care has to be taken to obtain meaningful results. To avoid artifacts caused by non-specific or cross-reactive effects, one aims to limit the amount of over-expression: ideally, the abundance of the expressed protein is small compared to the abundance of the endogenous protein. Artifacts might also be caused by the fluorescence tag masking a protein's functional sites or localization signals. Confidence can be increased when the results are consistent for both N- and C-terminal tag orientations (Wiemann et al., 2004).

In addition to probing for a phenotypic response, one can observe the tagged protein's localization in the cell by fluorescence microscopy (Simpson et al., 2000).

5.2.2   Loss of function assays

Loss of function assays are complementary to expression assays. Here, one aims to implicate proteins in cellular processes by observing the effect of their removal or partial removal. This can be achieved by knocking out a protein's encoding gene, disabling its mRNA by *RNA interference* (RNAi), or by inhibiting its activity through a small compound. RNAi has become popular (Meister and Tuschl, 2004) due to its ease and applicability to almost every gene. Genome-scale RNAi libraries for a number of model organisms are now available (Giaever et al., 2002; Kamath et al., 2003; Boutros et al., 2004), and they are quickly becoming a versatile and widely applicable tool of functional genomics. Among the challenges of this technology are the difficulty to monitor the success of each knock-down and to guarantee its specificity.

5.2.3   Monitoring the response

In principle any cellular process can be probed. This could be the (de-)activation of a certain pathway, differentiation, morphological changes, changes in viability, and so on. In some contexts, this is also called a *phenotype*. Phenotypes can be registered at various levels of detail: as a yes/no alternative, as a single quantitative variable, or as a more complex feature such as an image or a time series. There are three major technologies:

- plate reader

- flow cytometry
- automated microscopy

Each of these is able to deal with experiments in a 96- or 384-well format. The plate reader offers the highest throughput but also the lowest resolution. It measures the overall fluorescence of a population of cells, possibly at different wave-lengths, in each well. Each well may correspond, for example, to the silencing of a different transcript mediated by a specific siRNA. Such a plate reader may process a 96-well plate within a minute, and a whole eukaryotic transcriptome can be screened in a few hours.

Flow cytometry offers much more detail: here, fluorescence intensities as well as morphological parameters such as cell size and granularity are measured for each cell individually. This allows to make statements that go beyond population averages. Furthermore, it is possible to compare cells within the same well, for example, by correlating their individual level of over-expression to the strength of their response. This can be considerably more sensitive than just comparing the averages of treatment and control in different wells. The throughput is lower: it takes about four to five hours to screen one 96-well plate, depending on cell density and counting accuracy. The technology is rather robust and reliable, and commercially available flow cytometry machines have become standard laboratory equipment.

Automated microscopy offers the most detail but also has the lowest throughput. The technology is currently evolving rapidly (Liebel et al., 2003; Gerlich and Ellenberg, 2003; Huisken et al., 2004). The data that are available for each cell can be a 2D planar image, a 3D image obtained from confocal microscopy, or even a 4D movie with temporal as well as spatial resolution. The discussion of image analysis tasks, such as segmentation, tracking, and feature extraction is beyond the scope of this book. They are also currently not supported by Bioconductor packages. However, once quantitative features have been extracted for each cell, the further statistical analysis might proceed along similar lines as described below.

5.3 Reading data

Depending on the means of acquisition, data are stored in different formats. Usually, it is possible to transform the primary data from the recording instrument into a rectangular table that is amenable for reading through the function read.table or one of its relatives. In flow cytometry, a useful standard has been established, which we will describe in Section 5.3.3. First, we load the packages prada and facsDorit, which contains example data sets.

```
> library("prada")
> library("facsDorit")
```

5.3.1   Plate reader data

Let us look at a data set that was measured by Boutros et al. (2004). They used a library of 19,470 double-stranded (ds) RNAs targeting the corresponding set of genes in cultured *Drosophila* cells. By knocking out each gene in turn, their goal was to get a comprehensive list of genes that are essential for cell growth, and viability. The data are read from the file BoutrosKiger.tab, which we have provided in the facsDorit package.

```
> viab <- read.table(system.file("extdata", "BoutrosKiger.tab",
+     package = "facsDorit"), header = TRUE, as.is = TRUE,
+     sep = "\t")
> viab[1:2, ]

  LocationID   Plate Col Row AmpliconLen PeptideLen      ID
1 973108_A01 973108   A   1        1161        505 HDC00002
2 973108_A02 973108   A   2        1007        510 HDC00008
      Kc1     Kc2    S2R1   S2R2
1  471792  461639  945714 805464
2 1012685 1205919  647040 647415
```

The experiments were done on Kc_{167} and on $S2R^+$ cells, each in duplicate. The viability of the cells after dsRNA treatment was recorded by a plate reader measuring luciferase activity, which was indicative of ATP levels. Genes with a viability phenotype were found from those dsRNAs that showed low luciferase activity. The measurements are in columns Kc1, Kc2, S2R1, and S2R2. The 21312 rows of the data frame viab correspond to the 19470 dsRNAs, whose identifiers are in the column ID, plus 1842 controls and empty wells. The column Plate contains the identifier of the 96-well plate in which the dsRNAs are kept, and the columns Col and Row contain the coordinates within the plate. For the screening experiment, every set of four consecutive 96-well plates was combined into a 384-well plate, so we also calculate the identifier for the 384-well plates, using the integer division operator %/%:

```
> viab$Plate384 <- with(viab, (Plate - min(Plate))%/%4)
```

The boxplots in Figure 5.1a indicate locations and scales of the distributions of Kc1, one of the two series of measurements on Kc_{167} cells, grouped by 384-well plate.

```
> rg <- 1000 * c(250, 2000)
> boxplot(log(Kc1) ~ Plate384, data = viab, outline = FALSE,
+     col = "#A6CEE3", ylim = log(rg))
```

A further view of the data is provided by the scatterplot between the two replicates,

```
> fac <- factor(viab$Plate384)
> colors <- rainbow(nlevels(fac))[as.integer(fac)]
> perm <- sample(nrow(viab))
```

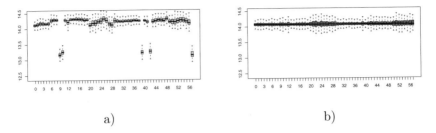

a)   b)

Figure 5.1. a) Boxplots of the `log(Kc1)` values grouped by `Plate384`. Different 384-well plates have different overall signal strengths. One can see long-range *trends* or *block effects* across plates. Furthermore, there are a few outlier plates with low signal. b) After normalization.

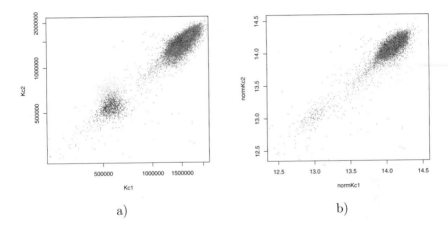

a)   b)

Figure 5.2. Scatterplots of `Kc1` versus `Kc2` values, colored by `Plate384`, a) before normalization, with logarithmic axis scaling, b) after logarithmic transformation and normalization.

```
> plot(viab[perm, c("Kc1", "Kc2")], col = colors[perm],
+      pch = ".", log = "xy", xlim = rg, ylim = rg)
```

The plot is shown in Figure 5.2a. We plot the points in randomly permuted order `perm` in order to avoid visual artifacts that derive from the order of the rows in `viab`. It appears that some *normalization* is necessary, similar to the normalization for microarray intensities discussed in Chapters 2 and 4. If we assume that the plate effects are multiplicative, we can adjust for them by fitting a robust linear model on the logarithmic scale,

```
> library("MASS")
> expts <- c("Kc1", "Kc2", "S2R1", "S2R2")
```

```
> allMedian <- log(median(as.matrix(viab[, expts])))
> for (ex in expts) {
+     lmRes <- rlm(log(viab[[ex]]) ~ fac)
+     viab[[paste("norm", ex, sep = "")]] <- residuals(lmRes) +
+         allMedian
+ }
```

The coefficients of `lmRes` absorb the plate effects, and the biologically interesting effects are contained in the residuals. We retain the overall median
`allMedian`, so the numerical values of the normalized intensities are in the
same range as the unnormalized ones. The result is shown in Figures 5.1b
and 5.2b.

Genes with a viability phenotype can now be found from those dsRNAs
that show low luciferase activity in both experimental repeats in either one
or both of the cell lines. For example, in Figure 5.2b, one might select the
genes in the lower left part of the scatterplot.

```
> score <- rowMeans(viab[, paste("normKc", 1:2,
+     sep = "")])
> sel <- order(score)[1:3]
> viab[sel, 1:7]

      LocationID  Plate Col Row AmpliconLen PeptideLen
20903 973331_H05 973331   H   5         719        425
12176 973238_A02 973238   A   2         301        455
21310 973331_H05 973331   H   5         719        425
            ID
20903 HDC08326
12176 HDC17664
21310 HDC08326
```

Note that the fluorescence data themselves do not suggest an unambiguous
cutoff for the selection; rather, such a cutoff has to be chosen according
to conventional trade-offs between false positives and false negatives. Control dsRNAs that hit known effectors and non-effectors are useful for the
calibration of this choice.

5.3.2 Further directions in normalization

Instead of invoking the linear modeling machinery of `rlm` in the code example above, it would have been possible to simply calculate and subtract
the mean or median of the logarithmic intensities for each plate and add
the overall median. The advantage of the presented approach is that it can
be extended to more complex normalizations that take into account further factors such as spatial effects within the plates or continuous-valued
covariates such as time.

In the code example above, we have called `rlm` separately for each of the four experiments. Alternatively, we can attempt to fit one overall model to the data, with a model formula such as

```
> lmRes <- rlm(log(luc) ~ expt * fac, data = rViab)
```

where `rViab` is a reshaped, "long" version of the data frame `viab`, obtained through the function `reshape` from the `stats` package, `luc` is its column containing the fluorescence intensities, and the factor `expt` codes the four experiments Kc1, Kc2, S2R1, S2R2. However, if approached in this naive manner, the calculation will be slow and memory-consuming due to the large matrices involved. It appears reasonable to employ the symmetries of the problem.

5.3.3 FCS format

FCS is a file standard for *flow cytometry* data now adopted by the major manufacturers of instruments and driver software. The current version is FCS 3.0 (Seamer et al., 1997). It contains the specifications needed to describe the flow cytometric measurements on one sample of cells and the accompanying meta-data in a single file. An FCS file is divided into segments: *header*, *text*, *data*, and an optional *analysis* segment. The *header* segment provides plain text numbers with byte offsets that point to start and end positions of the other segments. The *text* segment describes the experiment, the instrument and its settings, the specimen, specifications of the different fluorescence channels, plus any additional information the creator of the file wanted to include. These descriptions are provided in the form of keyword–value pairs. The *data* segment contains the actual fluorescence values in a binary format. The optional *analysis* segment can be used to describe subsequent analyses.

The function `readFCS` can be used to read data from FCS 3.0 files and transform them into an object of class `cytoFrame`. Here we use files from an experiment that screened for effectors of the MAP-kinase signaling pathway. The experiment was done at the German Cancer Research Center by Meher Majety (Wiemann et al., 2004).

```
> sampleDir <- system.file("extdata", "map", package = "facsDorit")
> B05 <- readFCS(file.path(sampleDir, "060304MAPK controls.B05"))
> B05
```

```
cytoFrame object with 1575 cells and 8 observables:
FSC-H SSC-H FL1-H FL2-H FL3-H FL2-A FL4-H Time
slot 'description' has 146 elements
```

The `cytoFrame` class has two slots `description` and `exprs`, which represent the *text* and *data* sections of the FCS file, respectively. The function `description` returns a named character vector, whose elements are the values and whose names are the keys. The current example file contains 146 key-value pairs, we print three of them:

```
> description(B05)[c(130, 137, 139)]

        $DATE &1Sample Vol &3Mixing Vol
  "03-Jun-04"        "200"        "180"
```

The function exprs returns a matrix which contains the fluorescence intensities. Its rows correspond to the cells and its columns to the different fluorescence channels.

```
> exprs(B05)[1:2, ]

      FSC-H SSC-H FL1-H FL2-H FL3-H FL2-A FL4-H Time
[1,]    621   454   973  1023   434  1023   566    3
[2,]    648   607   431   792   301   278   663    3
```

The experiments that we are interested in comprise more than one FCS file. We may be looking at samples of cells that were transfected with different genes or were fixed at different time points. Flow cytometry machines often allow the automated, serial processing of multiple samples in microtiter plate format, and their vendor software usually provides an automated naming for the resulting files. For example, the directory sampleDir contains 96 files, corresponding to the 96 wells of a microtiter plate. The files have extensions .A01, .A02, ..., .H12. In addition, the file plateIndex.txt keeps track of the samples contained in each well. It is a rectangular table with one row for each sample and a mandatory column named name with the names of the corresponding FCS files. The data can be read using

```
> mapk <- readCytoSet(path = sampleDir,
+       phenoData = "plateIndex.txt")
> pData(mapk)[1:2, ]

                     name clone wellnr
1 060304MAPK controls.A01  mock      1
2 060304MAPK controls.A02  mock      2
```

The resulting object is of class *cytoSet*. It has two slots: phenoData and frame. The phenoData slot is used to keep track of the covariates that are associated with each sample, and it functions in an analogous manner as for *exprSet* objects.

The slot frame is used to store the data from the individual wells, in particular, the fluorescence intensities. It is an R *environment* that contains a set of *cytoFrame* objects. The constructor and validity methods of the *cytoSet* class make sure that the component *cytoFrame* objects are compatible, in particular, that they have the same number of fluorescence channels and that these have the same names. We have defined subset operators [and [[for cytoSet objects that work similarly to those for lists:

```
> mapk[[1]]
```

```
cytoFrame object with 210 cells and 8 observables:
FSC-H SSC-H FL1-H FL2-H FL3-H FL2-A FL4-H Time
slot 'description' has 146 elements
```

```
> exprs(mapk[[1]])[1:2, ]
```

```
      FSC-H SSC-H FL1-H FL2-H FL3-H FL2-A FL4-H Time
[1,]    651   409    67   711   234   139   620   36
[2,]    235  1023   140   424    96     0   582   36
```

There is an important difference to lists, though, and users of the *cytoSet* are referred to the *Note on storage and performance* in its manual page.

5.4   Quality assessment and visualization

For cell-based assays, the quality of the data can be assessed on multiple levels: at that of single cells, entire wells, entire plates, all measurements from one transfection construct, and so on. We discuss some examples.

5.4.1   *Visualization at the level of individual cells*

Let us first have a closer look at flow cytometry data in general. For each individual cell, we can measure,

- Forward light scatter (FSC): this is a measure of a cell's size; the details depend on the experimental setup and the instrument.

- Sideward light scatter (SSC): this is a measure of a cell's granularity, that is, the appearance and structure of the cell surface as well as the amount of light-impermeable internal structures like lysosomes. A large value indicates high granularity.

- Several fluorescence channels that measure the abundance of fluorophores, which may be bound to specific antibodies for surface or intracellular markers or be encoded by and expressed from a tagged transcript.

The optical apparatus of a flow cytometry instrument is optimized for detecting fluorescent light. Measurements of morphological properties via the light scatter are usually not particularly accurate. Nonetheless they can be used for a rough segmentation of the cell populations.

Experimental cell populations are often contaminated by cell debris, cell conjugates, precipitates, or air bubbles. The instrument may not be able to discriminate these contaminants from single, living cells, and hence they may end up in the measured data. To a certain extent, it is possible to discriminate such contaminations by their size using the FSC signal and by their granularity using the SSC signal. We can look at the joint distribution of FSC and SSC by means of the scatterplot:

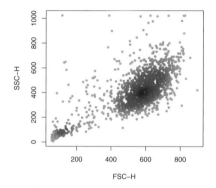

Figure 5.3. Scatterplot of flow cytometry data: FSC vs. SSC.

```
> x <- exprs(B05)[, c("FSC-H", "SSC-H")]
> plot(x, pch = 20, col = densCols(x))
```

The result is shown in Figure 5.3. We have used the function densCols to obtain a coloring of the points that is indicative of their local density.[1] The measurements with very small FSC and SSC values at the lower left corner of the plot are most likely due to debris, and we want to omit them from the subsequent analysis. This can be done manually, by defining certain thresholds ("gates") on the FSC and SSC values.

It is desirable to automate such a task. In Figure 5.3, we can observe that the main population of cells has roughly an elliptical shape. Consequently, we assume that the main population can be approximated by a bivariate normal distribution in the FSC–SSC space, and that outliers can be identified by having a low probability density in that distribution. Contours of equal probability density are ellipses, and we select those cells as outliers that lie outside a certain density threshold. This functionality is provided by the function fitNorm2. It fits a bivariate normal distribution into the data by robust estimation of its center and its 2×2 covariance matrix.

```
> nfit <- fitNorm2(x, scalefac = 2)
```

The parameter scalefac controls the probability density threshold. Its value corresponds to the distance from the center in the Mahalanobis metric. We can plot the result of this selection procedure together with the computed probabilities in false color coding using the function plotNorm2.

```
> plotNorm2(nfit, ellipse = TRUE)
```

---

[1]The visualization of scatterplots with many points is discussed in more detail in Section 10.3.

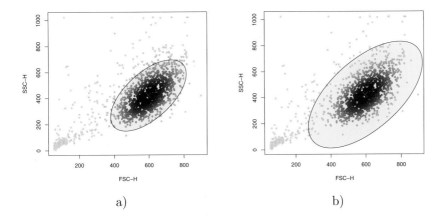

Figure 5.4. Outlier selection in the FSC-SSC plot: the points outside the ellipse, which is a contour of equal probability density in a bivariate normal distribution, can be considered outliers. The center of the distribution is marked by a red cross. The two panels correspond to two different values of the parameter `scalefac`: a) `scalefac`=2, b) `scalefac`=3.

The plot is shown in Figure 5.4. The return value `nfit` of the call to `fitNorm2` is a list. One of the list elements is the logical vector `nfit$sel`. It has the same length as the number of data points, and a value of `TRUE` indicates that the point lies within the ellipse.

```
> B05.sel <- B05[nfit$sel, ]
```

Figure 5.5 shows a comparison between the scatterplots of the two fluorescence channels FL1-H and FL4-H using the "clean" data `B05.sel` as well as the original data `B05`. In this example, FL4-H measures the effect of the perturbation in the MAP-kinase assay through the signal of a specific antibody, whereas FL1-H measures the overabundance of the respective effector protein via its YFP-tag.

```
> myPlot <- function(x) {
+     ex <- exprs(x)[, c("FL1-H", "FL4-H")]
+     plot(ex, pch = 20, col = densCols(ex))
+ }
> myPlot(B05)
> myPlot(B05.sel)
```

The convenience function `myPlot` allows us to produce the two similar plots in Figure 5.5 without repeatedly specifying the plot options.

The composition of a population of cells can also be used to get an overview of other properties. An unusually high proportion of contaminants may indicate problems during growth or experimental treatment; similarly, an unusual location or covariance of the fitted normal distribution. As

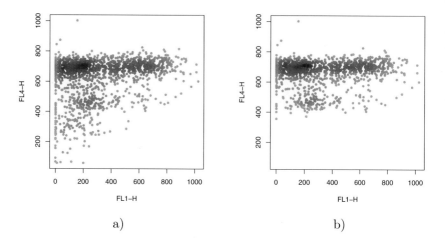

a)   b)

Figure 5.5. Scatterplots of FL1-H vs FL4-H. a) using all data (B05), b) using selected data only (B05.sel). Through the selection, we have reduced the proportion of data points with very small values of FL4-H, which apparently correspond to artifacts unrelated to the cell population of interest.

these quantities depend on instrument settings, cell line, and experimental treatment, such criteria will have to be adapted for each experiment.

Fitting a unimodal normal distribution as in Figure 5.4 is appropriate if the cell population is homogeneous and we screen for a phenotype that does not strongly affect the cell morphology. We note that flow cytometry is often also used to analyze more complex samples consisting of morphologically different sub-populations of cells, for example, whole blood samples. For these, more sophisticated clustering, mixture modeling, and classification algorithms are in order.

5.4.2   Visualization at the level of microtiter plates

Microtiter plates come in different formats, usually as a rectangular arrangement of $4 \times 6 = 24$, $8 \times 12 = 96$, or $16 \times 24 = 384$ wells. Each well may contain cells that were treated in a different manner. The wet lab handling of the plates is usually automated at least to some degree. A visualization of per-well statistics is provided by the function plotPlate. Figure 5.6 shows an example for a plate plot of a 96-well plate that indicates the number of cells that were found in each well.

```
> nrCells <- csApply(mapk, nrow)
> plotPlate(nrCells, nrow = 8, ncol = 12, main = "Cell number",
+       col = brewer.pal(9, "YlOrBr"), width = 6.3)
```

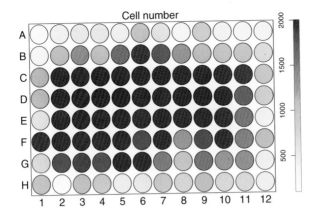

Figure 5.6. Plate plot indicating cell number. Cell number depends on factors like the condition of cells before plating, culture conditions, and purity of the reagents. Here, cell numbers at the edges of the plate are consistently low, which may indicate a handling problem.

Other quantities of interest include average fluorescences of each well, e.g., to monitor expression efficiency or artifactual shifts in the response.

The amount of information included in a plate plot can be expanded by decorating it with *tool-tips* and *hyperlinks*. When viewed in a browser, a tool-tip is a short textual annotation – for example, a gene name – that is displayed when the mouse pointer moves over a plot element. A hyperlink can be used to display more detailed information, even a graphic, in another browser window or frame. For example, underlying the values that are displayed in a plate plot such as Figure 5.9 can be a complex statistical analysis for each individual well, whose details can be displayed on demand by hyperlinking them to the corresponding well icons in the plate plot. For this purpose, the package geneplotter provides the function imageMap. From the output of the function platePlot and a list of hyper-links and tool-tips, it produces an HTML image map, which can be viewed with a web browser.

5.4.3 Brushing with Rggobi

The example data from Section 5.4.1 is four-dimensional: each cell is described by the four parameters FSC-H, SSC-H, FL1-H, and FL4-H. Although static scatterplots can only display two variables at a time, there are a number of interactive visualization techniques that can help to get insight into slightly higher-dimensional data sets. One of them is *brushing*. Here, one displays several two- or one-dimensional projections of the data. Interactive subset selections of the data are then simultaneously marked in all plots. Such a functionality can be achieved from within R; for example, see the package iSPlot. An alternative is to use an external program,

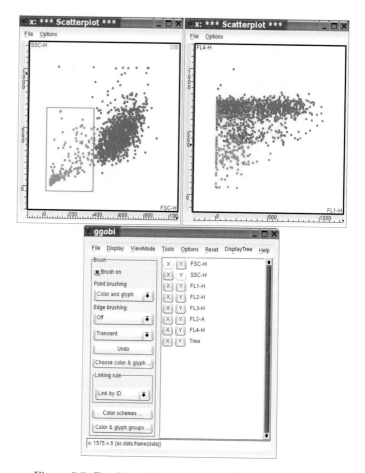

Figure 5.7. Brushing between scatterplots with Rggobi.

ggobi (Swayne et al., 2003), which has an elegant interface to R via the Rggobi package. Figure 5.7 shows an example for brushing between FSC-H vs. SSC-H and FL1-H vs. FL4-H scatterplots of the apoptosis assay data.

```
> library("Rggobi")
> x <- exprs(B05)
> gg <- ggobi(x)
> gg$setGlyphs(5, 1, 1:nrow(x))
> gg$setColors(rep(9, nrow(x)))
> gg$scatterplot("FL1-H", "FL4-H")
> gg$setBrushColor(5)
> gg$setMode("Brush")
```

5.5   Detection of effectors

We discuss two types of responses, which call for different methods of statistical analysis: discrete responses caused by on-off mechanisms and continuous responses that reflect a more gradual effect. Each type will be exemplified by a representative experiment.

5.5.1   Discrete Response

Apoptosis, or programmed cell death, is a strongly evolutionarily conserved cellular mechanism that plays a crucial role in development, growth, and tissue maintenance. One key player of this pathway is the enzyme caspase-3, that gets activated upon onset of apoptosis in most cell types. This activation is a rapid and irreversible step taking less than five minutes. Once the cell receives a signal to undergo apoptosis, most or all of its caspase-3 molecules are proteolytically cleaved, which then inevitably leads to the death of the cell. Thus caspase-3 activation can be used to measure the apoptotic state of a cell, and it can be considered an on-off switch.

   In a study that was performed by Mamatha Sauermann at the German Cancer Research Center, caspase-3 activation was monitored using a fluorochrome-coupled antibody specific for the activated form of caspase-3. The fluorescence was recorded in the FL4-H channel. Cells were transfected with expression vectors encoding different potential effector proteins and a fluorescent YFP tag whose fluorescence was recorded in the FL1-H channel. We first load and preprocess the data.

```
> preprocess <- function(x) {
+      for (i in 1:length(x)) {
+          dat <- exprs(x[[i]])
+          fn <- fitNorm2(dat[, c("FSC-H", "SSC-H")],
+              scalefac = 1.5)
+          x[[i]] <- dat[fn$sel, ]
+      }
+      return(x)
+ }
> apo <- readCytoSet(path = system.file("extdata",
+      "apoptosis", package = "facsDorit"),
+      phenoData = "plateIndex.txt")
> apoP <- preprocess(apo)
```

The `preprocess` function removes, for each well, the debris identified by `fit-Norm2`, as described in Section 5.4.1. The FL4-H (corresponding to caspase-3 activation) and FL1-H intensities (corresponding to YFP fluorescence) of mock transfected cells can now be used to define thresholds for approximately separating apoptotic from non-apoptotic cells and expressing from non-expressing cells.

```
> calcthr <- function(x) {
+      h <- hubers(x)
+      h$mu + 2.5 * h$s
+ }
> mock <- exprs(apoP[[1]])[, c("FL1-H", "FL4-H")]
> plot(mock, pch = ".", xlab = "protein expression",
+      xlim = c(0, 1023), ylab = "caspase-3 activation",
+      ylim = c(0, 1023))
> thrYFP <- calcthr(mock[, 1])
> thrCASP3 <- calcthr(mock[, 2])
> abline(v = thrYFP, h = thrCASP3, col = "red",
+      lty = 2)
```

The above code assumes that for mock transfected cells the fluorescence
values in both channels have a unimodal distribution, i.e., that almost
all cells are in a non-apoptotic state and that they do not express YFP.
Location and scale of these distributions can then be used for setting the
separation thresholds. These values are robustly estimated by the function
hubers from the package MASS. The plot is shown in Figure 5.8b.

We can now use the function thresholds to discretize the data into four
subsets and obtain a *contingency table*. The dependency between the counts
can be assessed using Fisher's exact test.

```
> cide <- exprs(apoP[[6]])[, c("FL1-H", "FL4-H")]

> ct <- thresholds(cide, xthr = thrYFP, ythr = thrCASP3)

        [,1] [,2]
[1,]     11  302
[2,]   8890 3957

> fisher.test(ct)

          Fisher's Exact Test for Count Data

data:  ct
p-value < 2.2e-16
alternative hypothesis: true odds ratio is not equal to 1
95 percent confidence interval:
 0.0080 0.0295
sample estimates:
odds ratio
      0.0162
```

Figure 5.9 shows results of such an analysis for a 96-well microtiter plate.
Not all wells were populated during the experiment, and these are omitted
from the plate plot. The false color coding represents the negative logarithm
of the odds ratios for those cases with p-values less than 0.01.

```
> calcOdds <- function(x) {
+      ct <- thresholds(x[, c("FL1-H", "FL4-H")],
```

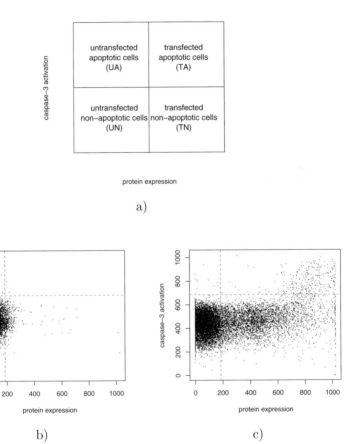

Figure 5.8. a) YFP fluorescence (FL1-H) can be used to approximately separate expressing from non-expressing cells, caspase-3 activation (FL4-H) to separate apoptotic from non-apoptotic cells. b) Scatterplot for a control population of mock-transfected cells, which can be used to determine the separation thresholds. c) Scatterplot for a population of cells that were transfected with an apoptosis-inducing CIDE3-YFP expression construct.

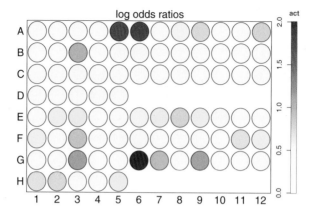

Figure 5.9. Plate plot classifying apoptosis pathway effectors. The color code represents the log odds ratio $\log[(TA/TN)/(UA/UN)]$. The numbers TA, TN, UA, and UN are defined in Figure 5.8.

```
+          xthr = thrYFP, ythr = thrCASP3)
+      f <- fisher.test(ct)
+      res <- -log10(f$estimate)
+      return(ifelse((f$p.value > 0.01 | is.infinite(res)),
+          0, res))
+ }
> odds <- csApply(apoP, calcOdds)

> cols <- brewer.pal(9, "Reds")[c(rep(1, 4), 2:9)]
> plotPlate(odds, nrow = 8, ncol = 12, main = "log odds ratios",
+      desc = c("act", ""), col = cols, width = 6.3,
+      na.action = "omit", ind = pData(apo)$wellnr)
```

5.5.2 Continuous response

Signaling in cells involves a system of proteins and small molecules that build up complex interacting pathways. Selective phosphorylation is one mechanism for passing signals along a cascade of proteins. Proteins with the ability to phosphorylate other molecules are called kinases. One prominent example of a kinase signaling cascade is the MAP-kinase pathway, which plays an important role in cell-cycle regulation. The activity of this pathway can be continuously regulated both in a positive and in a negative manner. In contrast to the previous section, where the response was essentially a "yes/no" decision, here the response is of a gradual nature. This calls for the use of regression analyses to assess the effect of overexpressed proteins. We return to the data set from Section 5.3. In that assay, an antibody against one of the cascade kinases, ERK, was used to monitor the activity of the MAP-kinase pathway. The over-expressed proteins were tagged with

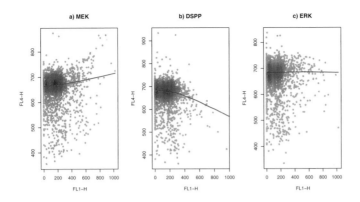

Figure 5.10. Regression analysis for MAP-kinase effectors: a) activator (MEK),
b) repressor (DSPP), c) neutral (ERK).

fluorescent YFP. In this data set, the wells within each row are replicates,
so we will combine the data from these wells. This is accomplished by the
function combineFrames. The function preprocess, which was defined above,
removes debris.

```
> mapkP <- preprocess(combineFrames(mapk, factor(pData(mapk)$clone)))
```

Figure 5.10 shows the fitted robust local regressions for a known activator, a repressor, and a protein with no effect on MAP-kinase activation. The
function locfit.robust is provided by the package locfit (Loader, 1999).

```
> groups <- c("MEK", "DSPP", "ERK")
> for (i in groups) {
+     dat <- exprs(mapkP[[i]])[, c("FL1-H", "FL4-H")]
+     lcft <- locfit.robust(x = dat[, 1], y = dat[,
+         2], deg = 1, alpha = 1, maxk = 512)
+     plot(dat, pch = 20, col = densCols(dat), main = i)
+     lines(lcft, col = "red", lwd = 2)
+ }
```

As a measure of the effect of the overexpressed protein, we use the estimated slope of the regression line for moderately expressing cells. To make
the slopes comparable across different experiments it can also be useful to
consider the z–score, that is the ratio of the estimated slope and its standard error. The following code computes local slope and z–score for the
example data in Figure 5.10 at YFP intensities of 600.

```
> sapply(groups, function(i) {
+     dat <- exprs(mapkP[[i]])[, c("FL1-H", "FL4-H")]
+     dlcft <- locfit.robust(x = dat[, 1], y = dat[,
+         2], deg = 1, alpha = 1, deriv = 1, maxk = 512)
+     pp <- preplot(dlcft, newdata = 600, band = "local")
```

```
+      c(delta = pp$fit, zscore = pp$fit/pp$se.fit)
+ })
```

```
              MEK      DSPP      ERK
delta   0.0339   -0.131   0.00406
zscore  5.3583  -16.659  0.68103
```

5.5.3   Outlook

We have shown three examples of cell-based assays: a genome-wide screen for viability in Section 5.3.1, a sensitive over-expression screen for a discrete response, apoptosis, in Section 5.5.1, and the monitoring of a continuous-valued response in MAP-kinase signaling in Section 5.5.2. We have discussed normalization, quality assessment, visualization, and statistical modeling of these data. The package prada provides data structures to manage these data. It also provides some specialized visualization and analysis routines; moreover, we have encouraged the reader to try out and adapt general-purpose statistical procedures, such as the local linear regression from the package locfit.

Besides the two response type discussed in this chapter, many other kinds of response or phenotype may be of interest, for example, time courses, microscopy images, particular features of the cells, even developmental features of small organisms such as worms or fish embryos. R and Bioconductor offer a large and steadily growing number of methods to import, process and statistically analyze this kind of data.

After the conversion of data from each individual well into quantities such as odds ratios, z-scores, the next step will usually be the summarization of data from multiple technical or biological replicates, further quality control, and the scoring and selection of "hits". Finally in order to interpret the results derived from our assays, we will want to assign our hits to known pathways, compare hit lists, investigate epistatic relationships, and integrate our data with other modes of experimentation such as expression profiles or protein interactions. For this, many of the methods described in the remainder of this book will be useful.

Acknowledgement

We would like to thank Dorit Arlt, Mamatha Sauermann, Meher Majety, Stefan Wiemann, and Michael Boutros for giving us access to their data and for many fruitful discussions about the analysis.

6

SELDI-TOF Mass Spectrometry Protein Data

X. Li, R. Gentleman, X. Lu, Q. Shi, J. D. Iglehart, L. Harris, and A. Miron

Abstract

The term proteome is used to denote the set of proteins encoded by a genome, and *proteomics* is the study of the expression and interactions of the proteins, which can depend on many factors such as cell type, treatment, tissue type, developmental state, and disease state. Conceptually, this is similar to the transcriptomics technologies discussed in Chapters 2–4; however, due to the more complicated chemistry of proteins, compared to RNA, the field has a different and diverse set of technologies and produces a wide range of specific challenges. Here we discuss one particular mass spectrometry technology.

6.1 Introduction

SELDI-TOF-MS is *surface enhanced laser desorption/ionization time-of-flight mass spectrometry*, a proprietary process of Ciphergen®,[1] a leading manufacturer of hardware and software for proteomic research. SELDI-TOF is a technology that can be used to profile protein markers from tissue or bodily fluids, such as serum. For ease of exposition, we will use the term protein, even when in some cases the term *polypeptide* would be more accurate.

Typically, biological samples from different patients or different conditions are compared. The sets of differentially expressed proteins are identified, and it is hoped that they will reveal biological processes or pathways that are involved in the different outcomes or different phenotypes

---

[1] Ciphergen is a registered trademark of Ciphergen Biosystems, Inc.

under study. This technology is often employed to identify biomarkers that can aid in diagnosis, prognosis or treatment, and the identified proteins could potentially serve as drug targets.

The technology has been used to study several different diseases using various bodily fluids. Cerebral spinal fluid was used to investigate markers for severe psychiatric disease (Johnston-Wilson et al., 2001). The urine of patients receiving radiocontrast media during cardiac catheterization provided markers of renal function (Hampel et al., 2001). In oncology the technology has been applied, for example, to transitional cell carcinoma (Vlahou et al., 2001), pancreatic cancer (Rosty et al., 2002), prostate cancer (Wang et al., 2001), ovarian cancer (Petricoin et al., 2002), and breast cancer (Vlahou et al., 2003).

We give a simple description of how the SELDI-TOF-MS technology works and refer the reader to Ciphergen's manual (Ciphergen, 2000) for more specific details. Biological samples are prepared and usually processed via some form of fractionation. *Fractionation* is the process of splitting the original sample into subsamples which contain proteins that are more homogeneous. The samples are placed on the active surface of an array, each in its own spot. Depending on the surface chemistry of the array, a specific subset of proteins in the sample are captured through chemical interactions. After binding and washing off weakly bound proteins, an *energy absorbing molecule* (EAM) solution is applied to the array surface and allowed to dry and crystallize. The array is then placed in a reader and queried with a laser beam. The laser ionizes the proteins, which become charged and gaseous. They fly away from a metal anode with a positive charge down a tube toward an ion detector, resulting in different times-of-flight (TOF) depending on the masses and charges of the polypeptides. The masses are then derived from a quadratic equation that relates TOF and the mass over charge ratio (m/z) of a polypeptide. The parameters of the equation are calibrated using known proteins or polypeptides under the same instrument conditions.

Charge z is quantized; it comes in multiples of the charge of the electron. Typically, the ionized polypeptides have unit charge, but some have higher charge numbers. For example, if a polypeptide has two elementary charges, it will have the same m/z value as a polypeptide half of its mass but with only one charge. In the m/z spectrum, such proteins' will be superimposed on each other; disentangling them is important. For each sample, we obtain a complete spectrum of m/z values with peaks indicating abundant proteins. The output data are often stored as comma delimited files with two columns, the mass over charge (m/z) values and the intensities. Each file represents one measured spectrum. Careful layout of replicate samples helps to reduce between run and spatial variations.

The proteins that are detected are dependent on many factors including the array surface, the EAM used and the fractionations that were used. In a sense, different values of these will yield *different* data sets and most of the parameters we discuss should be examined independently on the different data sets. We also note, that the output data from a single set of samples, given all of these options can be extremely large.

Before the output data can be analyzed to find biomarkers, a number of technical challenges in processing the SELDI-TOF-MS raw data (Fung and Enderwick, 2002; Baggerly et al., 2004; Yasui et al., 2004) need to be resolved. Although the specific details and methods are different from those used to process microarray data the general concepts are the same. For example, some of these fall in the realm of quality assessment while others carry out normalization and still others perform feature extraction. In the following sections, we shall explore those issues and discuss some possible solutions.

We have written a number of software routines to handle these data and have combined them into an Rpackage, PROcess. The routines remove baseline drift, detect peaks, normalize spectra, and align peaks to a set of proto-biomarkers. We also provide a routine for quality assessment of the spectra and a data-driven routine for selecting a cutoff point in spectra to reduce variability. This functionality helps to improve the quality of the processed data for subsequent machine learning efforts.

6.2   Baseline subtraction

We first load the required software and example data that are available from the PROcess package. We read in a raw spectrum and plot it.

```
> library("PROcess")
> fdat <- system.file("Test", package = "PROcess")
> fs <- list.files(fdat, pattern = "\\.*csv\\.*",
+      full.names = TRUE)
> f1 <- read.files(fs[1])
> plot(f1, type = "l", xlab = "m/z")
> title(basename(fs[1]))
```

In Figure 6.1, we see an elevated, non-constant baseline for m/z between about 2,000 and 10,000. At larger m/z, the baseline levels off to a plateau. This elevated baseline is mostly caused by the chemical noise in the EAM and by ion overload. Ideally a spectrum should rest more or less on the horizontal line $y = 0$. In order to make different spectra comparable, the baseline is subtracted from each raw spectrum. This step is similar to background subtraction in microarrays.

The removal of the baseline can be achieved by subtracting from a spectrum an estimate of its *bottom*. We estimate the bottom of a spectrum using local regression. A robust version of local regression should be used so that peaks do not inflate the estimate of the bottom. This can be achieved by fitting a local regression to the points below a certain quantile or to local

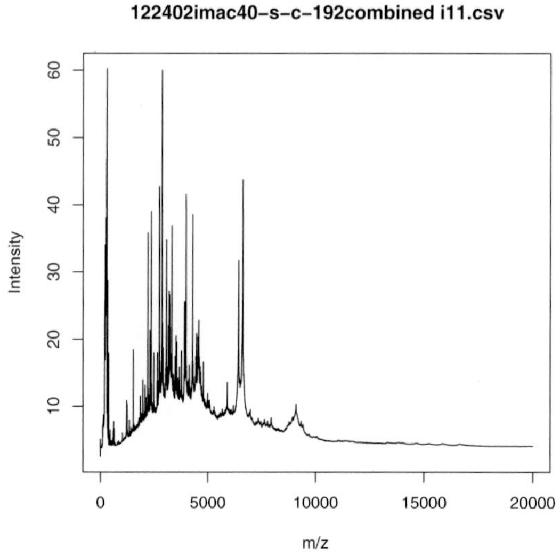
122402imac40-s-c-192combined i11.csv
Intensity
10
20
30
40
50
60
0
5000
10000
15000
20000
m/z

Figure 6.1. A raw spectrum. The x-axis corresponds to the m/z ratio, y-axis to ion intensity.

minima. Although we implemented both methods in PROcess, our experience suggests that using local minima to estimate the baseline yields better results. The process of baseline subtraction may introduce negative net intensity values, which are awkward for subsequent analyses. We have found that using local minima in the regression tends to yield fewer negative values. The algorithm we propose is as follows:

1. For each spectrum find local minima by either a moving window, or by segmenting the m/z range. If the latter, we split the m/z range on the log scale into n equally-spaced intervals and find a given quantile, e. g., the minimum, for each interval.

2. Fit a local regression to data below the local quantiles for each spectrum. A constant bandwidth can be used if the baseline is estimated from segmented data.

3. Subtract the estimated baseline from each spectrum.

The following code creates Figure 6.2, a spectrum with its baseline removed.

```
> bseoff <- bslnoff(f1, method = "loess", bw = 0.1,
+     xlab = "m/z", plot = TRUE)
> title(basename(fs[1]))
```

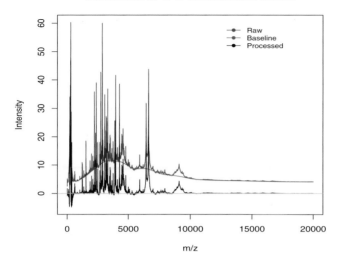

Figure 6.2. Spectrum baseline subtraction. Green: the raw spectrum; Red: the estimated baseline of the raw spectrum; Blue: the raw spectrum minus the estimated baseline.

Caution needs to be employed in choosing the bandwidth; too small a bandwidth may result in the removal of some peaks, especially wide short peaks, which may be potentially important.

6.3   Peak detection

Once we have the baseline-adjusted spectra, the next step is *peak detection*. The peaks represent specific, abundant polypeptides in the sample. Ultimately, interest is centered on the set of polypeptides that are differentially expressed between different samples.

   We locate peaks using the following algorithm. First, the spectrum is smoothed using moving averages of k_s nearest neighbors. Smoothing helps to remove spurious peaks. However, it should not be too vigorous because we want to retain short and wide peaks and also because we need precise estimates in peak locations. In the function isPeak, described below, k_s corresponds to the parameter sm.span. Local variability is computed as the median of the absolute deviations (MAD) of k_v nearest neighbors. The reason for using MAD instead of the standard deviation is that the standard deviation is not robust and tends to over-estimate local variability in the proximity of high peaks. Given that high peaks are sparse in the whole spectrum, MAD reduces the contribution of high peaks to the estimated noise. In `isPeak`, k_v corresponds to the parameter `span`. The use of a large value ensures that extended clusters of peaks are not treated as noise.

Once this has been done, local maxima of the smoothed spectrum are identified. Three thresholds are used in peak calling are:

1. the signal to noise ratio, which is calculated as the local smooth divided by the local estimate of variation;

2. the detection threshold for the whole spectrum: below which the intensities are considered as zero;

3. the shape ratio, which is computed as the area under the curve within a small distance (default 0.3%) of a peak candidate, as identified by the first two criteria, divided by the maximum of all such peak areas of a spectrum.

The first threshold is set to select peak candidates, for example, we retain peaks with signal to noise ratio greater than 2. The second threshold is used to keep out low peaks in relatively flat regions, where the local variation is nearly zero and the signal to noise ratio alone might select small peaks. The third threshold helps discard single spike peaks. Given the limited precision of the technology, a single spike is more likely to be a technical abnormality than a true signal. The parameters can all be set by the users and we suggest that some experimentation be used to find appropriate settings.

We use the following code to do peak detection on the baseline-subtracted spectrum `bseoff` obtained in the last section.

```
> pkgobj <- isPeak(bseoff, span = 81, sm.span = 11,
+      plot = TRUE, zerothrsh = 2, area.w = 0.003,
+      ratio = 0.2, main = "a)")
```

The result is shown in Figure 6.3a. Often it is helpful to inspect peaks in a particular range of m/z values, this is provided by the function `specZoom` (see Figure 6.3b).

```
> specZoom(pkgobj, xlim = c(5000, 10000), main = "b)")
```

6.4   Processing a set of calibration spectra

We have seen in the previous two sections the issues we need to resolve for a single spectrum. In this section, we discuss the additional problems encountered when a set of spectra are considered simultaneously. We presume that all individual spectra have been processed according to the recommendations given above.

Because of measurement variations, peaks in individual spectra that correspond to the same protein may not be in alignment. We compare peak locations are across spectra, and those which are close to each other (typically within a distance ϵ) are identified with a single m/z value. After

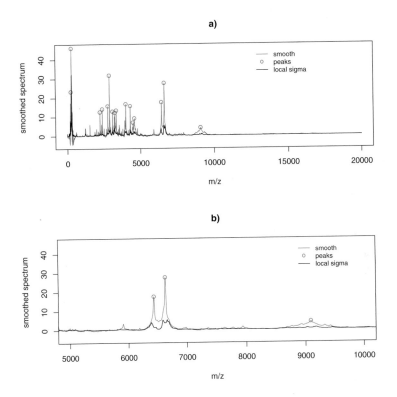

Figure 6.3. a) A spectrum with detected peaks marked by red circles. b) Zoom in produced by the function specZoom.

alignment, all spectra will use a set of common peaks and two peaks that correspond to the same protein should have the same m/z value.

Raw spectra exhibit not only variation in peak positions but also in amplitude. Hence some normalization in the vertical direction is generally also needed. There is often substantial noise observed for low values of m/z and excluding it before normalizing is preferred. So, we first find a suitable cutoff point. Once that is identified, a normalization procedure is applied to each spectrum for values of m/z larger than the selected cutoff value. In most cases, normalization is based on standardizing the area under the estimated (adjusted) spectra; we give details below.

For a given experiment, different combinations of biological samples, array types and EAM may call for different sets of parameters for processing the raw spectra. We recommend using bslnoff and isPeak on a few spectra from each run to tune the parameters to get satisfactory results in baseline subtraction and peak detection. But, in general, different values will be used for different experimental conditions.

We now demonstrate the batch functionality of this package using a set of 8 spectra from a calibration data set, where the same five proteins are present in the samples. Their masses are 1084, 1638, 3496, 5807, and 7034 atomic mass units (amu, often also called a Dalton). We use the following code to read in the eight spectra, plot them, and mark the protein positions by red vertical lines for each of them in Figure 6.4.

```
> amu.cali <- c(1084, 1638, 3496, 5807, 7034)
> plotCali <- function(f, main, lab.cali) {
+       x <- read.files(f)
+       plot(x, main = main, ylim = c(0, max(x[, 2])),
+             type = "n")
+       abline(h = 0, col = "gray")
+       abline(v = amu.cali, col = "salmon")
+       if (lab.cali)
+           axis(3, at = amu.cali, labels = amu.cali,
+                las = 3, tick = FALSE, col = "salmon",
+                cex.axis = 0.94)
+       lines(x)
+       return(invisible(x))
+ }
> dir.cali <- system.file("calibration", package = "PROcess")
> files <- dir(dir.cali, full.names = TRUE)
> i <- seq(along = files)

> mapply(plotCali, files, LETTERS[i], i <= 2)
```

We observe the following features of the eight spectra. First, there is baseline drift. Second, only two out of the five proteins, those with 3496 amu and 5807 amu are unequivocally present. The protein with 7034 amu is relatively short and of similar magnitude to other short peaks in the spectra. Third, the two smallest proteins, those with 1084 amu and 1638 amu are not visually discernible for all eight spectra. Finally, we observe that there is a lot of noise at low m/z values.

6.4.1 Apply baseline subtraction to a set of spectra

We subtract the baselines from this set of spectra by using the following command:

```
> Mcal <- rmBaseline(dir.cali)
```

The function rmBaseline simply calls bslnoff once for each spectrum. The baseline-subtracted spectra are stored column-wise in the matrix Mcal with m/z values as row-names and spectrum names as column-names.

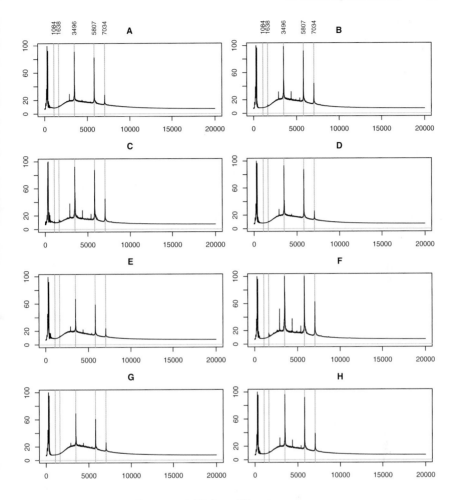

Figure 6.4. Eight calibration spectra.

6.4.2   Normalize spectra

Because of the variations in the amplitude of spectra, we normalize a set
of spectra using a procedure called *total ion normalization*. For each spectrum, we calculate its area under the curve (AUC) for m/z values greater
than the selected cutoff and then scale all spectra to the median AUC.
This step helps to reduce variation due to experimental noise, for example, systematic effects between samples due to varying amounts of applied
protein, degradation over time in the sample or change in the instrument
sensitivity. If intensities at m/z values are taken at equally spaced time
points, the sum of intensities of a spectrum can be seen as the AUC up to a
constant. This method of normalization relies on the assumptions that on
average, the number of proteins that are being over-expressed is approx-

imately equal to the number of proteins being under-expressed, and that the number of proteins whose expression levels change is small relative to the total number of proteins bound to the protein array surface. Ciphergen's internal studies (Fung and Enderwick, 2002) on replicates of the same sample run on a number of spots of the same ProteinChip Array surface type have shown improvements in the coefficient of variation on the order of 15%-20%. This normalization step is similar to those carried out when preprocessing microarrays, and is done for exactly the same reasons.

We normalize the baseline-subtracted spectra for the cutoff point $m/z = 400$ by issuing the following command.

```
> M.r <- renorm(Mcal, cutoff = 400)
```

The matrix M.r will be used later for identifying proto-biomarkers.

6.4.3   Cutoff selection

The observed spectra are not reliable throughout the whole m/z range. First, there are some negative m/z values reported by the instrument. These results from measurements outside the valid range of the quadratic equation relating TOF to mass and should be discarded. Second, we observe that the noise is very large at small m/z values. As mentioned earlier, chemical noise and ion overloading cause the baseline of a spectrum to elevate. They have a larger effect at smaller mass-over-charge (m/z) values (Fung and Enderwick, 2002; Baggerly et al., 2004). We want to choose a cutoff point such that the magnitude of the noise is relatively stable above that point. The cutoff should be chosen large enough to eliminate the initial noisy region but small enough to retain any peaks that correspond to real observable proteins. In order to compare spectra and to carry out the normalization the same cutoff should be used for all spectra.

Estimating a good cutoff is difficult. The existence of technical replicates can facilitate the computation. Once a group of spectra that will be used for common cutoff selection and a cutoff point have been determined, the following algorithm may be used:

1. Baseline-subtracted spectra within the group are normalized to the median of the sums of intensities of spectra.

2. The standard deviation of intensities at each m/z value is calculated.

3. The mean of those standard deviations is computed.

These steps should be carried out for different cutoff points, and a plot of the average standard deviations versus cutoff points should be examined. The following code carries out these calculations, and the resultant plot is shown in Figure 6.5.

```
> cts <- round(10^(seq(2, 4, length = 14)))
> sdsFirst <- sapply(cts, avesd, Ma = Mcal)
```

```
> plot(cts, sdsFirst, xlab = "cutpoint", pch = 21,
+      bg = "red", log = "x", ylab = "average sd")
```

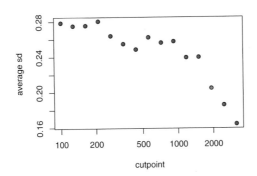

Figure 6.5. Average standard deviations versus cutoff points.

Figure 6.5 shows that the average of standard deviations first increases slightly and then decreases with the first local minimum at around 400. This indicates that 400 may be a suitable cutoff point because it seems to be outside of the noisy portion associated with low m/z values.

6.4.4   Identify peaks

For a single baseline-adjusted spectrum, we have shown in Section 6.3 how peaks can be located by using the function isPeak. For a set of spectra, we can use the function getPeaks, which runs isPeak on a batch of spectra. The following code identifies peaks in the eight individual calibration spectra.

```
> peakfile <- "calipeak.csv"
> getPeaks(M.r, peakfile, ratio = 0.1)
```

6.4.5   Quality assessment

Quality assessment is necessary because sometimes experiments can go wrong. Spectra of poor quality will not help in subsequent analysis in machine learning and may substantially affect estimates. We need to identify and eliminate them.

One type of failure results in a spectrum that has large amount of noise and very few peaks. In PROcess, we use three parameters, *Quality, Retain,* and *Peak* to assess the quality of a set of spectra; this is based on the approach proposed in Mani and Gillette (2004).

The function `quality` computes the three quantities for each spectrum. For a given baseline-adjusted spectrum, the algorithm is:

1. estimate the noise by subtracting from each spectrum its moving average with a window size of 5 points;

2. calculate the noise envelope as 3 times the standard deviation of the noise (as estimated from the previous step) in a 250–point window;

3. calculate the area under each spectrum A_0;

4. calculate the area after subtracting the noise envelope from the spectrum A_1;

5. obtain three statistics for each spectrum:
   (a) *Quality*: A_1/A_0,
   (b) *Retain*: the number of points with height greater than 5 times noise envelope over the total number of points in the spectrum,
   (c) *Peak*: the ratio of the number of peaks in each spectrum detected to the average number of peaks for all spectra in a run.

Quality is a measure of separation of signal from noise, *Retain* is the number of high peaks in a single spectrum, and *Peak* is the number of peaks in a spectrum relative to the average number of peaks of the whole set of spectra being considered. A spectrum is deemed of poor quality and might need to be excluded from subsequent analyses if it meets the following 3 conditions simultaneously: *Quality* < 0.4, *Retain* < 0.1, and *Peak* < 0.5.

For our set of calibration spectra, the following command carries out the quality assessment:

```
> qualRes <- quality(M.r, peakfile, cutoff = 400)
```

	Quality	Retain	peak
060503peptidecalib 1_128.csv	0.414	0.171	0.970
060503peptidecalib 1_16.csv	0.456	0.141	0.970
060503peptidecalib 1_2.csv	0.497	0.118	0.970
060503peptidecalib 1_256.csv	0.410	0.178	0.727
060503peptidecalib 1_32.csv	0.356	0.130	0.970
060503peptidecalib 1_4.csv	0.522	0.143	1.212
060503peptidecalib 1_64.csv	0.479	0.143	1.212
060503peptidecalib 1_8.csv	0.417	0.120	0.970

From the output, we see that none of the spectra in the calibration data set failed our quality control requirements.

6.4.6   Get proto-biomarkers

We now turn our attention to the problem of peak alignment – that is, identifying those peaks, across spectra, that are likely to represent the same protein. The observed variation in peak location is not simply a constant shift but is proportional to the value of m/z. Currently, the accuracy in the m/z position is believed to be within 0.3% of the observed m/z value.

Once the peaks are detected, we generate an interval around each peak that is centered at the m/z value for the peak (if a peak was observed to have m/z of x then the interval would be $[0.997\,x, 1.003\,x]$). We treat those intervals as a partially ordered set and use the locations of the maximal cliques to define the locations of the proto-biomarkers (Gentleman and Vandal, 2001). We call the peaks aligned across spectra *proto-biomarkers* and use the centers of the resulting intervals as the locations of the aligned peaks. For each spectrum, we determine which actual peaks are represented by a proto-biomarker and use the maximum value (within a spectrum) as the height of that proto-biomarker.

The function `pk2bmkr` uses this method to align the peaks and obtain the proto-biomarkers. The function has a parameter `binary` for outputting intensity or peak absence/presence (0/1) along with the m/z values. As discussed in Yasui et al. (2004) the absolute intensity values themselves may not be reliable for establishing proto-biomarkers because of the presence of experimental noise. Reducing the absolute intensity measures into peak absence or presence is an alternative. After the proto-biomarkers are obtained, statistical inference and machine learning algorithms can be applied to them.

Because the calibration data set contains eight homogeneous spectra, we used a criterion that a peak is retained if it is present for at least half of the spectra. The following code reads the results from the peak detection step, `peakfile`, aligns the peaks from the individual spectra and writes the proto-biomarkers into the file `calibmk.csv`.

```
> bmkfile <- "calibmk.csv"
> bmk1 <- pk2bmkr(peakfile, M.r, bmkfile, p.fltr = 0.5)
> mk1 <- round(as.numeric(gsub("M", "", names(bmk1))))
> mk1
```

```
[1] 2906 3498 5812 7036
```

The last three proto-biomarkers, 3498, 5812, and 7036, are within 0.3% of the m/z values of three proteins (3496, 5807, and 7034) of the five known to be present in the calibration samples. Hence they can be considered to be the proteins used for the calibration samples.

A peak at 2906 is detected in some of the samples. It may correspond to the protein with mass 5807 amu with two charges, as its m/z is close to $5807/2 = 2903.5$. To examine whether some peaks are a result of larger proteins with two charges, we can overlay, on each spectrum, the spectrum obtained by multiplying m/z values by 2. The following code does that for the eight calibration spectra. The function `plotCali` was defined above (page 98).

```
> plotCali2 <- function(...) {
+     x <- plotCali(...)
+     lines(x[, 1] * 2, x[, 2] + 25, col = "blue")
+ }

> mapply(plotCali2, files, LETTERS[i], i <= 2)
```

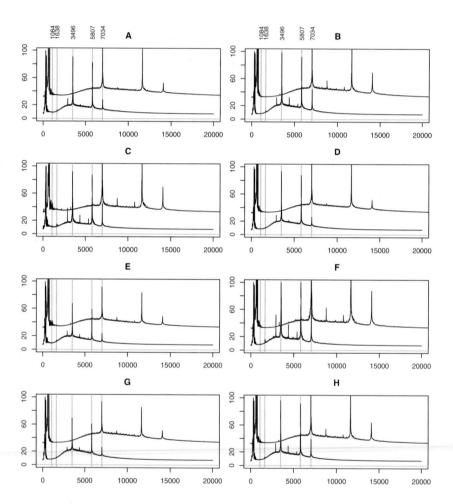

Figure 6.6. Eight calibration spectra, black is the original spectrum and blue is the spectrum with m/z values multiplied by 2, and shifted upwards by 25 to aid the visualization. Note the alignment of some black and blue peaks - indicating the possibility of doubly charged polypeptides.

As seen from Figure 6.6, the black and blue seem to overlap in two places, 5807 of black with 2×2906 of blue and 7034 of black with 2×3496 of blue. Because proteins with masses 3496 amu and 7034 amu are known to be present in the sample, we do not investigate further. Because no protein with mass around 2906 amu was supposed to be present in the calibration sample and because twice this m/z value is within 0.3% of 5807, the mass of a protein that was supposed to be present, the proto-biomarker 2906 may be the 5807 amu protein but with two charges. We tried other sets of parameter values but failed to detect peaks at $m/z = 1084$ and 1638 without finding other spurious peaks.

6.5   An example

One hundred sixty–seven samples were collected from 155 subjects in CPT tubes with plasma isolated and stored at -80°C until needed. Among the 167 samples, 55 were HER2 positive (A), 64 were normal healthy women (B), 35 were mostly ER/PR positive (C), and 13 samples were from a single healthy woman (D). Samples labeled D are the only ones from a single subject; all the other samples represent different individuals. Samples were thawed and aliquoted into 100 μl vials. The samples were *fractionated* so that proteins were separated into subsets of proteins based on biophysical characteristics. Fractions 4 and 5 (f45) were processed by the Ciphergen IMAC protocol with EAM of CHCA. This data set is stored in the ProData package also available from *Bioconductor*. The spectrum ID, the phenotype group, and the proto-biomarkers preprocessed by Ciphergen's software are stored in the *exprSet*: f45bmk of the same package.

We process these data using the tools and procedures described previously. We start with baseline subtraction.

```
> library("ProData")
> f45c <- system.file("f45c", package = "ProData")
> fs <- dir(f45c, full.names = TRUE)

> M1 <- rmBaseline(f45c)
```

Because the D samples are all from the same person, and hence represent biological replicates, they are ideally suited for use in selecting the cutoff as described in Section 6.4.3. We use the functions `regexp` and `match` to match the names of the spectra in `colnames(M1)` to the names in the sample annotation table `pData(f45cbmk)`.

```
> data(f45cbmk)
> SpecGrp <- pData(f45cbmk)
> fns <- colnames(M1)
> gi <- regexpr("i+[0-9]+", fns)
> specName <- substr(fns, gi, gi + attr(gi, "match.length") -
+     1)
> mt <- match(SpecGrp[, 2], toupper(specName))
> M2 <- M1[, mt]
> colnames(M2) <- SpecGrp[, 2]
```

Next we calculate the average standard deviations versus cutoff points plot, as explained in Section 6.4.3.

```
> sdsSecond <- sapply(cts, avesd, Ma = M2[, SpecGrp[,
+     1] == "D"])

> plot(cts, sdsSecond, xlab = "cutpoint", pch = 21,
+     bg = "red", log = "x", ylab = "average sd")
```

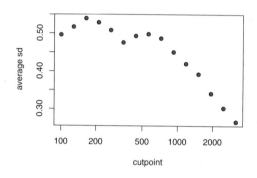

Figure 6.7. Average standard deviations versus cutoff points, calculated on the samples from group D.

From Figure 6.7, a m/z value of 1000 seems to be a reasonable choice for cutoff. We normalize the spectra after dropping all m/z values below 1000.

```
> nM <- renorm(M2, cutoff = 1000)
```

After normalization, we proceed to peak detection and quality control.

```
> peakfile <- "f45cpeak.csv"
> getPeaks(nM, peakfile, ratio = 0.1)
> qu <- quality(nM, peakfile, cutoff = 1000)

> bad <- qu[, 1] < 0.4 & qu[, 2] < 0.1 & qu[, 3] <
+     1/2
> sum(bad)

[1] 0
```

No failed spectra were detected. Before aligning the peaks, we exclude the control group, D from the following steps and proceed with 55 samples in group A, 64 in B, and 35 in C, a total of 154 spectra.

PROcess provides a visual tool gelmap for a quick discernment of patterns that may be present within different groups. Because the spectrum data are very large, we need to run a binning function to reduce the data before running `gelmap`. To best show the patterns, we restrict the m/z values to be under 10,000 and show the intensity on the log scale. Notice that a gelmap is very similar to a heatmap and represents a false color display with samples on one axis, features on the other, and the color of a cell determined by the experimental data. In this setting, our users preferred grayscale, but any set of colors could be used, and either rows or columns could be rearranged to identify interesting subsets.

```
> Ma <- binning(nM1, breaks = 300)
> colnames(Ma) <- SpecGrp[!drop, 1]
> par(xpd = TRUE)
> marks <- c(2666, 5055, 7560, 7934)
> sel <- as.numeric(rownames(Ma)) < 10000
> gelmap(log(Ma + 1)[sel, ], at.mz = marks, at.col = c(25,
+       90, 135), col = gray(10:0/10))
> segments(x0 = 1013 * c(1, 1), y0 = c(55, 119),
+       x1 = 10000 * c(1, 1), y1 = c(55, 119), col = "red")
> arrows(marks, c(-5, -5), marks, c(1, 1), length = 0.08,
+       angle = 20, col = "red")
```

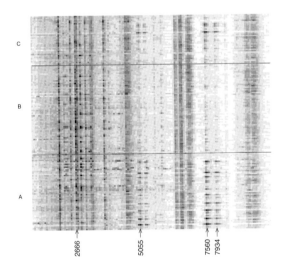

Figure 6.8. A gel-like image representing the spectrum data from 154 samples (rows) at 300 m/z values (columns).

Figure 6.8 shows that the spectrum intensities at m/z values 5055, 7560, and 7934 are higher in the cancer groups A and C than in the normal group

B. Furthermore, the spectrum intensities at m/z =2666 are higher in the normal group B than in the cancer groups A and C.

Next we align the peaks from individual spectra. In this step, we wish to drop those proto-biomarkers that appear in very few spectra. Because our smallest group is group C with 35 spectra, we shall set the filtering parameter p.fltr to be no greater than $35/154 = 0.21$. The following code gives us a list of aligned proto-biomarkers.

```
> peakfile1 <- "f45cpeak1.csv"
> getPeaks(nM1, peakfile1, ratio = 0.1)
> bmkfile <- "f45cbmk.csv"
> bmk <- pk2bmkr(peakfile1, nM1, bmkfile, p.fltr = 0.1,
+     eps = 0.003)

> mks <- round(as.numeric(gsub("M", "", names(bmk))))
> length(mks)

[1] 65

> mks[1:12]

 [1] 2024 2231 2360 2488 2556 2666 2669 2670 2671 2672 2735
[12] 2736
```

We have detected 65 proto-biomarkers and printed the first 12 of them. We can detect proto-biomarkers that are nearly exact multiples of each other, and hence potentially represent the same protein, using the function is.multiple. For example, we can find which proteins might have two to five charges by executing the following code.

```
> mults <- is.multiple(mks, k = 2:5)
> mults[[1]]

 2H+
2360
```

We have detected 13 proto-biomarkers that may have corresponding peaks with smaller m/z values that may represent the same protein. Taking the first component of the returned list as an example, the peak at 2360 may correspond to the same protein that caused a peak at around 4720, but some copies of it picked up two charges.

6.6 Conclusion

Measurements of protein abundance are likely to play an important role in the detection and diagnosis of disease. Although these tools are not currently as well developed as DNA microarrays, they nonetheless hold great promise. In this chapter, we have produced a flexible set of tools and a paradigm for analyzing SELDI-TOF data. We emphasize the need for, and the use of, good methods for quality control – the conclusions that can be drawn depend heavily on the quality of the data that have been collected. We have made our software freely available, and users can easily modify different aspects of our approach.

Part II

Meta-data: biological annotation and visualization

7

Meta-data Resources and Tools in Bioconductor

R. Gentleman, V. J. Carey, and J. Zhang

Abstract

Closing the gap between knowledge of sequence and knowledge of function requires aggressive, integrative use of biological research databases of many different types. For greatest effectiveness, analysis processes and interpretation of analytic results must be guided using relevant knowledge about the systems under investigation. However, this knowledge is often widely scattered and encoded in a variety of formats. In this section, we consider some of the different sources of biological information as well as the software tools that can be used to access these data and to integrate them into an analysis. Bioconductor provides tools for creating, distributing, and accessing annotation resources in ways that have been found effective in work-flows for statistical analysis of microarray and other high-throughput assays.

7.1 Introduction

In this monograph we will use the terms *annotation* and *meta-data* interchangeably. Both terms refer to the large body of information that is available, primarily through databases, on different aspects of the biological systems under study. This includes sequence information, catalogs of gene names and symbols, structural information, and virtually any relevant publication. We use the term *meta-data*, which means data about data, as well as the term *annotation*, because in many of the analyses existing meta-data are used to annotate analytic resources or results. Among the many challenges faced when using meta-data is to develop tools that adequately track the data as it evolves. The tools used to model these data should be

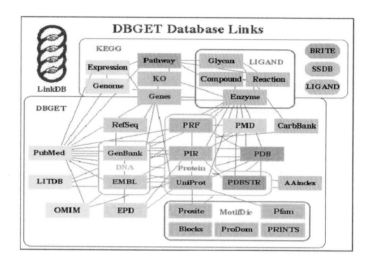

Figure 7.1. Kyoto Encyclopedia of Genes and Genomes DBGET schematic.

modular and should be capable of adapting to rapid changes in meta-data inputs.

We note the importance of version tracking. All persistent, curated meta-data resources should have version numbers that are incremented when new builds are made or when new meta-data are added to the set of curated data. Whenever version information for on-line Web services is made available, it should be propagated through to the consumer.

The relationship between biological annotation and a data-analytic workflow is inherently complex. Annotation may be used to perform dimension-reduction at an early stage, to introduce constraints on relationships between statistical model parameters in the model-building stage, or to interpret discovered patterns at the conclusion of data analysis; to name but a few of the potential uses.

Figure 7.1 (Kanehisa, 1997; Kanehisa et al., 2004) is a useful schematic of biological and biochemical research archives that may be queried interactively. Major classes of data resources cover genes and gene products (DNA, Protein, MotifDic components), pathways and gene clusters (KEGG component), biochemical pathway elements (Ligand component), and scientific literature (PubMed, LITDB, OMIM).

Complementary to the data resources catalogued in Figure 7.1 are assay-oriented annotation resources. These resources provide the links between assay probe identifiers and associated sequence catalog entries. Comprehensive assay-oriented annotation can be found at the Gene Expression Omnibus (GEO) and at TIGR's Resourcerer. Assay-oriented annotation is also typically provided by assay manufacturers.

Bioconductor provides access to annotation through two complementary mechanisms. One is based on curated, downloadable modules and the other is via direct real-time queries to Web services. The choice between real-time query resolution and use of curated packages introduces a dilemma between currency and reproducibility. On-line resolution will often lead to different results for a given query issued at different times. The alternative is to have locally curated and assembled data resources and to resolve all queries with respect to those data. In this case, some obsolete information may be used. However, two users resolving identifiers against the same version of a curated meta-data resource will get the same answers.

Much of what needs to be done to link different resources is essentially database technology. Our efforts in assembling annotation data for microarray experiments are represented by the AnnBuilder technology (Zhang et al., 2003). This software is used to construct all the meta-data packages distributed by Bioconductor.

The remainder of the section is as follows: this introductory chapter characterizes meta-data resources covered by Bioconductor, gives details of coverage (distinguishing curated data packages from interactive Web services), and provides examples of working with Gene Ontology, pathway databases, organism annotation, platform annotation, and platform archives. Subsequent chapters address concepts and tools for using R to query external annotation resources, tools for annotating analysis results, and visualization methods.

7.2   External annotation resources

An annual catalog of public databases relevant to molecular biology can be found in the first issue of each volume of the journal *Nucleic Acids Research*. The on-line category list covers nucleotide, RNA, and protein sequence, "structure" databases (small molecules, carbohydrates, nucleic acid and protein structures), vertebrate and non-vertebrate genome databases, pathways, human genes and diseases, microarray data, and proteomics resources. Brief comments on some of the resources that are commonly encountered in Bioconductor annotation follow. In certain cases, descriptive information is taken verbatim from web site content.

LocusLink is a catalog of genetic loci that connects curated sequence information to official nomenclature such as, sequence accessions, EC numbers (see below), UniGene clusters, homology, and map locations; to name but a few. During the writing of this monograph LocusLink was replaced by *EntrezGene* and we must apologize to the reader for the inconsistent and somewhat interchangable use of these two terms throughout the monograph.

UniGene defines sequence clusters. Large-scale sequencing of transcribed sequences has provided a large set of expressed sequence tags (ESTs). There are many more ESTs than there are genes and some method of grouping them is needed. UniGene focuses on protein-coding genes of the nuclear genome (excluding rRNA and mitochondrial sequences). Ideally there is a one-to-one mapping between UniGene clusters and gene loci.

RefSeq is a non-redundant data set of transcripts and proteins of known genes for a variety of species, including human, mouse, and rat.

Enzyme Commission (EC) numbers are assigned to different enzymes and linked to genes through their association with LocusLink identifiers.

Gene Ontology (GO) is a structured vocabulary of terms describing gene products according to relevant molecular function, biological process, or cellular component.

PubMed is a service of the National Library of Medicine. PubMed provides a very rich resource of data and tools for working with papers published in journals that are related to medicine and health. Genes are linked to published papers in different ways; perhaps the most commonly used linkage is via LocusLink identifiers. The data source, while large, is not comprehensive and not all papers have been abstracted.

The Protein Research Foundation curates *LITDB*, which covers all articles dealing with peptides from scientific journals accessible in Japan.

OMIM provides a keyword-driven search interface to narratives about inherited conditions that are heavily hyperlinked to relevant scientific literature.

Coordinated approaches to establishing public archives of descriptive information about microarray platforms have been undertaken by NCBI with *Gene Expression Omnibus (GEO)*, by TIGR with the *Resourcerer* database, and by EBI with *ArrayExpress*.

The NetAffx™ Analysis Center provides tools that correlate experimental data assayed using the AffymetrixGeneChip technology.

7.3   Bioconductor annotation concepts: curated persistent packages and Web services

The MetaData node of the Bioconductor portal provides an extensive catalog of R packages that encode annotation resources. These are distributed for installation and immediate use in R. In this section, we will illustrate interfaces to several of these annotation packages.

Each of the different biological data resources uses its own set of identifiers for distinguishing between the different entities that it documents, e.g., LocusLink, UniGene, EC, and so on. One of the roles of an annotation service is to provide links between these different sets of identifiers and to make it possible to use the different resources without having to understand their internal structure and labeling conventions.

For microarray-based assays we have taken the approach that for each distinct chip we will produce a collection of hash tables that provide access from the chip specific probe labels to the different identifiers associated with a reasonably large set of biological data. A hash table is simply a mapping from a set of keys, or known identifiers, to a set of values. Most of the Bioconductor meta-data packages use hash tables to provide mappings from one set of identifiers to the associated values. In R, hash tables are available through the environment class of objects. The annotate package contains much of the software infrastructure that is needed to simplify the user interface. The DPExplorer widget from tkWidgets provides a GUI for examining the contents of the different data packages.

7.3.1   Annotating a platform: HG-U95Av2

For our example we make use of one of the most widely used Affymetrix platforms, the HG-U95Av2 GeneChip. The annotation files provide mappings from the manufacturers identifiers to different targets, such as LocusLink identifier or chromosomal location. To access the data you must first load the package. We then list the contents of this package to see all the different targets for which mappings have been provided. It is important to note that the associations and data are likely to undergo constant change and that you will want to update the meta-data packages on a regular basis.

```
> library("hgu95av2")
> ls("package:hgu95av2")

 [1] "hgu95av2"            "hgu95av2ACCNUM"
 [3] "hgu95av2CHR"         "hgu95av2CHRLENGTHS"
 [5] "hgu95av2CHRLOC"      "hgu95av2ENZYME"
 [7] "hgu95av2ENZYME2PROBE" "hgu95av2GENENAME"
 [9] "hgu95av2GO"          "hgu95av2GO2ALLPROBES"
[11] "hgu95av2GO2PROBE"    "hgu95av2GRIF"
[13] "hgu95av2LOCUSID"     "hgu95av2MAP"
[15] "hgu95av2MAPCOUNTS"   "hgu95av2OMIM"
[17] "hgu95av2ORGANISM"    "hgu95av2PATH"
[19] "hgu95av2PATH2PROBE"  "hgu95av2PMID"
[21] "hgu95av2PMID2PROBE"  "hgu95av2QC"
[23] "hgu95av2REFSEQ"      "hgu95av2SUMFUNC"
[25] "hgu95av2SYMBOL"      "hgu95av2UNIGENE"
```

We can see in the output above that there are mappings to LocusLink, to KEGG, OMIM, gene symbols, chromosome, and chromosomal location. You can find out how many probes are mapped for each target by invoking the function hgu95av2; all meta-data packages have a function, with the same name as the package, that prints out some quality control information.

For some meta-data we also provide reverse mappings, for example from GO identifiers to all Affymetrixprobes (hgu95av2GO2PROBE). Other information is also provided such as the name of the organism (hgu95av2ORGANISM) and the lengths of the chromosomes (hgu95av2CHRLENGTHS).

7.3.2 An Example

Each of the chip-specific data packages provided by Bioconductor contains a hash table that maps from the probes on the array to KEGG pathways. Its suffix will be PATH, and so, for the HG-U95Av2 chip, the KEGG data are stored in hgu95av2PATH. The mapping from pathway IDs to all probes that are associated with the pathway are provided in the table with the suffix PATH2PROBE (i.e., hgu95av2PATH2PROBE).

In this example, we want to find the pathways that the gene BAD is involved in. The probeset associated with BAD is 1861_at. Now, equipped with this information we can find the pathways that BAD is associated with. Once we know these pathways, we can then find all the members of each pathway that are represented on the HG-U95Av2 GeneChip; but remember that in some cases a gene is represented by multiple different probesets.

```
> BADpath <- hgu95av2PATH$"1861_at"
> mget(BADpath, KEGGPATHID2NAME)

$"01510"
[1] "Neurodegenerative Disorders"

$"04210"
[1] "Apoptosis"

$"05030"
[1] "Amyotrophic lateral sclerosis (ALS)"

> allProbes <- mget(BADpath, hgu95av2PATH2PROBE)
> sapply(allProbes, length)

01510 04210 05030
   59   154    28
```

These can then be used to reduce an analysis to only those probes that are involved in a particular pathway, or to separate an analysis by pathway, perhaps by adding a term to a model. Once the pathway identifier has been obtained, it can be used together with the cMAP package and the graph package to carry out a variety of interesting analyses. See Section 7.6.2 for more details.

7.3.3 Annotating a genome

In the previous section, we consider assay- or platform-specific annotation packages. These are ideal for working with data from a single experiment. However they do have some limitations. Each contains only meta-data links for the probes that were assayed. In some situations, you will want to have access to meta-data information for all genes within a species. There are a few such packages available through Bioconductor, such as **YEAST** and **humanLLMappings**. However, these packages tend to be very large (and there is much redundancy).

In the next code chunk, we show the contents of the **YEAST** package. It is very similar to the assay specific packages in terms of the contents, except the primary keys are the systematic names for the yeast genes; such as YBL088C

```
> library("YEAST")
> ls("package:YEAST")

 [1] "YEAST"            "YEASTALIAS"
 [3] "YEASTCHR"         "YEASTCHRLENGTHS"
 [5] "YEASTCHRLOC"      "YEASTDESCRIPTION"
 [7] "YEASTENZYME"      "YEASTENZYME2PROBE"
 [9] "YEASTGENENAME"    "YEASTGO"
[11] "YEASTGO2ALLPROBES" "YEASTGO2PROBE"
[13] "YEASTMAPCOUNTS"   "YEASTORGANISM"
[15] "YEASTPATH"        "YEASTPATH2PROBE"
[17] "YEASTPMID"        "YEASTPMID2PROBE"
[19] "YEASTQC"
```

7.4 The annotate package

The **annotate** package is one of the foundational packages in Bioconductor. Functions provided in **annotate** cover various annotation processes including harvesting of curated persistent packages and simple HTTP queries to web service providers. Package **annotate** contains a certain amount of interface code that provides common calling sequences for the assay based meta-data packages provided by Bioconductor.

A variety of accessor functions have names beginning with `get`. These should be used when writing software as they are intended to define our API. For casual use, you can use any accessing methods you prefer.

The functions `getGI` and `getSEQ` perform Web queries to NCBI to extract the GI or nucleotide sequence corresponding to a GenBank accession number. GI numbers are a series of digits that the NCBI assigns to each sequence that it processes.

```
> ggi <- getGI("M22490")
> ggi
```

```
[1] "179503"
```

```
> gsq <- getSEQ("M22490")
> substring(gsq, 1, 40)
```

```
[1] "GGCAGAGGAGGAGGGAGGGAGGGAAGGAGCGCGGAGCCCG"
```

The functions, getGO, getSYMBOL, getPMID, and getLL simplify the decoding
of manufacturer identifiers:

```
> getSYMBOL("1000_at", "hgu95av2")
```

```
[1] "MAPK3"
```

```
> getGO("1000_at", "hgu95av2")[[1]][[1]]
```

```
[1] "GO:0005524"
```

The functions whose names start with pm work with lists of PubMed
identifiers for journal articles.

```
> hoxa9 <- "37809_at"
> absts <- pm.getabst(hoxa9, "hgu95av2")
> substring(abstText(absts[[1]][[1]]), 1, 60)
```

```
[1] "Members of the homeobox family of transcription factors are "
```

7.5   Software tools for working with Gene Ontology (GO)

An *ontology* is a structured vocabulary that characterizes some conceptual domain. The Gene Ontology (GO) Consortium defines three ontologies
characterizing aspects of knowledge about genes and gene products. These
ontologies are molecular function (MF), biological process (BP), and cellular component (CC). For explicit descriptions of these categories, you
should consult the GO Web page.

The *molecular function* of a gene product is what it does at the biochemical level. This describes what the gene product can do, but without
reference to where or when this activity actually occurs. Examples of
functional terms include "enzyme," "transporter," or "ligand."

A *biological process* is a biological objective to which the gene product
contributes. There is often a temporal aspect to a biological process. Biological processes usually involve the transformation of a physical thing. In
some sense they are larger than a molecular function, and several molecular
activities will be coordinated to carry out a biological process. The terms
"DNA replication" or "signal transduction" describe general biological processes. A *pathway* is not the same as a biological process in that pathways
include dependencies and dynamics that are not relevant to the notion of

a biological process. It is not always easy to distinguish between a molecular function and a biological process. The GO consortium suggests that a process must have more than one distinct step.

A *cellular component* is a part of a cell that is a component of some larger object or structure. Examples of cellular components include "chromosome," "nucleus," and "ribosome."

The GO ontologies are structured as directed acyclic graphs (DAGs) that represent a network in which each term may be a *child* of one or more *parents*. We use the expressions *GO node* and *GO term* interchangeably. Child terms are more specific than their parents. The term "transmembrane receptor protein-tyrosine kinase" is child of both "transmembrane receptor" and "protein tyrosine kinase." For each of the three different ontologies, there is a root node that has the ontology name associated with it, and that node is the most general node for terms in the respective ontology.

The relationship between a child and a parent can be either a *is a* relation or a *has a* (*part of*) relation. For example, "mitotic chromosome" is a child of "chromosome" and the relationship is an *is a* relation. On the other hand, a "telomere" is a child of "chromosome" with the *has a* relation. Child terms may have more than one parent term and may have a different relationship with different parents.

	Number of terms
BP	8578
CC	1335
MF	6891

Table 7.1. Number of GO terms per ontology.

Table 7.1 presents the number of terms associated with each of the three ontologies. For example, we see that for version 1.7.0 of the GO package there are 8578 terms in the biological process (BP) ontology.

Each term in the ontology is associated with a unique identifier. For example, the term "transcription factor complex" is in the molecular function ontology and has the GO label GO:0005667. The terms, and their parent–child relationships are provided by GO; but GO does not provide data on the mapping of gene products to terms. That process is carried out by the Gene Ontology Annotation project (GOA).

7.5.1 Basics of working with the GO package

Bioconductor's GO package encodes GO and simplifies navigation and interrogation of the structured vocabulary. The GO package also includes information on the mapping between GO terms and LocusLink identifiers.

For precision and conciseness, all indexing of GO resources employs the 7 digit tags with prefix GO:. Three very basic tasks that are commonly performed in conjunction with GO are

- navigating the hierarchy, determining parents and children of selected terms, and deriving subgraphs of the overall DAG constituting GO;

- resolving the mapping from GO tag to natural language characterizations of function, location, or process;

- resolving the mapping between GO tags or terms and elements of catalogs of genes or gene products.

To support these basic tasks, Bioconductor provides a number of different sets of mappings in the GO package (which provides encodings of the hierarchical structure of tags/terms and decodings of tags into natural language strings) and in assay-specific data packages such as the hgu95av2 data package (which provides mappings from Affymetrix probe identifiers to GO tags).

7.5.2   Navigating the hierarchy

Finding parents and children of different terms is handled by using the PARENT and CHILDREN mappings. These are available as R environments with names constructed using GO as a prefix, one of MF, BP, or CC to select subontology, and one of either PARENT or CHILDREN as a suffix. To find the children of "GO:0008094" we use:

```
> get("GO:0008094", GOMFCHILDREN)
```

```
[1] "GO:0004003" "GO:0008722" "GO:0015616" "GO:0043142"
```

We use the term *offspring* to refer to all descendants (children, grandchildren, and so on) of a node. Similarly we use the term *ancestor* to refer to the parents, grandparents, and so on, of a node. The sets of terms (but not their relationships) can be obtained from the OFFSPRING and ANCESTOR environments.

```
> get("GO:0008094", GOMFOFFSPRING)
```

```
[1] "GO:0004003" "GO:0008722" "GO:0015616" "GO:0043142"
[5] "GO:0017116" "GO:0043140" "GO:0043141"
```

7.5.3   Searching for terms

All GO terms are provided in the GOTERM environment. It is relatively easy to search for a term with the word chromosome in it using eapply and grep or agrep.

```
> hasChr <- eapply(GOTERM, function(x) x[grep("chromosome",
+       Term(x))])
> lens <- sapply(hasChr, length)
> hasChr <- hasChr[lens > 0]
> length(hasChr)
```

```
[1] 64
```

We see that there are 64 terms that have the word chromosome in them. To generalize this, we can write a function:

```
> GOTerm2Tag <- function(term) {
+       GTL <- eapply(GOTERM, function(x) {
+           grep(term, x@Term, value = TRUE)
+       })
+       Gl <- sapply(GTL, length)
+       names(GTL[Gl > 0])
+ }
```

Now we can apply this to find the term(s) in which some other phrase, for example, "transcription factor binding" is used:

```
> hasTFA <- GOTerm2Tag("transcription factor binding")
> hasTFA
```

```
[1] "GO:0003719" "GO:0008134"
```

The function GOTerm2Tag can be extended in several ways. More of the arguments to grep can be exposed. Users could then use quite general regular expressions (Friedl, 2002) allowing users to ignore capitalization, ignore multiple endings among many other things. The function agrep may be used in place of grep to allow for alternate spellings and approximate matches.

7.5.4 Annotation of GO terms to LocusLink sequences: evidence codes

The mapping of genes to GO terms is carried out separately by GOA (Camon et al., 2004). Both GO and GOA provide regular updates that account for both changes in GO and new mappings of genes to terms. The data from GOA consist of mappings between GO terms and LocusLink identifiers. But rather than map a LocusLink identifier to all terms that apply, only the most specific terms are used. The mapping to all less specific GO terms is implied.

In the GO package mappings from GO terms to specific genes are provided in the environment named GOLOCUSID. For each term, the set of LocusLink identifiers mapped to that term corresponds to the set of genes for which that term is a most specific mapping. To get all LocusLink IDs associated with a specific GO term use, GOALLOCUSID.

Term	Definition
IMP	inferred from mutant phenotype
IGI	inferred from genetic interaction
IPI	inferred from physical interaction
ISS	inferred from sequence similarity
IDA	inferred from direct assay
IEP	inferred from expression pattern
IEA	inferred from electronic annotation
TAS	traceable author statement
NAS	non-traceable author statement
ND	no biological data available
IC	inferred by curator

Table 7.2. GO evidence codes.

Four environments in the GO package address the association between LocusLink sequence entries and GO terms: GOLOCUSID, GOALLOCUSID, GOLO-CUSID2GO, and GOLOCUSID2ALLGO. The associations are labeled with the GO *evidence codes* characterizing the association. Definitions of evidence codes in use are briefly indicated in Table 7.2; more details are available on the GO Web site. We particularly want to draw your attention to the evidence code IEA, which stands for inferred from electronic annotation. This annotation is used when no curator has checked the annotation.

In some analyses, it will be important to make use of these annotation codes. For example, if you are analyzing data and want to study relationships between sequence similarity and molecular function, then you should remove all annotations made on the basis of ISS to avoid circularity in your arguments.

In the next code chunk, we find the GO identifier for "transcription factor binding" and use that to get all LocusLink identifiers that have that annotation. Then we look at a table of the different evidence codes that were used.

```
> gg1 <- get(GOTerm2Tag("^transcription factor binding$"),
+     GOLOCUSID)
> table(names(gg1))

IDA IMP IPI ISS NAS TAS
  9   1   8  17   4  28
```

In this next example, we consider the gene with LocusLink ID 7355, which is SLC35A2.

```
> lll <- GOLOCUSID2GO[["7355"]]
> length(lll)

[1] 9

> sapply(lll, function(x) x$Ontology)
```

```
GO:0000139 GO:0015785 GO:0005459 GO:0008643 GO:0006012
     "CC"       "BP"       "MF"       "BP"       "BP"
GO:0016021 GO:0015780 GO:0005338 GO:0005351
     "CC"       "BP"       "MF"       "MF"
```

We see that there are 9 different GO terms. We can get only those mappings for the BP ontology by using `getOntology`. We can get the evidence codes using `getEvidence`, and we can drop those codes we do not wish to use by using `dropEcode`. In the next code chunk, we first find all mappings and then remove those for which the evidence code is "IEA".

```
> getOntology(lll, "BP")

[1] "GO:0015785" "GO:0008643" "GO:0006012" "GO:0015780"

> getEvidence(lll)

GO:0000139 GO:0015785 GO:0005459 GO:0008643 GO:0006012
    "IEA"      "TAS"      "TAS"      "IEA"      "TAS"
GO:0016021 GO:0015780 GO:0005338 GO:0005351
    "IEA"      "IEA"      "IEA"      "IEA"

> zz <- dropECode(lll, code = "IEA")
> getEvidence(zz)

GO:0015785 GO:0005459 GO:0006012
    "TAS"      "TAS"      "TAS"
```

7.5.5   The GO graph associated with a term

We have shown how to determine the tag associated with a term. Starting with "transcription factor activity," we can construct the graph based on the GO relationships within the molecular function hierarchy. This is done by finding the parents (less specific terms), and then recursively finding their parents until the root node is reached. This graph is called the *induced graph* or the induced GO graph and an example is shown in Figure 19.2. More details on manipulating graphs are given in Part IV of this monograph.

7.6   Pathway annotation packages: KEGG and cMAP

A biological pathway can in some instances be modeled as a directed graph with labeled nodes and edges. In this section, we consider pathway data supplied by KEGG and cMAP. Here we concentrate on the data and its structure and refer the reader to Part IV of this monograph for more details on the graph manipulation tools that are available. Some interesting perspectives on modeling data of this form are presented in Bower and Bolouri (2001); Krishnamurthy et al. (2003); Sirava et al. (2002).

7.6.1 KEGG

The Kyoto Encyclopedia of Genes and Genomes (KEGG) provides a data resource that is primarily concentrated on pathways. KEGG associates each pathway with a number, and there are currently 236 available in the **KEGG** package. For each pathway there is a name and a KEGG identifier for that pathway (the relationships are stored in KEGGPATHID2NAME). Genes are associated with a pathway in a species specific manner and the resulting data are stored in KEGGPATHID2EXTID.

Data available in the **KEGG** package includes:

KEGGEXTID2PATHID which provides mappings from either LocusLink (for human, mouse, and rat) or Open Reading Frame (yeast) to KEGG pathways, a second environment, KEGGPATHID2EXTID contains the mappings in the other direction.

KEGGPATHID2NAME which provides mappings from the KEGG path ID to a name (textual description of the pathway). Only the numeric part of the KEGG pathway identifiers is used (and not the three letter species codes).

	ath	dme	hsa	mmu	rno	sce
Counts	108	108	135	127	118	101

Table 7.3. Pathway counts per species

The species and pathway counts for the current offerings in **KEGG** are given in Table 7.3. There are many more species specific pathways available from KEGG and users that would like to make use of them should do so through the SOAP interface discussed in Chapter 8.

If we consider one pathway, say 00140, then we can find its name and examine the different sets of gene identifiers for each species (in this case we will only examine human and yeast). Species specific mappings, from a pathway to the genes it contains, are indicated by gluing together a three letter species code, such as hsa for homo sapiens, to the numeric pathway code.

```
> KEGGPATHID2NAME$"00140"

[1] "C21-Steroid hormone metabolism"

> KEGGPATHID2EXTID$hsa00140

 [1] "1109" "1583" "1584" "1585" "1586" "1589" "3283" "3284"
 [9] "3290" "3291" "6718"

> KEGGPATHID2EXTID$sce00140

[1] "YGL001C"
```

And it is relatively easy to determine which pathways your favorite gene happens to be involved in. We look up PAK1, which has LocusLink ID 5058 in humans, and find that it is involved in three pathways, hsa04010, hsa04510 and hsa04810. For mice, the LocusLink ID for PAK1 is 18479.

```
> KEGGEXTID2PATHID$"5058"

[1] "hsa04010" "hsa04510" "hsa04810"

> KEGGEXTID2PATHID$"18479"

[1] "mmu04010" "mmu04810"
```

One of the case studies in Chapter 22 includes examples where gene expression data are related to KEGG pathways. The association of other data, such as gene expression data, with pathway membership clearly has some potential. The GenMAPP project (Doniger et al., 2003) provides some tools for linking gene expression data with pathway data.

7.6.2   cMAP

The cancer Molecular Analysis Project (cMAP) is a project that provides software and data for the comprehensive exploration of data relevant to cancer. cMAP provides pathway data in a format that is amenable to computational manipulation.

cMAP uses a graphical model of pathways in which molecular species and processes define graph nodes and associations of molecules with processes determine graph edges. A minimal pathway, which they term an interaction, consists of one process node and its adjacent molecule nodes. The model is basically a hierarchical one: interactions are sets of nodes (molecules and processes) and edges defining types of relationships between molecules and processes; pathways are sets of interactions.

cMAP has pathway data from both Biocarta and from KEGG, reorganized to allow for relatively easy access. In the cMAP package the data are stored in environments with the suffix PATHWAY; the environment cMAPKEGG-PATHWAY contains all KEGG pathways and cMAPCARTAPATHWAY contains those from BioCarta.

In the cMAP package each element of a PATHWAY is a list with five components. The components are source, process, reversible, condition, and component. The source element is a character string indicating whether the interactions between molecules are from a BioCarta or KEGG pathway. The process element is a character string describing the process the key molecule is involved. Potential values include *reaction, modification, transcription, translocation, macroprocess,* or a more specific subtype of macroprocess including any term from the GO biological process ontology. The reversible element is a Boolean value indicating whether the interaction is reversible. The condition element is a character string indicating the biological process the interactions take place. Potential values include any term from the GO biological process ontology.

Perhaps the most important piece of information comes from the `compo-nent` attribute. This attribute contains a set of keys for the `INTERACTION` database (for KEGG it would be `cMAPKEGGINTERACTION`). Each element of the interaction environments has the following components:

id the molecule ID of the interacting molecule, this can be used to extract further information from the cMAP molecule environments (for KEGG it would be `cMAPKEGGMOLECULE`),

edge indicating the way two molecules interact; possible values are `input`, `agent`, `inhibitor`, and `output`,

role the function of the key molecule; potential values include any term from the GO molecular function ontology,

location a GO cellular component ontology indicating the location of the interaction,

activity an abstract term that can be one of *inactive, active, active1, active2.*

We briefly explore some of the data that is available in the INTER-ACTION databases. In version 1.7.0 of this package all of the KEGG interactions are labeled as `reaction`, whereas the BioCarta interactions have a much more detailed set of descriptions.

```
> keggproc <- eapply(cMAPKEGGINTERACTION, function(x) x$process)
> table(unlist(keggproc))

character(0)

> cartaproc <- eapply(cMAPCARTAINTERACTION, function(x) x$process)
> length(table(unlist(cartaproc)))

[1] 0
```

In the cMAP package, molecules have different attributes; they are described briefly below, and elaborated further in the manual pages of the cMAP package.

type The type of molecule, this can be one of "protein," "complex," "compound," or "rna".

extid A set of external identifiers, so the molecule can be mapped to other annotation resources.

component A set of molecule identifiers, and other data, for the constituents of a "complex" molecule. Only "complex" molecules should have this attribute.

member A set of molecule identifiers for molecules that belong to the same protein family as the key molecule.

7.6.3 A Case Study

In many cases, working with the cMAP data will yield a set of interactions. These can be obtained by searching for specific pathways, for specific molecules, or any of a number of different actions. These interactions must then be combined into a single, possibly unconnected, graph. Interactions are connected on the basis of molecular identity (there should be a single node for each basic molecule with the same post-translational modifications located in the same cellular location).

You may want to label the different components of that graph in different ways. For example, if you use the software on the cMAP site they suggest that molecule labels can specify the location and any post-translational modifications while process labels can specify the generic nature of the process. cMAP makes use of GO vocabularies for some of the labeling. You can use edge labels to specify the general nature of each molecule's role in the process (i.e., input, output, agent, inhibitor). We note that both the graph package and the Rgraphviz package can facilitate these kinds of labeling (see Part IV of this monograph).

There are 85 pathways in the KEGG collection. Each pathway is identified by its KEGG ID (the three-letter species code followed by the five digit pathway code) and the pathway information is stored as a list. We can find the names of the KEGG pathways, and easily see which species are represented. We first find the names, then extract the first three characters, which should be the species code, and summarize these in a table. We can also obtain complete information on a pathway, by extracting it, and in the code chunk below we show how to obtain pathway information and use the pathway labeled hsa00020 as an example.

```
> cMK <- ls(cMAPKEGGPATHWAY)
> spec <- substr(cMK, 1, 3)
> table(spec)

spec
hsa map
 81   4

> cMK[[2]]

[1] "hsa00020"

> pw2 <- cMAPKEGGPATHWAY[[cMK[2]]]
> names(pw2)

[1] "id"        "organism" "source"    "name"
[5] "component"
```

```
> pw2$name

[1] "citrate cycle (tca cycle)"

> pw2$component

 [1]   63   71 3473   80   68   78   79   77   72   65   55
[12]   82   74   84   75   64   83   61   58   59   69   81
[23]   60   56   73   66   76 1367   62   54
```

We drill down into the KEGG-related environments to learn more about constituents of this pathway. We select the first element of the pw2$component; it is an interaction so we first extract it, and then explore its components.

```
> getI1 <- get("63", cMAPKEGGINTERACTION)
> unlist(getI1[1:4])

    source     process reversible   condition
    "KEGG"          NA     "TRUE"          NA

> unlist(getI1[[5]][[2]])

      id        edge        role location activity
       2          NA          NA       NA       NA
```

We find that ATP is an input to the citrate cycle:

```
> get("2", cMAPKEGGMOLECULE)[[2]][7:8]

                             AS                              AS
"adenosine 5'-triphosphate"                             "ATP"
```

As we have demonstrated these data are quite rich and can be used to develop a number of fairly sophisticated pathway analysis tools.

7.7   Cross-organism annotation: the homology packages

The HomoloGene project at the NCBI provides resources for the detection of homologs among annotated genes for a variety of organisms with sequenced genomes. Two genes are said to be *homologous* if they have descended from a common ancestral DNA sequence. The data have been abstracted into a collection of homology packages which we will briefly describe. There is one homology package for each species (the data are too large for a single combined package). Each package name begins with the three letter species name (i.e., for *Homo sapiens* it is hsa) and a suffix of homology. We load hsahomology in the next code chunk. There we can see that the mappings provided are between HomoloGene's identifiers and a variety of other commonly used identifiers, namely LocusLink and UniGene.

```
> library("hsahomology")
> ls("package:hsahomology")
```

```
 [1] "hsahomology"          "hsahomologyACC2HGID"
 [3] "hsahomologyDATA"      "hsahomologyHGID"
 [5] "hsahomologyHGID2ACC"  "hsahomologyHGID2LL"
 [7] "hsahomologyLL2HGID"   "hsahomologyMAPCOUNTS"
 [9] "hsahomologyORGCODE"   "hsahomologyQC"
```

The data linking genes is provided in hsahomologyDATA where the primary keys are the HomoloGene identifiers and the values are lists of homologous genes. Each element in the list represents a gene that is homologous to the key. There are three different types of homology that are recorded and represented in the data. A single letter is used (and it is stored as the homoType). The type can be either B (reciprocal best best between three or more organisms), b (reciprocal best match between two organisms), or c (curated homology relationship between two organisms). If the type is either B or b, then the percent similarity is given (it is the percent of identity of base pair alignment between the homologous sequences) and this value is stored as homoPS. If the type is c, then a URL giving the source of the homology relationship is given as homoURL.

Each species has a unique code assigned to it. For example the code for *Homo sapiens* is 9606 and this is used to label each homology. The Homologene project uses its own set of gene identifiers, and so users must map to these identifiers using the environments described above. To find the homologs for estrogen receptor 1 (ESR1) we use its LocusLink ID (2099) and find the corresponding HomoloGene ID. Then, we can find the likely homologs.

```
> esrHG <- hsahomologyLL2HGID$"2099"
> hesr <- get(as.character(esrHG), hsahomologyDATA)
> sapply(hesr, function(x) x$homoOrg)
```

```
10090 10090 10116 10090 10116  8022  8355  8364  9823  7955
"mmu" "mmu" "rno" "mmu" "rno" "omy" "xla" "xtr" "ssc" "dre"
 9913
"bta"
```

We see the different organisms that have known potential homologs with ESR1. Duplicated species names in this list indicate that there is more than one potential homolog for the gene of interest.

Another task that someone might often want to carry out is to find all potential homologs in one species, starting with the genes in a different species. Suppose, for example that we want to find all Xenopus Laevis homologs for human genes, then the following code carries out the necessary computations.

```
> hXp <- eapply(hsahomologyDATA, function(x) {
+     gd <- sapply(x, function(x) if (!is.na(x$homoOrg) &&
```

```
+          x$homoOrg == "xla")
+          TRUE
+      else FALSE)
+      x[gd]
+ })
> lh <- sapply(hXp, length)
> hXp2 <- hXp[lh > 0]
```

And we are left with 7021 human genes that have a potential homolog in Xenopus Laevis. The data in hXp2 tells us about the strength of the homology, what type of match was performed and the LocusLink ID (if there is one) or sequence identifier (either GenBank or RefSeq). These can then easily be used in any further analyses.

7.8   Annotation from other sources

The function, readGEOAnn, from annotate queries the Gene Expression Omnibus (GEO) for information about a microarray platform. The GEO Web site should be inspected for its structure, scope, and for details of its platform and archive enumeration system.

TIGR's Resourcerer project provides formatted descriptive information for many different microarray platforms. The Resourcerer package provides some tools to help Bioconductor users obtain and use these resources. The basic approach is to download the relevant files from the Resourcerer Web site, and then to transform them for use in Bioconductor. Users can access the data provided by Resourcerer in the form of a matrix by using the getResourcerer function. Alternatively, the resourcerer2BioC function in the Resourcerer package will transform any data set provided by TIGR Resourcerer into a platform annotation package for R. The resultant package takes the same form as all other Bioconductor meta-data packages and hence can be used with all of the tools that are discussed in this book. Consult the Resourcerer project for information on what platform descriptions are the describe.

To build packages based on Resourcerer data, users will need to have both the Resourcerer package and the AnnBuilder packages installed. The resourcer2BioC function accepts the name of a Resourcerer file, and downloads that file to produce the appropriate meta-data package. This can be quite time-consuming and resource-intensive as it involves downloading and manipulating large data files.

Another prominent public archive for microarray experimental resources is EBI ArrayExpress. Many of the experiments archived there are encoded in MAGE-ML, an XML dialect governed by the MAGE Object Model devised by MGED, the Microarray Gene Expression Data Society. The RMAGEML package can be used to deserialize these XML archives to R objects of the `marray` class. The RMAGEML package includes extensive examples and documentation illustrating this process.

7.9   Discussion

As we noted at the beginning of this chapter, analyzing genomic data relies heavily on the appropriate use of biological meta-data. Meta-data provide substantial opportunities for more sophisticated modeling of emerging large-scale experimental data resources. Expertise in the use and interpretation of diverse meta-data resources will become increasingly important over the coming years. It will be worthwhile to familiarize yourself with the different types of data and with methods for using them in different analytical agendas.

In this chapter, we have presented a number of different strategies that can be used to provide access to meta-data resources. For the most part, we have concentrated on creating small data packages that can be downloaded and used without access to other resources. However, that approach will need to evolve as the volume of relevant meta-data surpasses readily manageable limits. At the same time, database technology is improving, and powerful databases can be run on small computers (laptops) without extensive specialized knowledge. Since R has fairly seamless interactions with databases, our next moves in this area will be to develop database schemata that can be used interchangeably with the data packages that are now being supplied.

8

Querying On-line Resources

V. J. Carey, D. Temple Lang, J. Gentry, J. Zhang, and R. Gentleman

Abstract

Many different meta-data resources are available on-line, and several of these provide a Web services model for interactions. R and Bioconductor support the use of different technologies (including HTTP, SOAP, and XML-RPC) for accessing different Web services. In this chapter we describe the tools for accessing Web services and demonstrate their use in a number of examples.

Our view is very similar to that proposed by Stein (2002), who emphasized Web services as the basic computational resource for bioinformatics. Well-designed Web services will play an essential role in solving many bioinformatic problems and R has the capability of playing many different roles, both on the client and the server side.

8.1 The Tools

The familiar phrase *surfing the Web* connotes active travel between various information sources. In fact, one does not *visit* a Web site in any concrete sense. When the browser is directed to a given URL, it emits a request for information to the target server, and information returned by the server (typically HTML documents and images) is rendered locally in the browser. For integrating different software products across the Internet, commonly implemented standards are essential. Most of these standards are codified by the W3C consortium, and readers are referred to W3C Web sites for more details.

Web communication models support a wide variety of interactions beyond requesting and receiving HTML documents. A *web service* is a software application that is identified by a *uniform resource identifier* (URI) where the software interface, i.e., the services provided, can be discovered remotely. Definitions and descriptions are generally accessible through standardized requests to Web servers. A Web service supports direct interactions with external software, clients, using messages exchanged via Internet protocols. The client packages up the request, which may include data, in a pre-specified format, and sends the request to the web service over the Internet. The Web service carries out the requested computations and returns the result in a pre-specified format that the user then parses to extract the desired components.

In many cases, the data and requests are marked up in the eXtensible Markup Language (XML). XML is a standard mechanism for marking up data and text in a structured way. Servers and clients use XML dialects to negotiate queries on service availability, to submit requests, and to encode the results of requested computations. A tutorial discussion of XML concepts and tools for processing XML in R is provided in the on-line complements.

The Simple Object Access Protocol (SOAP) uses XML to structure requests and responses for Web service interactions. The SOAP protocol includes rules for encapsulating requests and responses (e.g., rules for specifying addresses, selecting methods, or specifying error handling actions), and for encoding complex data types that form parts of requests and responses (e.g., encoding arrays of floating point numbers). Examples are provided in the on-line complements.

Closely related to SOAP is the Web Services Description Language (WSDL) which provides a protocol for finding and describing Web services. XML and SOAP are tools that are used to interact with the web service, but it is also necessary to be able to locate particular services, and, given a service, to find out which queries it will respond to and the type of arguments that must be supplied. This is the role of WSDL. It allows for programmatic detection of services and construction of calls. This is a form of *reflectance* where one piece of software can locate another and determine the appropriate calling sequence.

Functionality for XML and SOAP in R is supplied through the Omegahat Project in the form of R packages XML and SSOAP. An additional supporting package, RCurl provides services for downloading files using URL syntax. Most of the tools that we describe in this chapter are built on top of these packages.

Using this software infrastructure the Bioconductor Project provides tools to support different Web services. These tools fulfill two roles, first they provide access to the specific services and second they act as prototypes that you can use to design and implement software to access Web services that interest you. In the remainder of this chapter, we discuss some of the different Web services available. We begin with a general discussion of the Entrez system. Subsequently we consider a set of tools that are tailored to querying PubMed. Our last example uses Kanehisa and Goto (2000) as it also provides a variety of services and a variety of interfaces.

8.1.1 Entrez

An interesting aspect of the Entrez system is that it supports the notion of a workspace. After certain queries have been executed, the user can obtain a `WebEnv` key, which can be used to restrict future operations to the results from the previous query. Thus, users can first select interesting PubMed articles (for example) and then operate only on the set that they have selected.

A brief description of some of the Entrez utilities follows.

EInfo Either lists all available databases or provides information on a specific database, depending on the supplied arguments.

ESearch This service allows users to perform a variety of searches of the different databases. If the `usehistory` argument is set, then Entrez will save the results (on their side) and return a `WebEnv` key that can be used by other Entrez services.

ESummary ESummary provides summaries of a list of primary IDs for a database or of the results of a previous search or action (using `WebEnv`.

ELink The ELink service provides links to external or related articles for primary IDs.

EFetch It returns the records for requested IDs (or those in the current Entrez environment) in a variety of formats.

EPost This services allows users to upload files to their current Entrez environment. These are then treated in the same manner as Entrez generated data.

EGQuery EGQuery is used to list the count across the NCBI databases of responses to a single query.

8.1.2 Entrez examples

In this example, we initiate a request to the EInfo utility to provide a list of all the databases that are available through the NCBI system. These can then be queried in turn to determine what their contents are.

```
> ezURL <- "http://eutils.ncbi.nlm.nih.gov/entrez/eutils/"
> library("XML")
> z <- xmlTreeParse(paste(ezURL, "einfo.fcgi", sep = ""),
+     isURL = TRUE, handlers = NULL, asTree = TRUE)
> dbL <- xmlChildren(z[[1]]$children$eInfoResult)$DbList
> dbNames <- xmlSApply(dbL, xmlValue)
> length(dbNames)

[1] 29

> dbNames[1:5]
```

```
   DbName        DbName        DbName        DbName
  "pubmed"     "protein" "nucleotide"   "structure"
   DbName
  "genome"
```

We see that at the time the query was issued, there were 29 databases. The names of five of them are listed, and the others are all available. Parsing of the XML is handled by fairly standard tools, and in particular we want to draw attention to the *apply*-like functions. Because XML document objects have complex R representations, the use of `xmlSApply` and other dedicated functions will generally simplify the code that needs to be written. For large documents, `xmlEventParse` (SAX style parser) will be preferred to `xmlTreeParse` (DOM style parsing). See the on-line complements for further details.

This example employs and HTTP request to activate CGI. The Entrez SOAP interface is very similar and provides no additional functionality. SOAP may be preferred in some instances, because WSDL provides an opportunity to auto-generate the software needed to make queries.

8.2   PubMed

The National Library of Medicine (NLM) provides support for a number of different Web services. In this section we describe a set of tools that we have developed to query PubMed. The software for carrying out these interactions is contained in the annotate package, and more details and documentation are provided as part of that package. Some of our earlier work in this area was reported in Gentleman and Gentry (2002).

There are four functions in annotate that provide the support for accessing on-line data resources. They are:

genbank Users specify GenBank identifiers and can request the related links to be rendered in the browser or returned in XML.

pubmed Users specify PubMed identifiers and can request them to be rendered in the browser or returned in XML. More details on parsing and manipulating the XML are given below.

locuslinkByID Users specify LocusLink identifiers, and the appropriate links are opened in the browser. LocusLink does not provide XML so there is currently no download option. The user can request that the URL be rendered or returned.

locuslinkQuery Users specify a string that will be used as the LocusLink query and the species of interest (there can be several). The user can request either that the URL be rendered or returned.

The function `locuslinkByID` takes a set of known LocusLink identifiers and constructs a URL that will have these rendered. The user can either save the URL (perhaps to send to someone else or to embed in an HTML page, see Chapter 9 for more details).

The function `locuslinkQuery` takes a character string to be used for querying PubMed. For example, this function call,

```
locuslinkQuery("leukemia", "Hs")
```

will find all human genes that have the word leukemia associated with them in their LocusLink records. Note that the R code is merely an interface to the services provided by NLM and NCBI and users are referred to those sites for complete descriptions of the underlying databases and algorithms.

8.2.1   Accessing PubMed information

In this section we demonstrate how to query PubMed and how to operate on the data that are returned. As noted above, these queries generate XML, which must then be parsed to provide the specific data items of interest. Our example is based on example data from the package Biobase. Users should be able to easily replace these data with their own.

```
> library("annotate")

> data(eset, package = "Biobase")
> affys <- geneNames(eset)[491:500]
> affys

 [1] "31730_at" "31731_at" "31732_at" "31733_at" "31734_at"
 [6] "31735_at" "31736_at" "31737_at" "31738_at" "31739_at"
```

Here we have selected an arbitrary set of 10 genes from our sample data for which abstracts will be sought. `eset` uses Affymetrix identifiers as gene names, but for the `pubmed` function, we need to use PubMed ID values. The following code performs the translation.

```
> library("hgu95av2")
> ids <- getPMID(affys, "hgu95av2")
> ids <- unlist(ids, use.names = FALSE)
> ids <- unique(ids[!is.na(as.numeric(ids))])
> length(ids)

[1] 56

> ids[1:8]

[1] "12477932" "12408966" "8325638"  "8175896"  "1889752"
[6] "12679040" "9315667"  "9199346"
```

We used `getPMID` to obtain the PubMed identifiers that are related to the selected probes. Duplicates and missing values are removed and we are left with 56 distinct PMIDs.

In the next code chunk, we show how to generate the query and store the results in a variable named x. This object is of class XMLDocument and to manipulate it we will use functions provided by the XML package.

```
> x <- pubmed(ids)
> a <- xmlRoot(x)

> numAbst <- length(xmlChildren(a))

[1] 56
```

Our search has obtained 56 abstracts from PubMed that can be processed locally. For easy review, the annotate package also provides a pubMedAbst class, which will take the XML tree from a call to pubmed and extract the interesting sections.

```
> arts <- xmlApply(a, buildPubMedAbst)
> class(arts[[7]])

[1] "pubMedAbst"
attr(,"package")
[1] "annotate"
```

The function xmlApply applies the function buildPubMedAbst to all children nodes of the root a, and returns the result in the list arts. The function buildPubMedAbst converts the *XML* representation of the abstract into an object of class *pubMedAbst*. One of the abstracts is printed below.

```
> arts[[7]]

An object of class 'pubMedAbst':
Title: Interference with the expression of a novel
     human polycomb protein, hPc2, results in
     cellular transformation and apoptosis.
PMID: 9315667
Authors: DP Satijn, DJ Olson, J van der Vlag, KM
     Hamer, C Lambrechts, H Masselink, MJ Gunster,
     RG Sewalt, R van Driel, AP Otte
Journal: Mol Cell Biol
Date: Oct 1997
```

The pubMedAbst class has a number of different slots. They are:

authors The list of authors (represented as a character vector).

pmid The PubMed record number.

abstText The actual abstract (in text).

articleTitle The title of the article.

journal The journal it is published in.

pubDate The publication date.

We can extract the text by applying the `absText` function to each of the R objects.

```
> absts <- sapply(arts, abstText)
> class(absts)
```

```
[1] "character"
```

Once the abstracts have been assembled, they can be searched in R using tools such as `grep` and `agrep`. Suppose, for example, that we wanted to know which abstracts have the term `cDNA` in them, then the following code chunk shows how to identify these abstracts.

```
> found <- grep("cDNA", absts)
> goodAbsts <- arts[found]
> length(goodAbsts)
```

```
[1] 18
```

So 18 of the articles related to our genes of interest mention the term `cDNA` in their abstracts. Although this approach is not as sophisticated as many text-mining procedures, it can be used to quickly identify papers of potential interest.

Next we use the same set of PubMed IDs with the `genbank` function. By default, the `genbank` function assumes that the ID values passed in are Genbank accession numbers so we use the `type` argument to indicate that we are using PubMed IDs.

```
> y <- genbank(ids[1:10], type = "uid")
> b <- xmlRoot(y)
> class(b)
```

```
[1] "XMLNode"
```

8.2.2   Generating HTML output for your abstracts

Rather than access the data and manipulate it directly, many users prefer to have static Web pages constructed. A fairly comprehensive treatment of Bioconductor tools and strategies is given in Chapter 9. In the following code chunk, we compute the pages and save the output into temporary files.

```
> fname <- tempfile()
> pmAbst2HTML(goodAbsts, filename = fname)
> fnameBase <- tempfile()
> pmAbst2HTML(goodAbsts, filename = fnameBase, frames = TRUE)
```

Figures 8.1 and 8.2 illustrate the results of directing the browser to the resulting files.

BioConductor Abstract List

Article Title	Publication Date
Cloning and characterization of hTAFII18, hTAFII20 and hTAFII28: three subunits of the human transcription factor TFIID.	Apr 1995
Generation and initial analysis of more than 15,000 full-length human and mouse cDNA sequences.	Dec 2002
[Expression, purification and characterization of human prorelaxin-like-protein H2 in Escherichia coli]	Sep 2002
Isolation and analysis of the 3'-untranslated regions of the human relaxin H1 and H2 genes.	Apr 2000
Two human relaxin genes are on chromosome 9.	Oct 1984
Relaxin gene expression in human ovaries and the predicted structure of a human preprorelaxin by analysis of cDNA clones.	Oct 1984
Expression of the human relaxin H1 gene in the decidua, trophoblast, and prostate.	Apr 1991
Molecular characterization and pharmacological properties of the human P2X3 purinoceptor.	Jul 1997
Characterization of recombinant human P2X4 receptor reveals pharmacological differences to the rat homologue.	Jan 1997
The HOXC11 homeodomain protein interacts with the lactase-phlorizin hydrolase promoter and stimulates HNF1alpha-dependent transcription.	May 1998
The human HOX gene family.	Dec 1989
Identification of the alternative splice products encoded by the human protein phosphatase inhibitor-1 gene.	Mar 2002
Identification and characterization of a novel protein inhibitor of type 1 protein phosphatase.	Nov 2000
Isolation and characterization of full-length cDNA clones for human alpha-, beta-, and gamma-actin mRNAs: skeletal but not cytoplasmic actins have an amino-terminal cysteine that is subsequently removed.	May 1983
Isolation and characterization of cDNA clones for human skeletal muscle alpha actin.	Jun 1983
Large-scale concatenation cDNA sequencing.	Apr 1997
Cloning and sequencing of the cDNA encoding human P5.	Oct 1995

Figure 8.1. pmAbst2HTML without frames.

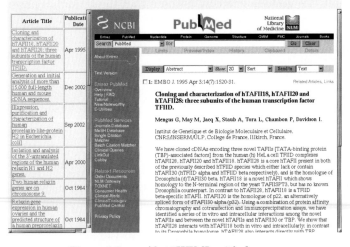

Figure 8.2. pmAbst2HTML with frames.

8.3   KEGG via SOAP

Kanehisa and Goto (2000) provides a rich resource of genomic and biological meta-data. As noted in Chapter 7, the Bioconductor provides mappings to KEGG pathways and other data through static, curated data packages.

However, this approach will not be sufficient for all needs. In addition to providing data for downloading over the Internet, KEGG supports a SOAP interface. The URI for their API is given in Appendix A.2. The API is quite general, and **KEGGSOAP** maintains bindings to a large subset of the KEGG functions. Users can extend this interface to achieve functionalities not currently provided in **KEGGSOAP**. Among the more interesting functions are:

- `get.genes.by.pathway`, which gets all genes in the named pathway.

- `get.enzymes.by.pathway`, which gets all enzymes in the named pathway.

- `get.motifs.by.gene`, which queries the Pfam, TIGRFAM, PROSITE pattern, or PROSITE profile databases for the motifs of a given gene.

- `get.genes.by.motifs`, which searches the databases implied by the motif IDs for genes containing the specified motifs.

Users can query KEGG according to pathway. In the example below we query KEGG for both the genes and the enzymes that are involved in the E. coli citrate cycle, pathway number 00020.

```
> KEGGPATHID2NAME$"00020"

[1] "Citrate cycle (TCA cycle)"

> genes <- get.genes.by.pathway("path:eco00020")
> enzymes <- get.enzymes.by.pathway("path:eco00020")

> enzymes[1:4]

[1] "ec:1.1.1.37" "ec:1.1.1.42" "ec:1.2.4.2"  "ec:1.3.99.1"
```

A motif is a locally conserved region of a sequence or a short sequence pattern shared by a set of sequences. In many situations, we are interested in finding genes that share a common motif, or in finding the motifs that are present in a particular gene. Users need to specify the databases that they would like to use, or the motifs in which they are interested. In general it will be necessary to look at the manual pages, or the KEGG API to determine the appropriate nomenclature needed to construct the call. We demonstrate a couple of simple calls in the code chunk below.

```
> motifs <- get.motifs.by.gene("eco:b0002", "pfam")

Type: SOAPStruct  & length =  5

> unlist(motifs[[1]][1:6])

               genes_id                    score
             "eco:b0002"                   "-1"
               motif_id             end_position
        "pfam:AA_kinase"                  "284"
```

```
           definition                  start_position
"Amino acid kinase family"                        "1"

> genes <- get.genes.by.motifs(c("pf:DnaJ", "ps:DNAJ_2"),
+     1, 10)

Type: SOAPStruct  & length =  10

> genes[1:3]

          chaperone DnaJ              chaperone DnaJ
            "aae:aq_1735"                "aae:aq_703"
curved DNA-binding protein
           "aci:ACIAD0406"
```

Note that these queries return SOAP structures that will require further processing.

8.4   Getting gene sequence information

The function getSeq4Acc can be used to obtain the nucleotide sequence for any provided GenBank or RefSeq accession numbers. The Biostrings package can then be used to do matching in nucleic acid sequences. Readers interested in pattern matching should consult the vignette and manual pages for the Biostrings package for more details.

In the code below, a single accession number is selected and then the nucleotide sequence is downloaded.

```
> ps <- ls(env = hgu95av2ACCNUM)
> myp <- ps[1001]
> myA <- get(myp, hgu95av2ACCNUM)
> myseq <- getSEQ(myA)
> nchar(myseq)

[1] 4839

> substr(myseq, 1, 10)

[1] "TCCGGTTTTT"
```

Alternatively, RefSeq identifiers, for either nucleic acid sequences or protein sequences, can be used.

```
> rsp <- getSEQ("NP_004327")
> substr(rsp, 1, 10)

[1] "MDTPENVLQM"

> rsn <- getSEQ("NM_004336")
> substr(rsn, 1, 10)

[1] "GAGCCGACTG"
```

One use for nucleotide sequence data is to check whether the probes on the microarray are actually in the the associated gene. The **hgu95av2probe** package provides probe sequence information for the AffymetrixHG-U95Av2 GeneChip. The probe data are stored as a *data.frame* with 6 columns:

sequence the sequence of the 25mer

x the x position of the probe on the array

y the y position of the probe on the array

Probe.Set.Name the AffymetrixID for the probeset

Probe.Interrogation.Position the location (in bases) of the 13th base in the 25mer, in the target sequence

Target.Strandedness whether the 25mer is a sense or an antisense match to the target sequence.

Using both the nucleotide sequence and the probe sequence information the **Biostrings** package can be used to do the matching. **Biostrings** has many different options and algorithms, including incomplete and partial matching, along with the ability to use the extended nucleotide alphabet (Cornish-Bowden, 1985).

```
> library("hgu95av2probe")

> wp <- hgu95av2probe$Probe.Set.Name == myp
> myPr <- hgu95av2probe[wp, ]
> library("Biostrings")
> mybs <- NucleotideString(myseq, "DNA")
> match1 <- matchDNAPattern(as.character(myPr[1,
+       1]), mybs)
> m1m <- as.matrix(match1)
> m1m

     [,1] [,2]
[1,] 4735 4759

> myPr[1, 5] == m1m[1, 1] + 13 - 1

[1] TRUE
```

And we can see that in this case the 13th nucleotide is indeed in exactly the place that has been predicted. It is relatively straightforward to check the other 25-mers.

8.5   Conclusion

The mechanics of issuing annotation requests to on-line providers will evolve along with Web protocols, provider populations, and contents of on-line databases. New technologies, such as those emerging in the semantic web initiative, may increase the efficiency with which on-line resources are harvested.

The Bioconductor approach to meta-data query resolution support described here emphasizes the requirements of flexibility and programmability. Developers supporting researchers with clearly patterned workflows related to annotation can create user-friendly, higher-level query processing facilities on the basis of the tools described here.

9

Interactive Outputs

C. A. Smith, W. Huber, and R. Gentleman

Abstract

In this chapter, we discuss creation of interactive outputs. We focus on the generation of reports, marked up in HTML, that link sets of genes with on-line resources, such as those supplied by the EBI or the NCBI, and which can be shared between different investigators. We discuss both the simple creation of these pages as well as some of the underlying software tools that can be used to construct new and different outputs. Although linked Web pages form the most commonly used outputs, we also consider some other tools that can be used to produce Web graphics that respond to the mouse in different ways.

9.1 Introduction

The advent of the World Wide Web revolutionized the way biologists, statisticians, and scientists from many other disciplines locate and disseminate research. Additionally, journals now afford authors the chance to deposit significant quantities of supplementary material on-line. This expands the amount of data that can be communicated far beyond that of a few figures and tables and generates new demands and opportunities. For many genomic experiments the outputs of the analysis are both large and complex. Conveying these outputs in a manner that is conducive to understanding and exploration suggests changes from publishing static printed materials to more interactive methods.

In this chapter, we discuss some methods for producing linked outputs, mostly of gene lists, that can be easily shared among investigators, regardless of their geographic locations. We also consider some other uses of interactive outputs such as the `imagemap` function that can be used to add *tool tips* to a graphic. A tool tip is a small window, containing text, that pops up when the mouse goes over a predefined location. Tool tips allow

developers to annotate their graphical outputs in a way that does not clutter them with text, but allows readers to interactively navigate and query
the graphic.

Bioconductor includes many tools that leverage the Web browser for
doing and reporting on microarray analyses. This chapter describes their
capabilities and use. There is a relatively primitive, but easy to use, function
(htmlpage) in the annotate package as well as a more sophisticated suite of
tools provided by the annaffy package. Both can be used to create annotated
lists of expressed genes. Those tabular lists are linked to numerous external
sources of information, including NCBI's LocusLink, GO, or the metabolic
pathway database from KEGG. annaffy takes advantage of Bioconductor's
pre-built annotation packages for most popular chips in the Affymetrix
platform. Users with other types of chips can use the tools in AnnBuilder, or
Resourcerer to build the appropriate annotation packages if they so desire.

Some of the Web interaction packages of Bioconductor are slanted towards Affymetrix analyses. This is a consequence of both the modularity
of the Affymetrix platform and the research demands that prompted the
development of the packages in the first place. Spotted arrays tend to be
less uniform, and the selection and layout of the spots can vary between
laboratories and between print-runs.

We will use data included in the annaffy package as the basis for our
examples and we first load that package. The data arise from a microarray
experiment which used the Affymetrix HG-U95Av2 GeneChip. There are
eight total samples in the set, four control samples and four experimental
samples. 250 expression measures were selected at random from the results,
and another 250 probe IDs were selected at random and assigned to those
expression measures.

We first discuss the simple interface provided by htmlpage. Following that
we discuss the annaffy package in some detail and finish this chapter with
some examples of using the imagemap.

9.2   A simple approach

The construction of a linked Web page requires substantial coordination.
First, the genes are identified by some analysis, and then the manufacturer's
identifiers must be linked to the other biological identifiers of interest (e.g.,
GO or KEGG). This can be done using the Bioconductor annotation packages. Next, the appropriate link text must be found and finally the HTML
must be generated and printed to a file.

In our example, we make use of the 10 probes in the aafExpr example
data set numbered from 41 to 50.

```
> library("annaffy")
> data(aafExpr)
```

```
> gN <- geneNames(aafExpr)[41:50]
> syms <- unlist(mget(gN, hgu95av2SYMBOL))
> lls <- unlist(mget(gN, hgu95av2LOCUSID))
> syms
```

```
   31848_at     31901_at     32010_at     32073_at     32155_at
   "CADPS"      "KCNAB2"      "EAN57"     "JMJD2A"     "TFAP2A"
   32230_at     32354_at     32443_at     32498_at   32525_r_at
   "EIF3S2"      "NPAS3"     "ZNF157"       "GRM2"           NA
```

Now we have obtained the Affymetrix IDs as well as LocusLink IDs and
the gene symbols for the probesets of interest. We want to create a Web
page with this information, and have the LocusLink identifiers linked to
the appropriate page at the NCBI, the Affymetrix identifiers linked to the
appropriate Web resources from Affymetrix and the symbols not linked.
The code is quite simple; linked identifiers are placed first, then the appropriate resources are indicated. Next we specify the other values that will
be output and these are followed by the column labels. Finally we specify
the name of the output file and call htmlpage to produce the output.

```
> gl <- list(gN, lls)
> repository <- list("affy", "ll")
> othernames <- list(syms)
> head <- c("Probe ID", "Symbol", "LocusLink")
> fName <- "out.html"
> htmlpage(gl, fName, title = "My Interesting Genes",
+     othernames, head, repository = repository)
```

Now the file, out.html can be opened in any Web browser and a table
is displayed. It has 10 rows and 3 columns. The Affymetrix identifiers are
linked to the appropriate location on the Affymetrix Web site and the
LocusLink identifiers are linked to the appropriate pages at the NCBI.
This file is available as part of the on-line complements.

9.3   Using the annaffy package

Now that we have seen a simple model for generating interactive Web
pages, we can turn our attention to the more sophisticated interface provided by the annaffy package. This package was designed to help facilitate
interactions between microarray analysis results and Web-based databases.
It provides classes and methods for accessing those resources both interactively, from the command line, as well as through statically generated
HTML pages.

annaffy uses a different class for each type of annotation data. This
approach was taken in order to take advantage of the object oriented programming paradigm in R and in particular the automatic dispatching based
on the class of the arguments to a function. Readers unfamiliar with R's

object-oriented programming paradigm are referred to Chambers (1998) for more details.

Some of the different classes defined and used are listed below and the others are documented in the annaffy package:

aafSymbol gene symbol

aafGenBank GenBank accession number

aafLocusLink NCBI Gene IDs (almost never more than one)

aafGO Gene Ontology identifiers, names, types, and evidence codes

aafPathway KEGG pathway identifiers and names

For each class, there is a constructor method with the same name as the class. This function takes as arguments a vector of probe identifiers as well as the name of the appropriate Bioconductor meta-data package. If the data package for the chip is not already loaded, the constructor will attempt to load it. The constructor finds the appropriate mapping from the supplied probe identifiers to the requested data and returns an object of class *aafList* which contains instances of the appropriate class. Note that annaffy handles missing annotation data in different ways, depending on the type of data. The value NA is used if there is no known mapping from the probe identifier to the requested output. For the purpose of demonstration, we will use the hgu95av2 meta-data package and probe IDs from the aafExpr data set (these have already been loaded).

In this next code chunk, we first access the Affymetrix probe IDs from the data, aafExpr and then use the aafSymbol method to find the gene symbols that are associated with those probe IDs. As mentioned above, symbols is an instance of the aafList class. R's subsetting functions provide access to the different mappings. In the code chunk below, we access the gene symbols, and note that here a missing value is encoded as a zero length string.

```
> probeids <- geneNames(aafExpr)
> symbols <- aafSymbol(probeids, "hgu95av2")

[1] "You have package hgu95av2 but the incorrect version"

> symbols[55:57]

An object of class "aafList"
[[1]]
An object of class "aafSymbol"
[1] "MRPS14"

[[2]]
An object of class "aafSymbol"
[1] "TDRD3"
```

```
[[3]]
An object of class "aafSymbol"
character(0)
```

All annotation constructors return their results as *aafList* objects, which act like normal lists but have special behavior when used with certain methods. One such method is `getText`, which returns a simple textual representation of most annaffy objects.

```
> getText(symbols[55:57])
```

```
[1] "MRPS14" "TDRD3"  ""
```

Other annotation constructors return more complex data structures, for example GO has a much richer structure and the object returned reflects that.

```
> gos <- aafGO(probeids, "hgu95av2")
> gos[[3]]
```

```
An object of class "aafGO"
[[1]]
An object of class "aafGOItem"
@id    "GO:0008083"
@name "growth factor activity"
@type "Molecular Function"
@evid "IEA"

[[2]]
An object of class "aafGOItem"
@id    "GO:0008083"
@name "growth factor activity"
@type "Molecular Function"
@evid "TAS"

[[3]]
An object of class "aafGOItem"
@id    "GO:0007399"
@name "neurogenesis"
@type "Biological Process"
@evid "TAS"
```

The GO constructor, `aafGO`, returns an *aafList* of *aafGO* objects, which are in turn lists of *aafGOItem* objects. Within each of those objects, there are four slots: `id`, `name`, `type`, and `evidence` code. The individual slots can be accessed with the `@` operator.

9.4   Linking to On-line Databases

One of the most important features of the annaffy package is its ability to create hyperlinks to various public on-line databases. Readers are referred to Chapter 7 for more details on the contents, structure, and use of the different meta-data resources. Most of the annotation classes in annaffy have a getURL method that returns single or multiple URLs, depending on the object type. These functions mimic other functions that provide similar functionality in other Bioconductor packages (most notably annotate). In the example code given below, we print out the URL for inspection. Making use of a URL involves initiating some form of Internet protocol request, for example using a url connection or opening the link in the Web browser using browseURL.

The simplest annotation class that produces a URL is *aafGenBank*. Most Affymetrix probes have corresponding GenBank accession numbers, even those missing other annotation data. The GenBank database provides information about the expressed sequence that was used to design the probes on the GeneChip. Additionally, it provides information about the functional parts of the sequence and the the authors that initially sequenced the gene fragment. The code chunk below demonstrates the steps needed to construct the URL, which is printed.

```
> gbs <- aafGenBank(probeids, "hgu95av2")
> getURL(gbs[[1]])
```

```
http://www.ncbi.nlm.nih.gov/entrez/query.fcgi?cmd=search&db=
    nucleotide&term=U41068%5BACCN%5D&doptcmdl=GenBank
```

NCBI Gene is a very useful on-line database that links to many other data sources not currently used in constructing Bioconductor meta-data packages. The NCBI Gene database was previously known as *LocusLink*. The class retains that name for historical reasons. Code demonstrating the construction of the appropriate URLs is given below. For the remainder of this section we suppress the printing of the resultant URLs.

```
> lls <- aafLocusLink(probeids, "hgu95av2")
> getURL(lls[[2]])
```

If you are interested in exploring the area of the genome surrounding a probe, the *aafCytoband* provides a link to NCBI's on-line genome viewer. It includes adjacent genes and other genomic annotations.

```
> bands <- aafCytoband(probeids, "hgu95av2")
> getURL(bands[[2]])
```

For primary literature information about a gene, use the *aafPubMed* class. It will provide a link to a list of abstracts on PubMed that describe the gene of interest. The list abstracts that can be accessed in this way are by no means exhaustive and will sometimes only include the paper in which the gene was cloned.

```
> pmids <- aafPubMed(probeids, "hgu95av2")
> getURL(pmids[[2]])
```

A number of interesting queries can be done with the GO class. You can display the GO family hierarchy for a set of GO IDs at once.

```
> getURL(getURL(gos[[1]]))
```

You can also show the family hierarchy for a single GO ID. See Chapter 22 for other methods of processing the GO hierarchy.

```
> getURL(gos[[1]][[4]])
```

```
http://godatabase.org/cgi-bin/go.cgi?open_0=GO:0005592
```

The last link type of note is that for *KEGG* pathway information. Most genes are not currently annotated with pathway data. However, for those that are, it is possible to retrieve schematics of the biochemical pathways a gene is involved in.

```
> paths <- aafPathway(probeids, "hgu95av2")
> getURL(paths[[5]])
```

9.5   Building HTML pages

In addition to using annaffy interactively through R, it may also be desirable to generate annotated reports summarizing your microarray analysis results. Such a report can be used by a scientist collaborator with no knowledge of either R or Bioconductor. Additionally, by having all the annotation and statistical data presented together on one page, connections between, and generalizations about, the data can be made in an efficient manner.

The primary role of these functions from the annaffy package is to produce such reports in HTML. Additionally, it can easily format the same report as tab-delimited text for import into a table, spreadsheet, or database. It supports nearly all the annotation data available through Bioconductor. Additionally, it has facilities for including and coloring user data in an informative manner.

9.5.1   Limiting the results

HTML reports generated by annaffy can grow to become quite large unless some measures are taken to limit the results. Multi-megabyte Web pages are unwieldy and should be avoided. Doing a ranked statistical analysis is one way to limit results, and will be shown here. We will rank the expression measures by putting their two-sample Welch t-statistics in order of decreasing absolute value.

The first step is to load the **multtest** package, which will be used to compute the *t*-statistics.

```
> library("multtest")
```

The `mt.teststat` function requires a vector that specifies which samples belong to the different observation classes. That vector can be produced directly from the first covariate of *pData*.

```
> class <- as.integer(pData(aafExpr)$covar1) - 1
```

Using the class vector, we calculate the *t*-statistic for each of the probes. We then generate an index vector that can be used to order the probes themselves in increasing order. As a last step, we produce the vector of ranked probe identifiers. Later annotation steps will only use the first 50 of those probes.

```
> teststat <- mt.teststat(exprs(aafExpr), class)
> index <- order(abs(teststat), decreasing = TRUE)
> probeids <- geneNames(aafExpr)[index]
```

9.5.2 Annotating the probes

Once there is a list of probes, annotation is quite simple. The only decision that needs to be made is which classes of annotation to include in the table. Including all the annotation classes, which is the default, may not be a good idea. If the table grows too wide, its usefulness may decrease. To see which columns of data can be included, use the `aaf.handler` function. When called with no arguments, it returns all of the annotation types **annaffy** can handle.

```
> aaf.handler()
 [1] "Probe"              "Symbol"
 [3] "Description"        "Function"
 [5] "Chromosome"         "Chromosome Location"
 [7] "GenBank"            "LocusLink"
 [9] "Cytoband"           "UniGene"
[11] "PubMed"             "Gene Ontology"
[13] "Pathway"
```

To avoid typing errors, you should subset the vector instead of retyping each column name.

```
> anncols <- aaf.handler()[c(1:3, 8:9, 11:13)]
[1] "Probe"          "Symbol"         "Description"
[4] "LocusLink"      "Cytoband"       "PubMed"
[7] "Gene Ontology"  "Pathway"
```

This may still be too many columns, but it is possible, at a later stage, to choose to either not show some of the columns or remove them altogether.

Example Table without Data

Probe	Symbol	Description	LocusLink	Cytoband	PubMed	Gene Ontology	Pathway
31521_f_at	HIST1H4J	histone 1, H4j	8363	6p22-p21.3	3		
36508_at	GPC4	glypican 4	2239	Xq26.1	4	transmembrane receptor activity morphogenesis cell proliferation membrane extracellular matrix integral to plasma membrane	
40614_at	SHB	SHB (Src homology 2 domain containing) adaptor protein B	6461	9p12-p11	5	SH3/SH2 adaptor protein activity intracellular signaling cascade	
1724_at	E2F4	E2F transcription factor 4, p107/p130-binding	1874	16q21-q22	17	transcription factor activity regulation of transcription, DNA-dependent regulation of cell cycle transcription factor complex nucleus	Cell cycle
38573_at	YAF2	YY1 associated factor 2	10138	12q12	2	regulation of transcription, DNA-dependent nucleus	
151_s_at	OK/SW-cl.56	beta 5-tubulin	203068	6p21.32	7	GTP binding structural molecule activity chaperone activity microtubule-based movement tubulin natural killer cell mediated cytolysis MHC class I protein binding	

Figure 9.1. HTML table with annotation links.

Now we generate the annotation table with the `aafTableAnn` function using the `hgu95av2` meta-data package.

```
> anntable <- aafTableAnn(probeids[1:50], "hgu95av2",
+     anncols)
```

To see what has been produced so far, use the `saveHTML` method to generate the HTML report. Using the optional argument `open=TRUE` will open the resulting file in your browser, as seen in Figure 9.1.

```
> saveHTML(anntable, "ex1.html", title = "Example Table without Data")
```

9.5.3 Adding other data

To add other data to the table, just use any of the other table constructors to generate your own table, and then merge the two. For instance, listing the *t*-statistics along with the annotation data is quite useful. annaffy provides the option of coloring signed data, making it easier to assimilate.

```
> testtable <- aafTable("t-statistic" = teststat[index[1:50]],
+     signed = TRUE)
> table <- merge(anntable, testtable)
```

After HTML generation, a one line change to the style sheet header will change the colors used to show the positive and negative values. In fact, with the use of cascading style sheets (CSS), it is possible to heavily customize the appearance of the tables very quickly, even on a column by column basis.

annaffy also provides an easy way to include expression data in the table. It colors the cells with varying intensities of green to show relative expression values. Additionally, because of the way `merge` works, it will always match probe ID rows together, regardless of their order. This allows a quick sanity check on the other statistics produced, and can help decrease user error. You can check, for example, that the t-statistics and ranking seem reasonable given the expression data.

```
> exprtable <- aafTableInt(aafExpr, probeids = probeids[1:50])
> table <- merge(table, exprtable)
> saveHTML(table, "example2.html", title = "Example Table with Data")
```

Producing a tab-delimited text version uses the `saveText` method. The text output also includes more digits of precision than the HTML; see Figure 9.2.

```
> saveText(table, "example2.txt")
```

Example Table with Data

Cytoband	PubMed	Gene Ontology	Pathway	t-statistic	Ctrl 1	Ctrl 2	Ctrl 3	Ctrl 4	Exp 1	Exp 2	Exp 3	Exp 4
3p22-p21.3	3			5.0313	10.1627	10.7319	10.149	10.5501	10.9595	11.652	11.782	11.5925
Xq26.1	4	transmembrane receptor activity morphogenesis cell proliferation membrane extracellular matrix integral to plasma membrane		4.99789	6.60766	6.3524	6.32759	6.59396	6.84562	7.02293	6.9051	6.85139
2p12-p11	5	SH3/SH2 adaptor protein activity intracellular signaling cascade		4.86627	2.64711	2.54982	2.62247	2.57495	2.77171	2.68999	2.75897	2.7301
16q21-q22	17	transcription factor activity regulation of transcription, DNA-dependent regulation of cell cycle transcription factor complex nucleus	Cell cycle	3.75708	8.02172	7.78865	9.21191	9.97504	6.87789	6.46994	6.88466	6.88744
12q12	2	regulation of transcription, DNA-dependent nucleus		2.94629	3.21518	3.2995	3.44363	3.36414	2.946	3.21611	3.17096	3.08585
5p21.32	7	GTP binding structural molecule activity chaperone activity microtubule-based movement tubulin natural killer cell mediated cytolysis MHC class I protein binding		2.92641	4.29144	4.29099	4.27513	4.21659	4.51866	4.36423	4.33972	4.37075

Figure 9.2. HTML table with data columns added.

9.6   Graphical displays with drill-down functionality

Visualization is an essential part of a virtually every analysis. More details on the visualization methods presented here are available in Chapter 10. With the large complex data sets that are common in functional genomics, it is often useful to choose a hierarchical approach to the presentation of the information, with one or a few overview plots, which then allow readers to interactively *drill-down* and reveal more detailed information on particular features of interest.

For example, one might want to augment the heatmap of Chapter 10, Figure 10.2, by adding clinical data on the tumor samples and patient, and bioinformatic annotations on the genes. Including these in a static display would be difficult and likely to result in a very cluttered visualization where the primary message (the heatmap) is largely obscured. A popular but decidedly cumbersome and error-prone approach is to copy and paste the sample and gene identifiers into a separate database or spreadsheet application and to use that to look up the information. A functionality that allows the data analyst to associate the appropriate meta-data directly with the graphical output but leaves the reader free to select and inspect those entities of interest seems to provide substantial benefits and overcomes many of the stated difficulties.

There are a number of graphics formats and associated viewers that support this kind of drill-down, and we discuss two of these: HTML image maps and scalable vector graphics (SVG).

9.6.1 HTML image maps

An HTML *image map* consists of (at least) two files: the actual image file itself, in some graphic format suitable for rendering in a Web browser or other similar tool and an HTML file that contains the image annotation. The following HTML fragment shows a schematic example.

```
<html><body>
  <!-- further HTML here -->
  ...
<img src="myImage.png" usemap=#foo/>
<map name="foo">
<area shape="rect" coords="100,20,110,30"
   href="http://...." title="alpha-helical ferredoxin"/>
<area shape="rect" coords="210,300,280,310"
   href="http://...." title="RNase III endonuclease domain"/>
  ...
</map>
  <!-- further HTML here -->
</body></html>
```

The image is called `myImage.png` and is included using standard HTML. When the file that contains these commands is rendered in a browser, the image is rendered and the series of `area` statements are evaluated. The example assumes that the image file `myImage.png` exists and has dimensions of at least 280 pixels in the x-direction and 310 pixels in the y-direction. The text in the `title` parameter of the `area` tag will be used as a *tool-tip* for the corresponding area of the image, and the `href` parameter defines a *hyperlink*.

Tool-tips are short bits of explanatory text that are displayed when the mouse pointer moves into a certain region and disappear when the pointer leaves the region. The mouse-sensitive areas are specified by their shape and coordinates. In the example, the first area is the rectangular region with x-coordinates between 100 and 110 and y-coordinates between 20 and 30. The area tag allows many more parameters, for which we refer the reader to the documentation of the HTML language.

The `imageMap` from the package `geneplotter` can be used to generate an HTML fragment such as shown above from a specification of coordinates, tool-tips, and hyperlinks. In many cases, the hard part in using it with a plot produced by R will be the determination of the coordinates of the plot elements that are supposed to be mapped. For two specialized types of plots, support for this is already provided: for plate plots (see Figure 5.6), the function `plotPlate` in the package `prada` returns a list that contains the coordinates of the relevant plot elements, namely the circles that represent the wells. These coordinates then simply need to be passed to the `imageMap` function. For graphs, the package `Rgraphviz` defines a generic function `imageMap`, which takes a graph layout as input. Users can use this function to add tool-tips and hyper-links to the graph's nodes (see Section 21.4.5).

Image files can be produced, in a variety of formats, using built-in R commands. For example, the `png` device of R will often be a convenient choice. In addition, R has the devices `jpeg` and `bitmap`, the latter is a unified interface to many different bitmap file formats.

The interactivity of an image map cannot be adequately reproduced in a printed book; for an example we refer the reader to the on-line complements for this volume for an example. Or, they can simply try the example for the `imageMap` function and view the results on their own computer.

9.6.2 Scalable Vector Graphics (SVG)

SVG is a modularized language for describing two-dimensional vector and mixed vector/raster graphics in XML (W3C, 2003). Version 1.1 is dated January 2003. This is a very general and powerful graphics language, and among other things, it allows the annotation of graphics elements with tool-tips and hyper-links.

The R package `RSvgDevice`, which is available from *CRAN*, provides a graphics device that produces SVG output. Viewers for SVG, while not plentiful, are available for most platforms. A freely available basic viewer that can also be used as a browser plug-in is provided by Adobe Systems Incorporated.

9.7   Searching Meta-data

We now turn our attention to other programmatic interactions with the biological meta-data. Readers are referred to Chapter 8 for a detailed discussion of some of the other tools and capabilities in both R and Bioconductor. In order to facilitate the formulation and testing of such different hypotheses, annaffy includes functions to search annotation meta-data using various criteria. These search functions return character vectors of Affymetrix probe IDs that can be used to subset data and annotation.

9.7.1   Text searching

Currently, there are two simple search functions available in annaffy. The first is a text search that matches against the textual representation of biological meta-data. Recall that textual representations are extracted using the getText method. For complex annotation structures, the textual representation can include a variety of information, including numeric identifiers and textual descriptions. For the purposes of demonstration, we will use the hgu95av2 annotation data package available through Bioconductor.

The textual search is most straightforwardly applied to the Symbol, Description, and Pathway meta-data types. A specialized *GO* search will be discussed subsequently. For instance, to find all annotated kinases on a chip, a user can simply perform a text search of Description for kinases. It is important to emphasize that the search performed is based entirely on data supplied as part of the Bioconductor annotation packages.

```
> kinases <- aafSearchText("hgu95av2", "Description",
+     "kinase")
> kinases[1:5]

[1] "1000_at"   "1001_at"   "1008_f_at" "1010_at"
[5] "1015_s_at"

> length(kinases)

[1] 696
```

One can search multiple meta-data types with multiple queries all with a single function call. For instance, to find all genes with "ribosome" or "polymerase" in the Description or Pathway annotation, use the following function call.

```
> probes <- aafSearchText("hgu95av2", c("Description",
+     "Pathway"), c("ribosome", "polymerase"))
> print(length(probes))

[1] 198
```

When doing searches of multiple annotation data types or multiple terms, by default, the search returns all probe IDs matching any of the search criteria. That can be altered by changing the logical operator from *or* to *and* using the `logic="AND"` argument. This is useful because `aafSearchText` does not automatically split a search query into separate words (this is often referred to as *tokenizing*) as Google and many other search engines do. For example, "DNA polymerase" finds all occurrences of that exact string. To find all probes whose description contains both "DNA" and "polymerase," use the following function call.

```
> probes <- aafSearchText("hgu95av2", "Description",
+     c("DNA", "polymerase"), logic = "AND")
> print(length(probes))

[1] 42
```

Another useful application of text searching is to map a vector of Gen-Bank accession numbers onto a vector of probe ids. This comes in handy if you wish to filter microarray data based on the output of a BLAST query. Here we take a list of GenBank IDs and find every probe ID that contains one of those GenBank IDs in its annotation.

```
> gbs <- c("AF035121", "AL021546", "AJ006123", "AL080082",
+     "AI289489")
> aafSearchText("hgu95av2", "GenBank", gbs)

[1] "1954_at"    "32573_at"    "32955_at"    "34040_s_at"
[5] "35581_at"    "38199_at"
```

The text search is always case insensitive. Individual search terms are treated as Perl compatible regular expressions so you should be cautious of special regular expression characters. See Friedl (2002) for further information about how to use regular expressions.

9.8   Concluding Remarks

In this section we have provided a brief tour of some of the tools that can be used to construct interactive outputs. Our own use of them is largely as a method for communicating results to colleagues that are not necessarily computer savvy. We have found that they are an effective mechanism for communicating, especially at the early stages of manuscript preparation. They ensure that all authors are looking at, and talking about, the same outputs. We have also found them useful for providing on-line complements to published papers, seminars and workshops.

10

Visualizing Data

W. Huber, X. Li, and R. Gentleman

Abstract

Visualization is an essential part of exploring, analyzing, and reporting data. Visualizations are used in all chapters in this monograph and in most scientific papers. Here we review some of the recurring concepts in visualizing genomic and biological data. We discuss scatterplots to investigate the dependency between pairs of variables, heatmaps for the visualization of matrix-like data, the visualization of distance relationships between objects, and the visualization of data along genomic coordinates.

10.1 Introduction

Visualization has long been recognized as an essential part of *exploratory data analysis* practice and of reporting the results of an analysis. It is also instrumental in quickly answering goodness-of-fit and data quality questions. A well-designed plot has the ability to reveal things that we did not anticipate, and in many cases it should permit the reader to easily contrast what was observed with what should have been observed if certain assumptions held.

Given the importance of visualization methods for the comprehension of microarray and other high-throughput experiments, it is similarly important that good tools for producing plots and diagnostic displays be readily available. In this chapter, we will consider some of the basic tools. Visualizations are used in many places throughout this monograph, and in most cases specialized tools are described alongside the examples. *Silhouette plots* are discussed in Chapter 13, while plotting methods for *graphs and networks* are covered in detail in Part IV, specifically in Chapters 21 and 22. *Spatial plots* are useful for detecting spatial patterns, and applications to microarray quality control and to the analysis of data measured on microtiter-plates have been shown in Chapters 3, 4, and in Section 5.4.2.

High-throughput biological data from functional genomics, proteomics, etc., create many challenges for visualization methods. The data sets are large, and in most cases it is essential that the data be related to known biological information, for example the mapping of the probe sequences to genomic loci.

One of the strengths of R is the fact that plots are highly customizable. You can easily create your own visualizations and then make tools available to others so they can create similar plots. The grid and lattice packages offer new and expanded capabilities compared to the base graphics system. Many of the contributed packages on Bioconductor and CRAN extend the graphical capabilities of R or provide specialized graphical outputs for particular applications. From Bioconductor, we mention the arrayMagic, arrayQuality, and aCGH packages for microarray data visualization, from CRAN, gclus for the ordering of panels in scatterplot matrices and parallel coordinate displays by a merit index, scatterplot3d for 3D scatter plots, vcd for the visualization of categorical data.

10.2 Practicalities

In statistics and in other areas of science, there have been many good books about the visualization of data. We recommend the following monographs: Cleveland (1993, 1994), Bertin (1973), Tufte (2001), and Tufte (1990). Merely drawing a plot does not constitute visualization. Visualization is about conveying *important* information to the reader *accurately*. It should *reveal* information that is in the data and should not impose structure on the data. These are not trivial requirements, and there are many examples of plots that are full of irrelevant clutter or that distort the facts.

Color is an important aspect of visualization. We particularly note the work of Brewer (1994a,b), some of which has been implemented in the R package RColorBrewer. Brewer showed that specific color schemes are appropriate for specific tasks. For example, a color scheme that is suitable for comparing ordinal data is different from one suitable for the display of nominal data.

An important notion is that of *distance* in color space. For quantitative data, distances in color space should reflect an appropriate distance measure in data space. A related requirement is that the available room in color space should be used to convey as much information as possible. Color schemes should be intuitive, consistent, and ergonomic. A now famous example for an unergonomic color scheme is the red-green color scheme that has been widely used for the display of microarray data. That color scheme is undesirable because a sizeable proportion of the audience is red-green color-blind and hence unable to interpret these plots.

The complexity and richness of genomic data argues persuasively for some form of interactivity. However, the interactive visualization capabilities in R itself are somewhat limited. The functions `locator` and `identify` can be used in conjunction with other functions to provide some interactive graphics but their use is not widespread. More comprehensive capabilities are provided by add-on packages such as tcltk and RGtk (available from Omegahat). Both offer the potential for interactive graphical interfaces. The package iSPlot, which is based on RGtk, provides tools for creating interactive, linked plots (brushing). Options for the production of graphical displays with *drill-down* functionality have been discussed in Chapter 9.6. Another option for interactive visualization that includes *brushing* and 3D scatterplots is the visualization program *GGobi* (Swayne et al., 2003), which has an elegant interface to R via the Rggobi package. An example for this is described in Section 5.4.3.

10.3   High-volume scatterplots

Scatterplots are a powerful tool for the visualization of data from two variables with a small or intermediate number of observations. However, it is difficult to get a good impression of the distribution of the data when the number of observations gets large and the points become so dense that they form a featureless blot. Methods for plotting large data sets have been studied systematically for many years. The work of Carr et al. (1987) is of particular relevance, and their method of dividing the data into a two-dimensional histogram of hexagonal bins has been implemented in the hexbin package.

   In Figure 10.1a, we show a scatterplot of the data from two Affymetrix GeneChips with 409,600 probes each. The code below loads an example data set, extracts the intensity vector of the first two arrays, transforms them to the logarithmic scale, and calculates the sum `A` and difference `M` of the two vectors. The matrix multiplication operator `%*%` provides a compact and fast way to this.

```
> library("affydata")
> data("Dilution")
> x <- log2(exprs(Dilution)[, 1:2])
> x <- x %*% cbind(A = c(1, 1), M = c(-1, 1))

> plot(x, pch = ".")
```

   Above a certain threshold density of points it is not possible to visually distinguish between more or less dense regions: the plot becomes a featureless black shape. Some relief from this problem is provided by the hexagonal binning algorithm in the hexbin package.

```
> library("hexbin")
> library("geneplotter")
> hb <- hexbin(x, xbins = 50)
> plot(hb, colramp = colorRampPalette(brewer.pal(9,
+     "YlGnBu")[-1]))
```

In the call to hexbin we need to specify the parameter xbins, which is the number of bins that are used for the x-dimension, and determines the hexagon size. Note that the hexagons are equilateral, which implies that the variables along the x and y dimensions are on comparable scales. We use the YlGnBu colormap from the RColorBrewer package, which is appropriate for sequential variables. The map extends in nine steps from light yellow to dark blue. We remove the first step, which is too close to the white background color and interpolate into a smooth color ramp with the colorRampPalette. The result is shown in Figure 10.1b.

An alternative approach is provided by the function smoothScatter in the package prada.

```
> library("prada")
> smoothScatter(x, nrpoints = 500)
```

The result is shown in Figure 10.1c. A variation on this theme is the function densCols, which calculates the local density of data points for each observation, and returns a false color representation (Figure 10.1d).

```
> plot(x, col = densCols(x), pch = 20)
```

10.3.1 A note on performance

Scatterplots like the one in Figure 10.1a can take a long time to render, and when saved in a vector graphics format such as PDF or Postscript, they tend to produce large files. More importantly, they are not informative and hide rather than reveal the structure of the data. Many of these problems can be ameliorated by the use of a binning method such as the hexagon binning shown in Figure 10.1b. For such plots, the file size and the drawing time, is proportional to the number of bins and, once the binning has been carried out, independent of the number of data points.

We note that the reduction to hexagon bins can also be used to speed up regression or smoothing algorithms, which can be applied to observations located at the bin centers using weights that correspond to the bin counts. This makes the complexity of these algorithms independent of the number of data points. If the bin size is chosen appropriately there will be no noteworthy loss in the precision of the fitted curves.

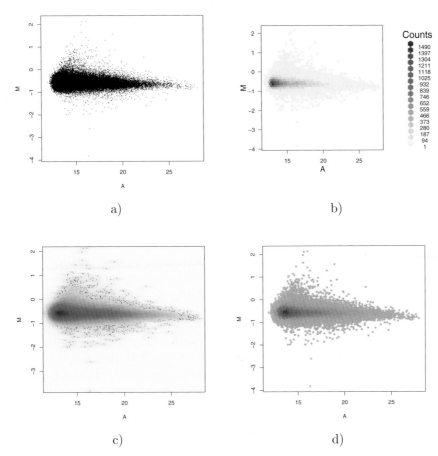

a)   b)

c)   d)

Figure 10.1. Four different visualizations of the same data, the mean-difference plot of the unprocessed probe intensities from a pair of microarrays. Panel a) shows the usual scatterplot. Because of the large number of points, it is a rather featureless black blot. Panel b) shows the result of a hexagon binning procedure. The color code at each hexagon represents the number of data points that it contains. In panel c) we see a color representation of a smooth density on the (x, y)-plane calculated from the data using a kernel density estimator. In the sparse regions of the density, the plot is augmented by black dots that represent individual data points. In the denser regions, these are omitted. Note that at the boundary between sparse and dense region (as assigned by the algorithm), a visual artifact is created. Panel d) shows the usual scatterplot, however with points colored according to the local density.

10.4   Heatmaps

Heatmaps, or false color images have a reasonably long history, as has the notion of rearranging the columns and rows to show structure in the data. They were applied to microarray data by Eisen et al. (1998) and have become a standard visualization method for this type of data.

A heatmap is a two-dimensional, rectangular, colored grid. It displays data that themselves come in the form of a rectangular matrix. The color of each rectangle is determined by the value of the corresponding entry in the matrix. The rows and columns of the matrix can be rearranged independently. Usually they are reordered so that similar rows are placed next to each other, and the same for columns. Among the orderings that are widely used are those derived from a hierarchical clustering, but many other orderings are possible. If hierarchical clustering is used, then it is customary that the dendrograms are provided as well. In many cases the resulting image has rectangular regions that are relatively homogeneous and hence the graphic can aid in determining which rows (generally the genes) have similar expression values within which subgroups of samples (generally the columns).

The function heatmap is an implementation with many options. In particular, users can control the ordering of rows and columns independently from each other. They can use row and column labels of their own choosing or select their own color scheme. They can also add a colored bar to annotate either the row or the column data (e.g., to show the association with some phenotype). And perhaps most importantly they can take the standard R implementation and extend it in any way they would like.

We return to the ALL example to provide a demonstration of a heatmap. We select two small subgroups of patients for this examination, the ALL1/AF4 group and the E2A/PBX1 group.

```
> library("ALL")
> data("ALL")
> selSamples <- ALL$mol.biol %in% c("ALL1/AF4",
+       "E2A/PBX1")
> ALLs <- ALL[, selSamples]
> ALLs$mol.biol <- factor(ALLs$mol.biol)
> colnames(exprs(ALLs)) <- paste(ALLs$mol.biol,
+       colnames(exprs(ALLs)))
```

There are 15 samples and they are stored in the *exprSet* object ALLs. Among the total of 12625 probes in this data set, we select those with mean expression larger than 100 in at least one of the two groups, and a p value of the two-sample t test of less than 0.0002.

```
> library("genefilter")
> meanThr <- log2(100)
> g <- ALLs$mol.biol
```

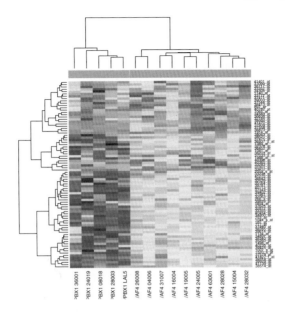

Figure 10.2. A heatmap comparing the ALL1/AF4 group (brown) to the E2A/PBX1 group (light blue).

```
> s1 <- rowMeans(exprs(ALLs)[, g == levels(g)[1]]) >
+     meanThr
> s2 <- rowMeans(exprs(ALLs)[, g == levels(g)[2]]) >
+     meanThr
> s3 <- rowttests(ALLs, g)$p.value < 2e-04
> selProbes <- (s1 | s2) & s3
> ALLhm <- ALLs[selProbes, ]
```

This results in a set of 81 probes. We can now draw the heatmap for this data set, Figure 10.2. The results are quite striking and we can easily separate the two groups. We have chosen to use a set of colors from the RColorBrewer package that come from a *diverging palette*. Diverging palettes are designed to put equal emphasis on mid-range values and the extremes at both ends of the data range. You may want to try a number of choices, different palettes are given in the RColorBrewer package.

```
> hmcol <- colorRampPalette(brewer.pal(10, "RdBu"))(256)
> spcol <- ifelse(ALLhm$mol.biol == "ALL1/AF4",
+     "goldenrod", "skyblue")
> heatmap(exprs(ALLhm), col = hmcol, ColSideColors = spcol)
```

10.4.1   Heatmaps of residuals

Statistical models try to explain the observed data in terms of systematic effects and residual random variation. One example is the functional model with additive error

$$Y = f(x) + \varepsilon, \tag{10.1}$$

where Y is the vector of observed data, x represents a set of explanatory variables, the function f models the systematic dependence of the observed quantities on the explanatory variables, and ε represents the random error. Often, interest centers on the estimation of f and there are many different methods that can be employed to obtain an estimate, $\hat{f}$. Equipped with $\hat{f}$, we can compute the residuals, $\hat{\epsilon} = Y - \hat{f}$.

The residuals often play an important role in assessing the fit of the model, and a failure of $\hat{f}$ to adequately capture the structure that is present in Y is usually reflected in the residuals. A similar approach can be used for gene expression data; in many cases we estimate some model to explain the observed data, and hence we can use that model to obtain a matrix of residuals.

To demonstrate the approach, we will use the estrogen data. These are explained in Appendix A.1.3. We preprocess the data, define a linear model, and obtain the resulting object fit in the same way as described in Chapter 14 on pages 241 and 244.

For each probeset, we have eight measurements, and the model has four coefficients: an overall baseline, the treatment effect [estrogen stimulation yes (+) or no (-)], the time effect (10h or 48h), and the interaction between treatment and time, that is, the difference in the treatment effect between the 10h and 48h time points. That leaves four residual degrees of freedom for each probeset.

We can now compare the expression values that would be predicted by this model to actual expression values. For this, we first define a function predict.MArrayLM that produces a synthetic expression matrix according the fitted model, and then calculate the difference between these values and the values in the *exprSet* esEset.

```
> predict.MArrayLM <- function(f, design = f$design) {
+     return(f$coefficients %*% t(design))
+ }
> esFit <- predict(fit)
> res <- exprs(esEset) - esFit
```

It is difficult to visualize the full data set with 12625 probesets all ot once, hence let us focus on the 50 probesets with the largest estimated values of the treatment–time interaction:

```
> sel <- order(fit$coefficients[, "ES:T48"], decreasing = TRUE)[1:50]
> four.groups <- as.integer(factor(colnames(exprs(esEset))))
```

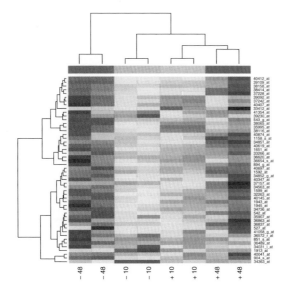

Figure 10.3. Heatmap of the estrogen data for the 50 probesets with the highest treatment–time interaction. The horizontal color bar corresponds to the 2×2 factor levels for treatment and time.

```
> csc <- brewer.pal(4, "Paired")[four.groups]
> heatmap(exprs(esEset)[sel, ], col = hmcol, ColSideColors = csc)
```

The resulting heatmap is shown in Figure 10.3.

```
> heatmap(res[sel, ], col = hmcol, ColSideColors = csc)
```

In Figure 10.4, we see that the array in the leftmost column has predominantly positive (red) residuals for the 50 probes shown here. Further inspection reveals that the overall distribution of all 12625 residuals for that array is approximately symmetric about 0. Also, the size of the residuals for this array is much larger than those of, say, the second array. The dependence of the residuals between different probes and the heteroskedasticity point to problems with data quality or normalization. Note that the residuals are, by the nature of the linear model, not independent across arrays: the fit is such that the residuals within each factor level sum up to zero. In this example, each factor level corresponds to two arrays, so the residuals from one array are -1 times those of the other. Due to the way that the column reordering in the heatmap function works, this leads to the antisymmetry of the plot about the vertical line between the fourth and fifth column.

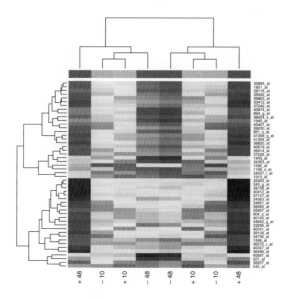

Figure 10.4. Heatmap of the residuals of the linear model for the estrogen data described in Section 14.3.1. In such a plot, we can look for patterns in the residuals that might indicate problems with the model fit.

10.5   Visualizing distances

The data-analytic and modeling aspects of distances are covered in some detail in Chapter 12. In this section, we will discuss some of the tools that are used to visualize distances and expose some of their strengths and weaknesses.

While the *dendrogram* has been widely used to represent distances between objects, it cannot really be considered to be a visualization method. Dendrograms do not necessarily expose structure that exists in the data. In many cases they impose structure on the data, and when that is the case it is dangerous to interpret the observed structure. Hierarchical clustering creates a new set of between-object distances, corresponding to the path lengths between the leaves of the dendrogram. It is interesting to ask whether these new distances reflect the distances that were used as inputs to the hierarchical clustering algorithm. The cophenetic correlation [e.g. Sneath and Sokal (1973, p. 278)], implemented in the function `cophenetic`, can be used to measure the association between these two different distance measures.

As an example, we compute between sample distances for the ALL data. We first standardize the gene expression values across samples and then use Euclidean distance on the standardized values as the metric between samples (so the distance measured is equivalent to a correlation-based distance).

```
> standardize <- function(z) {
+       rowmed <- apply(z, 1, median)
+       rowmad <- apply(z, 1, mad)
+       rv <- sweep(z, 1, rowmed)
+       rv <- sweep(rv, 1, rowmad, "/")
+       return(rv)
+ }
> ALLhme <- exprs(ALLhm)
> ALLdist1 <- dist(t(standardize(ALLhme)))
> ALLhc1 <- hclust(ALLdist1)

> plot(ALLhc1, xlab = "", sub = "", main = "ALLhc1")
```

In Figure 10.5a, we see that there is a substantial difference between two groups. If instead of selecting probes based on their t-statistic from the two-sample comparison, we had selected probes simply on their overall variability, a different picture emerges.

```
> ALLsub2 <- exprs(ALLs[(s1 | s2), ])
> rowMads <- apply(ALLsub2, 1, mad)
> ALLsub2 <- ALLsub2[rowMads > 1.4, ]
> ALLdist2 <- dist(t(standardize(ALLsub2)))
> ALLhc2 <- hclust(ALLdist2)

> plot(ALLhc2, xlab = "", sub = "", main = "ALLhc2")
```

The resulting dendrogram is shown in Figure 10.5b. The two groups still separate, but the differences are not as strong and the within group variability is larger.

In the next code chunk, we compute the cophenetic distances for both data sets and then compare them to the original distances. High correlations indicate good agreement between the original distances and those assigned by hierarchical clustering. We note that the correlations are very good, and indicate that the dendrogram is a better match for the t-test selected genes than for those selected on the basis of overall variability. The scatterplots are shown in Figure 10.6.

```
> ALLcph1 <- cophenetic(ALLhc1)
> cor(ALLdist1, ALLcph1)

[1] 0.99

> plot(ALLdist1, ALLcph1, pch = "|", col = blue)

> ALLcph2 <- cophenetic(ALLhc2)
> cor(ALLdist2, ALLcph2)

[1] 0.877

> plot(ALLdist2, ALLcph2, pch = "|", col = blue)
```

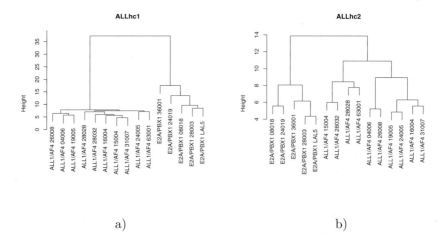

Figure 10.5. Dendrograms for the ALL1/AF4 and E2A/PBX1 samples. The clustering was obtained a) using the 81 probes in ALLhme that were selected in Section 10.4 by the t-statistic, b) using the 58 probes in ALLsub2 that were filtered for their overall variability.

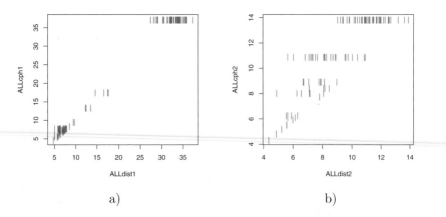

Figure 10.6. Scatterplots of actual distances versus cophenetic distances. a) Distances calculated with t-test selected probes, b) variability-selected probes. Each pair of distances is shown by a vertical bar. Note that the cophenetic distances only take on a discrete set of values.

10.5.1 Multidimensional scaling

Another useful tool for examining distances is *multidimensional scaling* (MDS). Starting from a matrix of all pairwise distances or dissimilarities between n objects, the aim of MDS is to arrange n points in a k-dimensional Euclidean space such that the distances between the points are as much like the given distances as possible. There are a variety of ways in which "like" can be strictly defined, and these lead to different flavors of MDS. See the references (Cox and Cox, 2001; Ripley, 1996a; Borg and Groenen, 1997) for more details on MDS.

In R there are three different MDS functions that are readily available, cmdscale in the stats package, and isoMDS and sammon in MASS.

cmdscale computes classical metric MDS, which uses a least-squares definition of "like." Its solution can be found by computing the eigendecomposition of a suitably defined matrix, the so-called doubly centered matrix of squared distances. A nice property of classical MDS is that the dimensions are nested, that is, the first two dimensions of the $k = 3$ solution are the same as the $k = 2$ solution.

To assess the goodness-of-fit of a classical MDS solution, cmdscale returns two statistics. One is the sum of the eigenvalues for the components S divided by the sum of the absolute value of all eigenvalues, and the other is S divided by the sum of all positive eigenvalues. To decide how many dimensions are necessary to adequately represent your data, it is useful to look at the *scree plot*, that is, the plot of the goodness-of-fit statistic as a function of k. A criterion for the choice of k is to pick a solution for which adding more dimensions does not significantly improve the goodness-of-fit.

isoMDS provides one form of non-metric MDS. It chooses a k-dimensional configuration to minimize the loss-function

$$s^2 = \frac{\sum_{i \neq j} [f(p_{ij}) - d_{ij}]^2}{\sum_{i \neq j} d_{ij}^2}, \tag{10.2}$$

where p_{ij} is the original distance matrix, f is a monotonic transformation, and d_{ij} are the distances between the MDS points. s is also called the *stress*.

Yet another variant of non-metric MDS is provided by sammon, which uses a different kind of loss-function. The different variants of MDS lead to different relative importances of large versus small distances to the fitted MDS solution. In some cases, one may be more interested in preserving the local similarities at the expense of more distant relationships, whereas in other cases, it may just be the global structure that one is interested in and that should be preserved in the low-dimensional reduction.

In the code below, we use both metric MDS (cmdscale) and the Sammon version of non-metric MDS (sammon). We request the computation of eigenvalues for `cmdscale` so that we can explore the goodness-of-fit question. The results are shown in Figure 10.7.

```
> library(MASS)
> cm1 <- cmdscale(ALLdist1, eig = TRUE)
> cm1$GOF

[1] 0.908 0.908

> samm1 <- sammon(ALLdist1, trace = FALSE)
> cm2 <- cmdscale(ALLdist2, eig = TRUE)
> cm2$GOF

[1] 0.646 0.646

> samm2 <- sammon(ALLdist2, trace = FALSE)

> ALLscol <- c("goldenrod", "skyblue")[as.integer(ALLs$mol.biol)]
> plot(cm1$points, col = ALLscol, ...)
```

We note that there is an opportunity for circularity to enter into your analysis; if you select genes, as we did, based on a two-sample *t*-test, then it should come as no surprise that the samples fall in two groups.

In our last example of a visualization method for distances, we make use of the `heatmap` function described in the previous section. Here we call `heatmap` with both `sym=TRUE` and we specify our own distance function. The resulting plot is shown in Figure 10.8. Where we can see that it is symmetric and again there is a strong indication that there are two (or possibly three) groups of samples.

```
> heatmap(as.matrix(ALLdist2), sym = TRUE, col = hmcol,
+        distfun = function(x) as.dist(x))
```

10.6   Plotting along genomic coordinates

Some of the genetic defects that are associated with cancer such as deletions and amplifications induce correlations in expression that are related to chromosomal proximity. Genomic regions of correlated transcription have also been identified in normal tissues (Su et al., 2004). This motivates the development of tools that relate gene expression to chromosomal location.

Genomic DNA is double stranded: one strand is called the *sense strand*, the other the *antisense strand*. The sense strand is sometimes also called the *Watson* or "+" strand, and the antisense strand is sometimes called the *Crick* or "-" strand. Both strands can contain coding sequences for genes, and the visualization methods we consider reflect this.

The plotting functions that we consider in this section are available in the **geneplotter** package. We first build the object `chrLoc`,

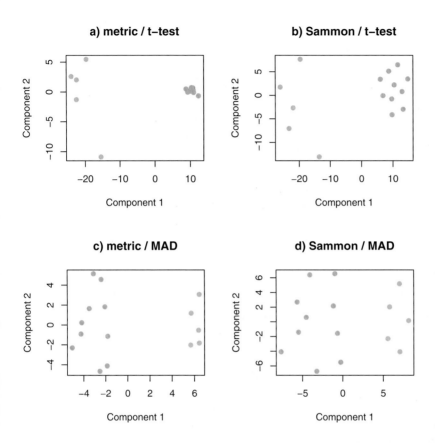

Figure 10.7. MDS plots of the ALL data. The patients with E2A/PBX are colored red and those with ALL1/AF4 are colored blue. The four panels compare different methods of MDS and feature selection. a) metric MDS using *t*-test selected features, b) Sammon MDS using *t*-test selected features, c) metric MDS using MAD selected features, d) Sammon MDS using MAD selected features.

```
> library("geneplotter")
> chrLoc <- buildChromLocation("hgu95av2")
```

which is of class *chromLocation*. It contains the location of all genes that were assayed on the HG-U95Av2 chip, on which the example data set ALL was measured. We select the highly expressing genes using s1 and s2 from above and compute the mean expression for each probe separately for the two groups of patient samples, ALL1/AF4 and E2A/PBX1.

```
> ALLch <- ALLs[s1 | s2, ]
> m1 <- rowMeans(exprs(ALLch)[, ALLch$mol.biol ==
+      "ALL1/AF4"])
```

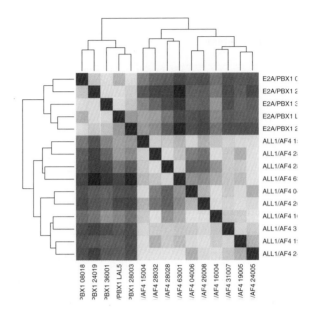

Figure 10.8. A heatmap of the between-sample distances.

```
> m2 <- rowMeans(exprs(ALLch)[, ALLch$mol.biol ==
+     "E2A/PBX1"])
```

Next, we compute the deciles of the combined data. We will color the genes in each decile differently.

```
> deciles <- quantile(c(m1, m2), probs = seq(0,
+     1, 0.1))
> s1dec <- cut(m1, deciles)
> s2dec <- cut(m2, deciles)
> gN <- names(s1dec) <- names(s2dec) <- geneNames(ALLch)
```

In the following code, we select a sequential color palette from the RColor-Brewer package and define a plot layout with three panels. We then use the function cPlot to plot horizontal lines, one for each chromosome, on which the gene locations are marked by vertical ticks. The function cPlot is then used to color the selected probes in gN according to the decile in which they are in.

```
> colors <- brewer.pal(10, "RdBu")
> layout(matrix(1:3, nr = 1), widths = c(5, 5, 2))
> cPlot(chrLoc, main = "ALL1/AF4")
> cColor(gN, colors[s1dec], chrLoc)
> cPlot(chrLoc, main = "E2A/PBX1")
> cColor(gN, colors[s2dec], chrLoc)
> image(1, 1:10, matrix(1:10, nc = 10), col = colors,
```

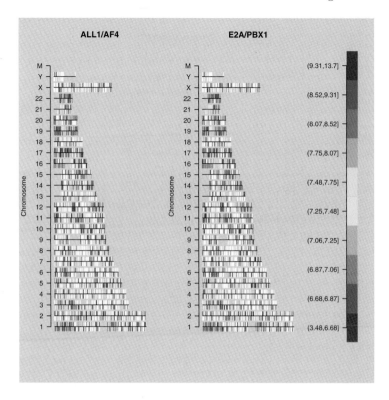

Figure 10.9. Side-by-side whole genome plots comparing expression levels in ALL1/AF4 and E2A/PBX1.

```
+       axes = FALSE, xlab = "", ylab = "")
> axis(2, at = (1:10), labels = levels(s1dec), las = 1)
```

The result is shown in Figure 10.9. Genes are represented by short vertical lines which go up if the gene is on the sense strand and down if it is on the anti-sense strand. Although not all differences are obvious, some inspection reveals that there appear to be some real differences on Chromosomes 20 and 22. We will explore Chromosome 22 further using a related, but more detailed, chromosomal plotting mechanism, provided by the function plotChr.

plotChr produces one plot per chromosome. Each sample has two smooth lines. The one in the top half of the plot represents genes on the sense strand and the line in the bottom half of the plot represents expression for genes encoded on the anti-sense strand. Low expression values are near the center line, and high expression values are toward the edge of the plot.

```
> par(mfrow = c(1, 1))
> msobj <- Makesense(ALLs, "hgu95av2")
```

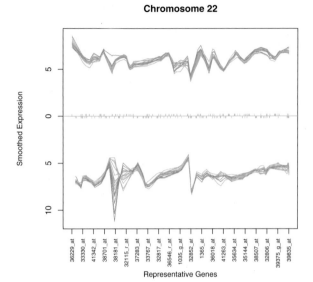

Figure 10.10. Plot of the expression data of individual ALL1/AF4 and E2A/PBX1 samples. Each sample is plotted as a continuous line, the x-values are determined by the location of the gene on the chromosome and the y values are expression values. To remove some of the noise, the lines have been smoothed.

```
> plotChr("22", msobj, col = ifelse(ALLs$mol.biol ==
+    "ALL1/AF4", "#EF8A62", "#67A9CF"), log = FALSE)
```

We can see in the plot of Chromosome 22 in Figure 10.10 that there are a few loci of interest, two on the anti-sense strand and one of the sense strand. Next we could try to extract the relevant genes and try to understand whether these differences in expression might be related to the outcome.

10.6.1   Cumulative Expression

The function `alongChrom` plots gene expression with the genes ordered by their chromosomal location. It can sometimes be useful to look at cumulative expression, that is, the running sum of expression values along the chromosome. The motivation for this is that on the level of individual loci, the technical and biological variability between samples can be large enough to obscur systematic differences due to copy number changes, while the cumulative values are more precise.

J.-P. Bourquin compared gene expression profiles between children with Down's syndrome (trisomy 21) and a transient myeloid disorder and

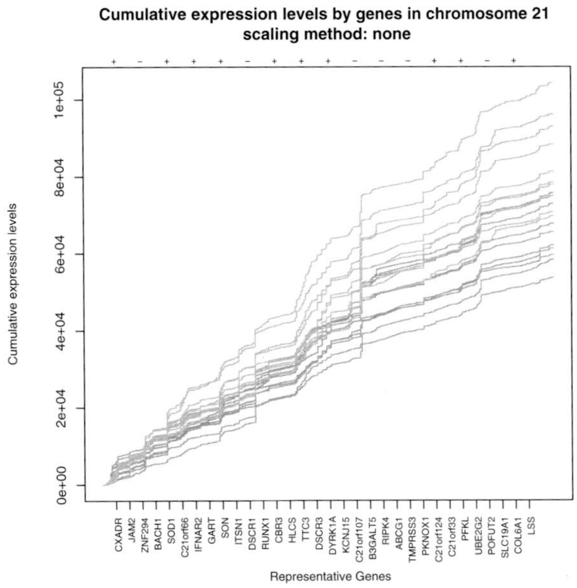

Figure 10.11. Cumulative expression profiles along Chromosome 21 for samples from 10 children with trisomy 21 and a transient myeloid disorder, colored in red, and children with different subtypes of acute myeloid leukemia (M7), colored in blue.

children with different subtypes of acute myeloid leukemia (M7). In Figure 10.11, we show cumulative expression for patient samples (10 from each group). Those with Down syndrome are colored red. There is a fairly good divergence between the two groups.

10.7   Conclusion

We have shown suggestions for the visualization of different aspects of microarray expression data. There are many more visualizations that might also be useful to comprehend the data. In particular, for plots along genomic coordinates, the increasing density and resolution of microarray probes, with applications such as chromatin immuno precipitation (ChIP) arrays, genomic tiling arrays, SNP chips, and DNA copy number arrays (Solinas-Toldo et al., 1997; Snijders et al., 2001), and the desire to integrate and report these data are generating many new challenges for visualization.

Part III

Statistical analysis for genomic experiments

11

Analysis Overview

V. J. Carey and R. Gentleman

Abstract

Chapters in this part of the book address tasks common in the downstream analysis (after preprocessing) of high-dimensional data. The basic assumption is that preprocessing has led to a sample for which it is reasonable to make comparisons between samples or between feature-vectors assembled across samples. Most examples are based on microarray data, but the principles are much broader and apply to many other sources of data. In this overview, the basic concepts and assumptions are briefly sketched.

11.1 Introduction and road map

Chapters in this section address approaches for deriving biological knowledge and formally testing biological hypotheses on the basis of experimental data. We concentrate on DNA microarray data, but other high-throughput technologies such as protein mass spectrometry, array comparative genomic hybridization (aCGH), or chromatin immuno-precipitation (ChIP) are also relevant.

The major focus of this section is on the application of unsupervised and supervised machine learning technologies to the analyses of these large complex data sets. We begin by considering distance measures, as these play an important role in machine learning. We next consider supervised and unsupervised machine learning in some detail and consider multiple testing methodologies and their application to the problems considered in the earlier chapters. The final chapter reviews a Bioconductor approach to browser-based workflow support for downstream analysis.

11.1.1   Distance concepts

It is both common and fruitful to invoke metaphors of spatial organization when discussing high-dimensional data structures arising in various disciplines. Thus while it is sometimes physically sensible to speak of the distance between two genes as a quantity measured in base pairs along a chromosome, it is also sometimes appropriate to speak of the distance between two genes as a quantity measured by the correlation of expression values obtained on a series of samples. The former concept of distance is precise but breaks down for genes present on different chromosomes, whereas the latter concept of distance can be made meaningful in a wide variety of settings. Chapter 12 describes conceptualizations and formalisms of distances for general structures represented in mathematical models of multidimensional spaces. The implications for microarray data analysis are numerous. The definition of a gene cluster in expression space over a series of samples is crucially dependent on selection of a distance definition. Cluster structures and inferences on co-regulation may change when aspects of the underlying distance model are altered. Distances among samples defined in terms of sample phenotype or clinical features are also of interest, but the mathematical construction of a distance function for such features can be complex.

11.1.2   Differential expression

A very common objective of microarray studies is the identification of sets of genes that are consistently expressed at different levels under different conditions. Chapter 14 illustrates this activity with data on leukemia, kidney cancer, and estrogen responsiveness.

11.1.3   Cluster analysis

Identification of shared patterns of expression across samples is basic to exploratory reasoning about co-regulation. Chapter 13 describes new developments in hierarchical clustering based on intensive resampling and evaluation of strength of cluster membership based on the silhouette function. This function, defined formally in Section 13.2.7, measures the relative magnitudes of within- and between-cluster proximities.

11.1.4   Machine learning

The volume of information in high-throughput bioinformatics gives rise to some doubts that traditional approaches to exploratory and confirmatory statistical inference can discover the latent patterns from which new biological understanding can be developed. Machine learning theory and methods attack problems of pattern recognition in voluminous noisy data with minimal human input. Chapter 16 describes basic concepts of computational learning theory and illustrates a number of applications of such tools as neural nets and random forests to microarray data. Chapter 17 specializes the focus to learning procedures based on weighted voting results among ensembles of learners.

11.1.5   Multiple comparisons

Effective use of statistics with data involves recognizing the inherent trade-off between sensitivity and specificity of inference procedures. In the context of differential expression studies, the sensitivity of a procedure is related to its tendency to identify differential expression when it is actually present. The specificity of a procedure is related to its tendency to refrain from identifying differential expression when it is in fact not present. When large numbers of inferences are attempted, as is common in microarray studies, the calibration of sensitivity and specificity of procedures is complex and requires understanding of statistical dependencies among the test statistics used for inference. Chapter 15 reviews the key concepts and illustrates fundamental tools available in Bioconductor.

11.1.6   Workflow support

Inference with Affymetrix microarray data proceeds from capture of CEL files with minimally processed intensities, specification of sample-level covariates, selection of analysis strategies, calibration of multiple test procedures, and interpretation of resulting gene sets using biological meta-data. The webbioc package (Chapter 18) is a browser-based interface that guides users through these steps.

11.2   Absolute and relative expression measures

One of the main differences between the Affymetrix and cDNA array technologies is that Affymetrix arrays are typically used to measure the *overall* abundance of a probe sequence in a target sample, whereas cDNA arrays typically measure the *relative* abundance of a probe sequence in two target samples. That is, the expression measures for Affymetrix arrays are typically *absolute* (log) intensities, whereas they are (log) *ratios* of intensities for cDNA arrays. In many cases, one of the samples in a cDNA array hybridization is a common reference used across multiple slides and whose sole purpose is to provide a baseline for direct comparison of expression measures between arrays.

For Affymetrix arrays, a direct comparison of expression measures between genes is problematic because the measurement units are not the same across genes. The measured fluorescence intensities are roughly proportional to mRNA abundance, but the proportionality factor is different for each gene. When using short oligonucleotide arrays, it is a function of the probes used and in particular of the frequencies of the different nucleotides in each probe. What we mean specifically is that between-sample, within-gene comparisons are valid and sensible, but within-sample, between-gene comparisons are not easy to make. If for gene X patient i has an estimated expression measure of 100, while for gene Y that same patient has an expression value of 200, these observed data tell us nothing about the real relative abundance of the mRNAs for these two genes. There could, in fact be more copies of the mRNA for gene X. On the other hand, if a second patient, j, say has an expression measure of 200 for gene X, we would conclude that the abundance of mRNA for X in patient j is likely higher than that observed in patient i.

For cDNA arrays, abundance is not measured directly but rather relative to some standard reference. Consider a patient with estimated relative abundance of 1 for gene X and 2 for gene Y. Then, we infer that gene X is expressed at approximately the same level in patient i as in the reference sample, while gene Y has approximately twice the abundance in patient i as in the reference sample. Note that we still do not know if the absolute abundance is the same or not since the we do not know the true abundance of either mRNA in the reference sample.

In some sense, the distinction between the two types of expression measures is artificial, as one could always select a particular Affymetrix array to use as a reference and take ratios of all expression measures to this referent. This will not be quite as successful as it is for cDNA arrays, because with cDNA arrays both the sample of interest and the reference sample are co-hybridized to the same slide. Co-hybridization is a form of blocking, and blocking in experimental design can provide substantial increases in precision.

We would like to stress that whether there is any real difference between the use of absolute and relative measures depends on the distance being considered, as demonstrated below.

A secondary consideration, that will not be explored here, is that differences in the probes can affect the variability of the expression measures. For short oligonucleotide arrays, there is some evidence that the variability of the estimated expression levels can be quite different across probes for the same gene. The variability may be a function of the mean level of expression (Rocke and Durbin, 2001), but there can be other substantial sources of variation as well. It is unlikely that this phenomenon is restricted to short oligonucleotide data. In particular, one would expect cDNA array data to exhibit similar behavior. Taking ratios with respect to a common reference (that was subjected to the same hybridization and scanning conditions) may provide some relief.

12

Distance Measures in DNA Microarray Data Analysis.

R. Gentleman, B. Ding, S. Dudoit, and J. Ibrahim

Abstract

Both supervised and unsupervised machine learning techniques require selection of a measure of distance between, or similarity among, the objects to be classified or clustered. Different measures of distance or similarity will lead to different machine learning performance. The appropriateness of a distance measure will typically depend on the types of features being used in the learning process.

In this chapter, we examine the properties of distance measures in the context of the analysis of gene expression data from DNA microarray experiments. The feature vectors represent transcript levels, i.e., mRNA abundance or relative abundance, either across biological samples (if comparing genes) or across genes (if comparing samples).

We consider different aspects of distances that help address the heterogeneity of the data and differences in interpretation depending on the source of the data (cDNA arrays versus short oligonucleotide arrays). Traditional measures, such as Euclidean and Manhattan distances as well as correlation-based distances, are considered. Other dissimilarity functions, which involve comparisons of distributions based on the Kullback-Leibler and mutual information criteria, are also examined.

12.1 Introduction

Genomic experiments generate large and complex multivariate data sets. Machine learning approaches are important tools in microarray data analysis, for the purposes of identifying patterns in expression among genes and/or biological samples, and for predicting clinical or other outcomes using gene expression data. Chapters 13, 16, and 17 consider different aspects of machine learning in more detail. We briefly review some of the concepts here as motivation for the discussion in this chapter.

Inherent in every machine learning approach is a notion of a distance or similarity between the objects to be clustered or classified. In general, any distance measure can be used with any machine learning algorithm. The choice of distance measure is probably more important than the choice of machine learning algorithm, and some attention should be paid to the selection of an appropriate measure for each problem. In this chapter, we describe distances in quite general terms and consider both their mathematical properties as well as their implementation in different R packages.

The notion of distance is explicit in clustering procedures that operate directly on a matrix of pairwise distances between the objects to be clustered, e.g., partitioning around medoid (PAM) and hierarchical clustering (Kaufman and Rousseeuw, 1990). Certain supervised learning methods, such as nearest neighbor classifiers, also involve explicitly specifying a distance. Although the choice of distance may not be as transparent for other supervised approaches, observations are in fact assigned to classes on the basis of their distances from objects known to be in the classes. For instance, linear discriminant analysis is based on the Mahalanobis distance [Mardia et al. (1979); Ripley (1996a); see Equation (12.3) below] of the observations from the class means. The weighted gene voting scheme of Golub et al. (1999) is a variant of a special case of linear discriminant analysis, also known as naive Bayes classification. In addition, the distance and its behavior are intimately related to the scale on which measurements are made. The choice of a transformation and distance should thus be made jointly and in conjunction with the choice of a classifier or clustering procedure.

In this chapter, we consider the impact of distance selection on the analysis of genomic data. We assume that the data have been preprocessed using appropriate techniques and normalization methods and that the researcher is presented with an array containing G features (genes) for I samples. For microarray data there are potentially two values per feature: an estimate of the abundance of mRNA for that gene and a standard error of estimated abundance.

Our development goes as follows. In the next section, we give a general introduction to distances and discuss specific classes of distances. We provide formal definitions and discuss the relevant resources available in R. Then in Section 12.3, we focus on gene expression data from Affymetrix and two-color cDNA microarray experiments and discuss standardization and issues specific to these two platforms. We provide some examples of the use of different distance measures, in particular we make use of literature co-citation data, in Section 12.4. Visualization methods for distance data are described in Chapter 10.

12.2   Distances

Distances, metrics, dissimilarities, and similarities are related concepts. We provide some general definitions and then consider specific classes of distance measures.

12.2.1   Definitions

Any function d that satisfies the following five properties is termed a *metric*:

(i) non-negativity $d(\mathbf{x}, \mathbf{y}) \geq 0$;

(ii) symmetry $d(\mathbf{x}, \mathbf{y}) = d(\mathbf{y}, \mathbf{x})$;

(iii) identification mark $d(\mathbf{x}, \mathbf{x}) = 0$;

(iv) definiteness $d(\mathbf{x}, \mathbf{y}) = 0$ if and only if $\mathbf{x} = \mathbf{y}$;

(v) triangle inequality $d(\mathbf{x}, \mathbf{y}) + d(\mathbf{y}, \mathbf{z}) \geq d(\mathbf{x}, \mathbf{z})$.

A function that satisfied only properties (i)-(iii) is termed a *distance*. For many of the techniques we will consider, distances are sufficient. Hence, we will generally refer to distances (which include metrics) and only mention metrics specifically when properties (iv) and (v) are relevant.

A *similarity function* S is more loosely defined and satisfies the three following properties

(i) non-negativity $S(\mathbf{x}, \mathbf{y}) \geq 0$;

(ii) symmetry $S(\mathbf{x}, \mathbf{y}) = S(\mathbf{y}, \mathbf{x})$;

(iii) $S(\mathbf{x}, \mathbf{y})$ increases in a monotone fashion as objects $\mathbf{x}$ and $\mathbf{y}$ are more and more *similar*.

A *dissimilarity* function satisfies (i) and (ii), but for (iii), $S(\mathbf{x}, \mathbf{y})$ increases as objects $\mathbf{x}$ and $\mathbf{y}$ are more and more dissimilar. It is worth noting that there is, in fact, no need to require symmetry although some adjustments generally need to be made if the measures are not symmetric. The airplane flight time between two cities is an example of an asymmetric distance.

Many options are available in selection of a distance for machine learning tasks. Because there are many different types of data (e.g., ordinal, nominal, continuous) and approaches for analyzing these data, the literature on distances is quite broad. References that consider the application of distances in either clustering or classification include: Duda et al. (2001, Section 4.7); Gordon (1999, Chapter 2); Kaufman and Rousseeuw (1990, Chapter 1); (Mardia et al., 1979, Chapter 13).

As noted above, we are most concerned with a situation where G features have been measured for I observations, or samples. There is substantial interest in applying some form of machine learning to both the samples (e.g., to identify patients with similar patterns of mRNA expression) and the features (e.g., to identify genes with similar patterns of expression).

We distinguish between two main classes of distance measures. Consider computing the distance between the expression profiles of two genes across I samples. In the first approach, we view the gene expression profiles as two I-vectors in some space and compute distances in a pairwise (within-sample) manner (Section 12.2.2). In contrast, the second approach ignores the natural pairing of observations and instead, views the two gene expression profiles as two different samples generated from underlying probability density functions for mRNA expression measures. In this case, distances between densities or distribution functions are relevant (Section 12.2.3). Of course, one is certainly not limited to an either-or approach. It may, in fact, be quite sensible to devise measures that combine the two. Genes with expression patterns that are similar in both aspects are possibly more interesting than those that are close in only one.

12.2.2   Distances between points

For m-vectors $\mathbf{x} = (x_1, \ldots, x_m)$ and $\mathbf{y} = (y_1, \ldots, y_m)$ consider distances of the form

$$d(\mathbf{x}, \mathbf{y}) = F[d_1(x_1, y_1), \ldots, d_m(x_m, y_m)], \qquad (12.1)$$

where the d_k are themselves distances for each of the $k = 1, \ldots, m$ features. We refer to these functions as *pairwise distance functions*, as the pairing of observations within features is preserved. This representation is quite general: there is no need for the d_k to be the same. In particular, features may be of different types (e.g., the data may consist of a mixture of continuous and binary features) and may be weighed differentially (e.g., weighted Euclidean distance).

Common metrics within this class include the Minkowski metric, with $z_k = d_k(x_k, y_k) = |x_k - y_k|$ and $F(z_1, \ldots, z_m) = (\sum_{k=1}^{m} z_k^\lambda)^{1/\lambda}$. Special cases of the Minkowski metric considered in this chapter are the Manhattan and Euclidean metrics corresponding to $\lambda = 1$ and $\lambda = 2$, respectively.

EUC Euclidean metric

$$d_{euc}(\mathbf{x}, \mathbf{y}) = \sqrt{\sum_{i=1}^{m} (x_i - y_i)^2}. \qquad (12.2)$$

MAN Manhattan metric

$$d_{man}(\mathbf{x}, \mathbf{y}) = \sum_{i=1}^{m} |x_i - y_i|.$$

Correlation-based distance measures have been widely used in the microarray literature (Eisen et al., 1998). They include one minus the standard

Pearson correlation coefficient and one minus an uncentered correlation coefficient (or cosine correlation coefficient) considered by Eisen et al. (1998), Spearman's rank correlation, and Kendall's τ (Conover, 1971).

COR Pearson sample correlation distance

$$d_{cor}(\mathbf{x}, \mathbf{y}) = 1 - r(\mathbf{x}, \mathbf{y}) = 1 - \frac{\sum_{i=1}^{m}(x_i - \bar{x})(y_i - \bar{y})}{\sqrt{\sum_{i=1}^{m}(x_i - \bar{x})^2 \sum_{i=1}^{m}(y_i - \bar{y})^2}}.$$

EISEN Cosine correlation distance

$$d_{eisen}(\mathbf{x}, \mathbf{y}) = 1 - \frac{\mathbf{x}'\mathbf{y}}{\|\mathbf{x}\|\|\mathbf{y}\|} = 1 - \frac{|\sum_{i=1}^{m} x_i y_i|}{\sqrt{\sum_{i=1}^{m} x_i^2 \sum_{i=1}^{m} y_i^2}}$$

which is a special case of Pearson's correlation with $\bar{x}$ and $\bar{y}$ both replaced by zero.

SPEAR Spearman sample correlation distance

$$d_{spear}(\mathbf{x}, \mathbf{y}) = 1 - \frac{\sum_{i=1}^{m}(x_i' - \bar{x}')(y_i' - \bar{y}')}{\sqrt{\sum_{i=1}^{m}(x_i' - \bar{x}')^2 \sum_{i=1}^{m}(y_i' - \bar{y}')^2}}.$$

where $x_i' = rank(x_i)$ and $y_i' = rank(y_i)$.

TAU Kendall's τ sample correlation

$$d_{tau}(\mathbf{x}, \mathbf{y}) = 1 - \tau(\mathbf{x}, \mathbf{y})| = 1 - \frac{|\sum_{i=1}^{m}\sum_{j=1}^{m} C_{x_{ij}} C_{y_{ij}}}{m(m-1)}$$

where $C_{x_{ij}} = sign(x_i - x_j)$ and $C_{y_{ij}} = sign(y_i - y_j)$.

Note that we have transformed the correlations by subtracting them from one. This is done so that two vectors that are strongly positively correlated are regarded as close together. Using this transformation, data that exhibit a strong negative correlation will be far apart. In some cases, you might want to treat negative and positive correlations similarly, and that can be achieved by using the absolute value of the correlation. Correlation-based measures are in general invariant to location and scale transformations and tend to group together genes whose expression patterns are linearly related. While correlation-based distances have many nice properties, they tend to be adversely affected by outliers and then the non-parametric versions (SPEAR or TAU) are preferred.

When the data are standardized using the mean and variance, so that both $\mathbf{x}$ and $\mathbf{y}$ are m-vectors with zero mean and unit length, there is a functional relationship between the Pearson correlation coefficient $r(\mathbf{x}, \mathbf{y})$ and the Euclidean distance. The relationship is

$$d_{euc}(\mathbf{x}, \mathbf{y}) = \sqrt{2m[1 - r(\mathbf{x}, \mathbf{y})]}.$$

We note that expression values are generally measured with error. The standard deviation of measurement errors can be estimated and is sometimes available along with intensity measures in the form of "standard errors." This variability information can be exploited in *errors-in-variables* models. This is the approach taken by Tadesse et al. (2005) for modeling survival data. Estimated standard errors can also be used when considering Kullback-Leibler distances, as is shown below.

Finally, we mention the *Mahalanobis distance*. Consider a situation where a pair of vectors, $\mathbf{x}$ and $\mathbf{y}$, are generated from some multivariate distribution with mean vector μ and variance-covariance matrix $\mathbf{\Sigma}$. Then the Mahalanobis distance between them is defined as

$$(\mathbf{x} - \mathbf{y})'\mathbf{\Sigma}^{-1}(\mathbf{x} - \mathbf{y}). \tag{12.3}$$

When $\mathbf{\Sigma}$ is unknown, it is generally replaced with the sample variance-covariance matrix. In general terms, the Mahalanobis distance reflects the notion that the data are more variable in some directions than in others.

Distances and transformations. Distances and data transformations are closely related. If g is an invertible, possibly non-linear, transformation $g : x \rightarrow x'$, then this can be used to induce a new metric d' via

$$d(\mathbf{x}, \mathbf{y}) = d[g^{-1}(\mathbf{x}'), g^{-1}(\mathbf{y}')] = d'(\mathbf{x}', \mathbf{y}').$$

The metric d operates on the original variables $\mathbf{x}$, whereas d' works on the transformed variables $\mathbf{x}'$, and the two are equivalent, even though they can have quite different functional forms. Conversely, the same distance function, say d_{euc} from Equation (12.2), can lead to quite different distances, between the same data points, when applied on different scales. Hence the choice of the scale is important. For microarray data, at least three different scales are generally considered: that of the original scanned fluorescence intensities, the logarithmically transformed scale, or the generalized logarithmic (variance-stabilized) scale proposed by Huber et al. (2002) and Durbin et al. (2002). A more general discussion of transformations in regression can be found, for example, in Ryan (1997).

Practicalities. Many pairwise distances can be computed in R using the `dist` function, including `euclidean`, `manhattan`. The function returns an object of class *dist*, which represents the distances between the rows of the input argument (which can be either a matrix or a dataframe). This function assumes that distances are symmetric and saves storage space by using a lower-triangular representation for the returned value.

The function `daisy` in the `cluster` package also provides distance computations. This function returns an object of class *dissimilarity* which contains the distances between the rows of the input matrix or dataframe. This class *inherits* from the *dist* class so that it will automatically use methods appropriate to that class. When some of the input variables are categorical, such as sex, then it makes no sense to compute distances between the numerical encodings and `daisy` has functionality to compute appropriate between-observation distances.

The package bioDist has implementations of the various correlation distances, such as `spearman.dist` and `tau.dist`. These functions return objects of class *dist*.

The functions in cluster take either a data matrix or a dissimilarity matrix as input. Other machine learning algorithms are less flexible and may require that the user manipulate the data in order to alter the distance measure that is used. An approach toward standardization is considered in Chapter 16.

12.2.3   Distances between distributions

The distances enumerated in the preceding section treat the expression measurements as points in some metric space, where each observation (gene or sample, depending on the problem) contributes one point and the coordinates are given by the corresponding expression measures. Distances are computed in a pairwise manner within features (samples when genes are being compared, and vice versa). A different approach is to consider the data for each feature as an independent sample from a population. In this case, we are interested in questions such as whether the shape of the distribution of features is similar between two genes. For example whether they are bimodal or, perhaps have long right-tails. Other authors have also considered using distances between distributions as a means of analyzing genomic data. For example, Butte and Kohane (2000) suggest binning the data and then using a mutual information distance. Quite a different approach to the comparison of distributions is taken in Gentleman and Carey (2003, Section 2.4.3); see also Section 16.4.6 below.

Alternatively, for each gene, across samples, we can consider the data as random I-vectors from some distribution. The simplest case is to assume that the expression measures for a particular gene follow an I-dimensional multivariate normal distribution with diagonal variance-covariance matrix. Using this approach, each gene provides a multivariate observation. Each of the I measurements for a given gene come from different samples, which are assumed to be independent, and hence the estimated variance-covariance matrix is diagonal. This approach can be used when both expression levels and their associated standard errors are available. The observed expression values are used to estimate the mean vector and the observed standard errors are used to estimate the variance-covariance matrix.

Many different distance measures can be used to assess the similarities between two densities. We consider two measures that are not actually distances: the Kullback-Leibler information and Hamming's mutual information.

Kullback-Leibler Information. The *Kullback-Leibler Information* (KLI) measure between densities f_1 and f_2 is defined as

$$
\begin{aligned}
KLI(f_1, f_2) &= E_1\{\log[f_1(X)/f_2(X)]\}\\
&= \int \log[f_1(x)/f_2(x)]f_1(x)dx, \quad (12.4)
\end{aligned}
$$

where X is a random variable with density f_1, and E_1 denotes expectation with respect to f_1. This ratio can be infinite and hence so can the KLI. The KLI is not a distance because it is not symmetric. KLI does not satisfy the triangle inequality either.

The KLI can be symmetrized in a number of ways, including the approach described in Cook and Weisberg (1982, p. 163). They define the *Kullback-Leibler Distance* (KLD) to be,

$$
2d_{KLD}(f_1, f_2) = KLI(f_1, f_2) + KLI(f_2, f_1).
$$

The measure is symmetric and positive if f_1 and f_2 are different, however, it still does not satisfy the triangle inequality.

In the special case where $f_1 = N_m(\mu_1, \Sigma_1)$ and $f_2 = N_m(\mu_2, \Sigma_2)$, and assuming that Σ_1 and Σ_2 are positive definite, the expression for $d_{KLD}(f_1, f_2)$ simplifies and we get:

$$
\begin{aligned}
2d_{KLD}(f_1, f_2) &= (\mu_1 - \mu_2)^T \Sigma_2^{-1}(\mu_1 - \mu_2)\\
&\quad + \log(|\Sigma_1|/|\Sigma_2|) + \mathrm{tr}(\Sigma_1 \Sigma_2^{-1}) - m. \quad (12.5)
\end{aligned}
$$

However, this simplification involves making a strong assumption and requires knowledge of both variance-covariance matrices. Note that if Σ_1 and Σ_2 are identical, this is a form of Mahalanobis distance. However, we should emphasize that the treatment here is slightly different.

To compute between gene distances from microarray data, the expression measures for a given gene, across samples, can be treated as a single observation from an I-dimensional multivariate normal distribution. For each gene, we estimate the mean in each coordinate (sample) by the observed expression measure for that sample, and we estimate the variances using, for example, the Li and Wong estimated standard errors for Affymetrix data (Li and Wong, 2001a). When viewed from this perspective, KLD (Equation 12.5) is more similar to the distances in Section 12.2.2 than it is to either KLI or the mutual information distance described below. This is a model that accounts for measurement error, though not as explicitly as an errors-in-variables approach.

Mutual Information. Closely related to the KLI is the *mutual information* (MI). The MI measures the extent to which two random variables X and Y are dependent. Let $f(\cdot, \cdot)$ denote the joint density function and $f_1(\cdot)$ and $f_2(\cdot)$ the two marginal densities for X and Y, respectively. Then the MI is defined as

$$
MI(f_1, f_2) = E_f\left\{\log\left[\frac{f(X, Y)}{f_1(X)f_2(Y)}\right]\right\}, \quad (12.6)
$$

and is zero in the case of independence. We note that like KLI, MI is not a distance although we will sometimes refer to it as if it were. This can easily be determined by noticing the relationship between the MI distance and the KLI. The MI is basically the KLI between $f(x, y)$ and $g(x, y) = f_1(x)f_2(y)$, where $g(x, y)$ is the joint distribution obtained by assuming that the two marginals are independent,

$$
\begin{aligned}
KLI(f, g) &= \int_x \int_y \log[f(x, y)/g(x, y)]f(x, y)dxdy \\
&= E_f \left\{ \log \left[\frac{f(X, Y)}{f_1(X)f_2(Y)} \right] \right\} \\
&= MI(f_1, f_2).
\end{aligned}
$$

MI and KLD focus on very different aspects of distributions and that is reflected in their performance. MI is large when the joint distribution is quite different from the product of the marginals. Thus, it attempts to measure the distance from *independence*. KLD, on the other hand, measures how much the shape of one distribution resembles that of the other.

Joe (Joe, 1989) considers MI and its role as a multivariate measure of association. He shows that if the transformation,

$$
\delta^* = [1 - \exp(-2MI)]^{1/2} \tag{12.7}
$$

is used, then δ^* takes values in the interval $[0, 1]$ and can be interpreted as a a generalization of the correlation. He further notes that δ^* is related to Kendall's τ. We will make the further transformation to $1 - \delta^*$ so that our measure has the same interpretation as the other correlation-based distance measures discussed in this chapter.

Practicalities. The distances being considered are functionals of the underlying probability density functions. Given the observed data, there are many different methods for providing the appropriate estimates. We consider three of the more commonly used methods in this chapter. The simplest method is to assume some parametric distribution for the data and to estimate the parameters for that distribution; these can then be used, together with the functional form of the density, to estimate the mutual information. A second approach is to roughly group the data and to then treat it as discrete. A third approach is to use density estimation followed by either numerical integration or explicit calculation. The second and third approaches involve some form of smoothing, and this should be dealt with explicitly. Much work remains to be done before any method can be recommended for general use.

To apply the binning approach, the samples are separately divided into k common bins and then each sample is treated as if it were data from a discrete distribution. This approach can be problematic, as the estimated KLI will be infinite whenever a bin has an observation from f_1 but not one from f_2. In our experience, this occurs quite often. We note that there are other problems with the binning approach; a straightforward calculation shows that the binned version of MI distance tends to the logarithm of the number of sample points as the number of bins goes to infinity, since in the limit every point will end up in a bin of its own.

An alternative procedure is to employ a density estimation procedure followed by numerical integration. One could standardize the data (shift so that a measure of central location is approximately zero and scale so that a measure of dispersion is approximately unity), estimate the densities and then apply numerical integration (using the range -3 to 3) to estimate KLI in Equation (12.4). This approach could be extended to MI as well. There are many good density estimation routines available in R, and one-dimensional integration is straightforward. In our examples for MI, we used the binning approach because density estimation followed by numerical integration proved too computationally expensive.

We have created the bioDist package, which contains code for some of the methods described here. It is used in the examples given later in this chapter. bioDist contains an implementation of the KL distances that rely on binning; `KLdist.matrix` and one that uses density estimation followed by numerical integration, `KLD.matrix` . For mutual information there are two functions, `mutualInfo` that computes the distance from independence and `MIdist` that computes the transformation in Equation (12.7). We note that the computations are not terribly fast computing these distances on very large data sets is time consuming.

12.2.4   Experiment-specific distances between genes

The between-gene distances considered thus far do not take into account the *structure* or *design* of the microarray experiment. Such distance measures may be appropriate for situations where there is no particular structure of interest among the arrays, e.g., when the target samples hybridized to the arrays are viewed as a random sample from a particular population. However, microarray experiments can be highly structured, as in time-course and multifactorial experiments. It is desirable to derive between-gene distances that reflect the design of the experiment under consideration. Such distances may serve to produce a more vivid visualization of the data and to permit focus on more meaningful patterns in gene expression. In this section, we consider some modifications that are more suitable for data arising from designed experiments or other situations where the samples have specific relationships to one another.

Instead of computing distances directly on the genes-by-arrays data matrices, one may use covariate information (e.g., treatment, cell type, dose, time) to derive suitable transformations of this matrix. Linear models and extensions thereof (e.g., generalized linear models) can be used to estimate experiment specific effects for each gene and hence produce new gene profiles. For factorial experiments studying the simultaneous gene expression response to two treatments, say, the new profiles could be based on main effects and interactions. In time-course experiments, it makes sense to consider distances that are not time-exchangeable and use the time index in an essential way. This could be done by penalizing for non-smoothness as in Sobolev metrics, where the squared Sobolev distance between two functions is based on the sum of squared distances, in some standard metric (e.g., $\mathcal{L}_2$), between the two functions, their first derivatives, second derivatives, etc., up to some order p. For time-course data with a large enough number of equally spaced time points, one of the standard wavelet decompositions could be used to decompose expression profiles into potentially interpretable quantities corresponding to local frequency components.

The use of covariate information as described above produces new profiles for each gene. Distances can then be computed for the new profiles, and genes can be clustered based on these distances. A preliminary application of such an approach can be found in Lin et al. (2004), for a study of spatial differential expression in the mouse olfactory bulb experiment. Distances on the new profiles can also be used to match profiles to a library of profiles of interest for a particular experiment, by ranking projections of the new gene profiles along specified directions in an appropriate geometric representation of the problem. For instance, in factorial experiments across time, interesting reference profiles for main effects and interactions might include: cyclical, early, or late effects, or the effects over time for a known gene.

12.3 Microarray data

For our purpose, gene expression data on G genes for I mRNA samples may be summarized by a $G \times I$ matrix $X = (x_{gi})$, where x_{gi} denotes the expression measure of gene g in mRNA sample i. The expression levels might be either absolute (e.g., Affymetrix oligonucleotide arrays) or relative to the expression levels of a suitably defined common reference sample (e.g., cDNA microarrays.)

12.3.1 Distances and standardization

The behavior of the distance is closely related to the scale on which the observations have been made. Standardization of features is thus an important issue when considering distances between objects and is one method of making the features comparable. However, standardization also has the effect of removing some of the potentially interesting features in the data. Thus, in some cases it will be sensible to explore other approaches to obtaining comparability across features.

In the context of microarray data, one may standardize genes and/or samples. When standardizing genes, expression measures are transformed as follows

$$x_{gi} = \frac{x_{gi} - \text{center}(x_{g.})}{\text{scale}(x_{g.})}$$

where center$(x_{g.})$ is some measure of the center of the distribution of the set of values x_{gi}, $i = 1, \ldots, I$, such as mean or median, and scale$(x_{g.})$ is a measure of scale such as the standard deviation, interquartile range, or MAD (median absolute deviation about the median). Alternatively, one may want to standardize arrays (samples) if there is interest in clustering or classifying them (rather than clustering or classifying the genes). Now we use

$$x_{gi} = \frac{x_{gi} - \text{center}(x_{.i})}{\text{scale}(x_{.i})},$$

where the centering and scaling operations are carried out across all genes measured on sample (or array) i.

We now consider the implications of the preceding discussion on standardization in the context of both relative mRNA expression measurements (cDNA) and absolute (Affymetrix) mRNA expression measurements. Consider the standard situation where x_{gi} represents the expression measure on a log scale for gene g on patient (i.e., array or sample) i. Let $y_{gi} = x_{gi} - x_{gA}$, where patient A is our reference. Then, the relative expression measures y_{gi} correspond to the standard data available from a cDNA experiment with a common reference. The use of relative expression measures represents a location transformation for each gene (gene centering). Now, suppose that we want to measure the distance between patient samples i and j. Then, for the classes of distances considered in Equation (12.1) of Section 12.2.2,

$$d(\mathbf{y}_{.i}, \mathbf{y}_{.j}) = \sum_{g=1}^{G} d_g(y_{gi}, y_{gj}) = \sum_{g=1}^{G} d_g(x_{gi} - x_{gA}, x_{gj} - x_{gA}).$$

When the $d_g(x, y)$ are functions of $x - y$ alone, then $d(\mathbf{y}_{.i}, \mathbf{y}_{.j}) = d(\mathbf{x}_{.i}, \mathbf{x}_{.j})$, and it does not matter if we look at relative (the $\mathbf{y}$'s) or absolute (the $\mathbf{x}$'s) expression measures.

Suppose that we are interested instead in comparing genes and not samples. Then the distance between genes g and h is

$$d(\mathbf{y}_{g.}, \mathbf{y}_{h.}) = \sum_{i=1}^{I} d_i(y_{gi}, y_{hi}) = \sum_{i=1}^{I} d_i(x_{gi} - x_{gA}, x_{hi} - x_{hA}).$$

If $d(\mathbf{x}, \mathbf{y})$ has the property that $d(\mathbf{x} - \mathbf{c}, \mathbf{y}) = d(\mathbf{x}, \mathbf{y})$ for any c, then the distance measure is the same for absolute and relative expression measures.

Thus, for Minkowski distances, the distance between samples is the same for relative and absolute expression measures. This does not hold for the distance between genes. On the other hand, distances based on the Pearson correlation yield the same distances between genes for both relative and absolute measures. This does not hold for the distance between samples. Arguments can be made in favor of either approach: invariance of (i) gene distances or (ii) sample distances, for absolute and relative expression measures. The data analyst will have to weigh these and other biological considerations when selecting a distance measure.

12.4   Examples

For our examples in this chapter we make use of the data reported in Chiaretti et al. (2004) and described in Appendix A. We consider only the subset of patients that have a reciprocal translocation between the long arms of Chromosomes 9 and 22 that has been causally related to chronic and acute leukemia (Cilloni et al., 2002). They are labeled BCR/ABL.

We select genes for our distance measurements by first carrying out a non-specific filtering (as described in Chapter 14) where we imposed three requirements: the gene must have an expression level greater than log(100) in at least 25% of the samples, it must have an IQR that is larger than 0.5, and it must have median expression level greater than log(300). Genes that passed all three filters will be referred to as *expressed*. We then adjusted each gene across samples by subtracting the median and dividing by the MAD (median absolute deviation from the median). This step makes computations between genes and across different distance measures more comparable. By standardizing the genes, we have made the four distances, EISEN, COR, EUC, and MAN, more similar than they would be if we worked with untransformed data.

The code below shows how we constructed our filters, using genefilter and then the resulting manipulations, to restrict the data to those selected genes. Finally, we standardize the genes, across samples, as described above.

```
> library("genefilter")
> data(ALL)
> Bsub <- (ALL$mol == "BCR/ABL")
> Bs <- ALL[, Bsub]

> f1 <- pOverA(0.25, log2(100))
> f2 <- function(x) (IQR(x) > 0.5)
> f3 <- function(x) (median(2^x) > 300)
> ff <- filterfun(f1, f2, f3)
> selected <- genefilter(Bs, ff)
> sum(selected)

[1] 637

> BSub <- Bs[selected, ]
> eS <- exprs(BSub)
```

```
> mads <- apply(eS, 1, mad)
> meds <- apply(eS, 1, median)
> e1 <- sweep(eS, 1, meds)
> e2 <- sweep(e1, 1, mads, FUN = "/")
> BSubStd <- BSub
> exprs(BSubStd) <- e2
```

We now show how some of the distance measures we have discussed can
be applied to the ALL data. In order to have a small set of genes to work
with, we select genes that are in the GO BP category GO:0006917, which
corresponds to the induction of apoptosis.

```
> library("GO")
> library("annotate")
> GOTERM$"GO:0006917"
```

```
GOID = GO:0006917
Term = induction of apoptosis
Synonym = apoptosis signaling
Synonym = positive regulation of apoptosis
Definition = A process that directly activates any
      of the steps required for cell death by
      apoptosis.
Ontology = BP
```

```
> library("hgu95av2")
> apop <- hgu95av2GO2ALLPROBES$"GO:0006917"
> inboth <- apop %in% row.names(e2)
> whsel <- apop[inboth]
> exprApop <- e2[whsel, ]
> unlist(mget(whsel, hgu95av2LOCUSID))
```

36199_at	39020_at	2031_s_at	39723_at	1635_at
1611	10572	1026	8454	25
1636_g_at	39730_at	34740_at	41763_g_at	38050_at
25	25	2309	7073	9774

Next we load the bioDist package and compute some pairwise distances
between probesets.

```
> library("bioDist")
> man <- dist(exprApop, "manhattan")
> MI <- MIdist(exprApop)
> KLsmooth <- KLD.matrix(exprApop)
> KLbin <- KLdist.matrix(exprApop)
```

False color representations of the distance matrices are shown in Figure 12.1. We have used the transformation of mutual information distance
described in Equation (12.7). The KL distances are small the more similar the shape of the two densities and are larger if the shapes are quite
different.

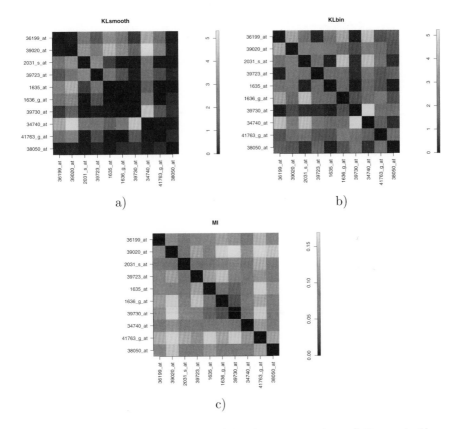

Figure 12.1. False color representation of the distance matrices. a) KLsmooth, b) KLbin, c) MI.

To further compare the distances, we produced pairwise scatterplots of the different distances in Figure 12.2. From that we can see the general positive correlation of the KL based distances and note that, as expected, there is little relationship between the MI distance and the KL distances – they are measuring different things.

12.4.1   A co-citation example

We now consider an example that relates distances and co-citation in the medical literature as measures of biological similarity. This approach can be contrasted with the one taken in Chapter 22. Two or more genes that share a common reference (i.e. were written about in the same paper) are more likely to be meaningfully biologically related than genes that are never jointly mentioned in any paper. Joint mention of genes A and B does not imply that these genes are strongly or even remotely biologically related.

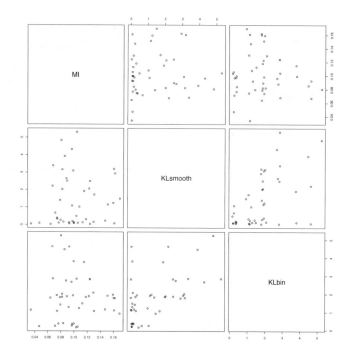

Figure 12.2. Pairwise scatterplots of different distances computed between the same set of points.

However, the data resource is large, there have been a number of studies that make use of co-citation data, and it is an active research area (Jenssen et al., 2001; Masys et al., 2001). Our approach is simple but could easily be extended to make use of other data sources as they become available.

Data on co-citation were obtained from PubMed. See Chapter 7 for more details on this reference database. We mapped Affymetrix identifiers to their corresponding LocusLink values and from there to PubMed identifiers (PMIDs).

We distinguish two relationships between genes that can be identified from these data. Any two genes that directly share a citation are called *adjacent*. Two genes will be called *accessible* if they can be connected, possibly by other genes, through a co-citation path. To be specific, suppose that genes X and Y are co-cited and that X and Z are also co-cited. Then we would say that X and Y are adjacent (as are X and Z) and that Y and Z are accessible. In the literature co-citation context, accessibility is too weak a relationship to explore further.

Next we selected a target gene and then determined the 100 genes closest to that target using the different distance measures under consideration, namely COR, SPEAR, TAU, EUC, MAN, KLD, and MI. We first look at

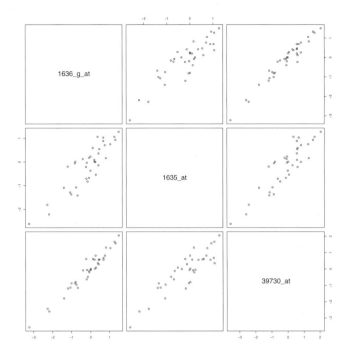

Figure 12.3. Scatterplot of multiple expressed probesets for ABL1

the agreement among distances in terms of percentages of common genes chosen.

As an example, we selected ABL1 as the target gene, which has 8 probesets on the HG-U95Av2 GeneChip. The filtering steps left us with 3 probesets that satisfied our selection criterion (see Figure 12.3). We emphasize the fact that these three probesets should be measuring the same thing and certainly the correlations, Figure 12.3, do appear quite strong. So we anticipate that starting with any one of them, we would find that the other two were close to the one we started with. To test this, we carried out the following experiment.

For each of the 3 expressed probesets a list of the 100 closest probesets, using each of the 7 distance measures, was computed. The between-list agreement, in terms of whether a probeset was selected or not, is shown next.

```
$"1636_g_at"
      cor spear  tau  euc  man  kld
spear 0.73
tau   0.72 0.94
euc   0.74 0.66 0.67
man   0.65 0.65 0.66 0.81
```

```
kld    0.19   0.15 0.13 0.10 0.09
mi     0.23   0.26 0.28 0.26 0.32 0.13
```

$"1635_at"

```
       cor spear  tau  euc  man  kld
spear 0.85
tau    0.85  0.93
euc    0.75  0.71 0.69
man    0.69  0.70 0.69 0.84
kld    0.15  0.17 0.18 0.15 0.15
mi     0.35  0.36 0.36 0.39 0.41 0.21
```

$"39730_at"

```
       cor spear  tau  euc  man  kld
spear 0.77
tau    0.75  0.95
euc    0.78  0.71 0.71
man    0.69  0.74 0.74 0.81
kld    0.22  0.19 0.20 0.23 0.23
mi     0.29  0.31 0.30 0.31 0.34 0.22
```

As we can see, the agreement among the correlational distances (i.e. COR, SPEAR and TAU) and Minkowski metric distances (i.e., EUC and MAN) are good, especially those between the nonparametric correlational distances (i.e., SPEAR and TAU). There is not much commonality between the distributional distances (i.e., KLD and MI) and other distances and the agreement within the distributional distances is also low. For each of the multiple probesets, we further look at the rank of the other two probesets chosen by the various distance measures.

```
1636_g_at
     cor spear tau euc man kld mi
[1,]   2     2   2   2   2  NA 10
[2,]   1     1   1   1   1  NA  1
1635_at
     cor spear tau euc man kld mi
[1,]   2     1   2   1   1  NA 23
[2,]   1     2   1   2   2  55  1
39730_at
     cor spear tau euc man kld mi
[1,]   1     1   1   1   1  NA  1
[2,]   2     2   2   2   2  NA  3
```

We can see that whenever one of the multiple probesets for ABL1 was chosen as target probeset, the other two probesets for the same gene were always the top two probesets using correlational or metric distances. MI also captured the other multiple probesets in the top list although the ranks tended to be larger. Again we see that KLD fared worst. Note that here we used discrete versions of KLD (symmetrized) and MI. The missing values reported for `kld` arise because the other probesets for ABL1 were not always chosen.

12.4.2   Adjacency

We now compare co-citation and distance. Recall that two genes are called adjacent if they are cited together in any article, and in a sense this is a measure of similarity. So we might then ask whether any of the distance measures under consideration reflect this same measure of similarity. To answer that question we examine how many of the co-cited genes are within the top 100 list for each of the different distances. We first subset the original data set to include probesets that have been cited, this reduced the number of probesets to 632. There were 38 genes that were co-cited with ABL1 in the `humanLLMappings` package, out of which 2 were in the filtered *ALL* data set. We then used the 7 distance measures to generate the top 100 probesets and computed the number of genes among those 100 closest that had co-citations with the target gene. The results are shown below

	1636_g_at	1635_at	39730_at
cor	0	1	1
spear	1	1	1
tau	1	0	1
euc	1	2	1
man	1	1	1
kld	0	1	0
mi	0	1	0

Notice that both EUC and MAN did quite well and that KLD seemed to fare the worse. The Hypergeometric distribution can be used to assess the significance of the results given above. Consider an urn with 629 balls in it. Of these 2 are colored white, the remainder are black. Under the null hypothesis that there is no relationship between co-citation and being close, as computed by one of the distances, then each selection of the 100 nearest genes is like drawing 100 balls from the urn and counting how many white ones were selected. The computation of p-values is easily carried out.

```
P(X >= 1) P(X >= 2) P(X >= 3)
   0.4068     0.0678     0.0040
```

Note that the assumption of independence among the enumerated events required for applicability of the Hypergeometric model is not tenable for these data. Thus this Hypergeometric computation should be regarded as a rough approximation to the truth.

12.5   Discussion

Distances are an integral part of all machine learning algorithms and hence play a central role in the analysis of most experimental data. The distance that is used for any particular task can have a profound effect on the output of the machine learning method, and it is therefore essential that users ensure that the same distance method is used when comparing machine learning algorithms.

It is also important that the investigator be able to select, and use, a distance that is appropriate for the task at hand. There is no single distance that is always relevant, and similarity can be measured in many ways. We find R to be a good platform for these sorts of analyses, as it has a wealth of built-in distance functions, and supports the addition of new distance functions straightforwardly.

13

Cluster Analysis of Genomic Data

K. S. Pollard and M. J. van der Laan

Abstract

We provide an overview of existing partitioning and hierarchical clustering algorithms in R. We discuss statistical issues and methods in choosing the number of clusters, the choice of clustering algorithm, and the choice of dissimilarity matrix. We also show how to visualize a clustering result by plotting ordered dissimilarity matrices in R. A new R package hopach, which implements the Hierarchical Ordered Partitioning And Collapsing Hybrid (HOPACH) algorithm, is presented (van der Laan and Pollard, 2003). The methodology is applied to a renal cell cancer gene expression data set.

13.1   Introduction

As the means for collecting and storing ever larger amounts of data develop, it is essential to have good methods for identifying patterns. For example, an important goal with large-scale gene expression studies is to find biologically important subsets of genes or samples. Clustering algorithms have been widely applied to microarray microarray data analysis (Eisen et al., 1998).

Consider a study in which one collects on each of I randomly sampled subjects (or more generally, experimental units) a J-dimensional gene expression profile X_i, $i = 1, \ldots, I$: for example, X_i can denote the gene expression profile of cancer tissue relative to healthy tissue within a randomly sampled cancer patient. To view clustering as a statistical procedure it is important to consider X_i as an observation of a random vector with a population distribution we will denote with P. These I independent and identically distributed (*i.i.d.*) observations can be stored in an observed $J \times I$ data matrix X. Genes are represented by I-dimensional vectors

$[X_i(j) : i = 1, \dots, I]$, while the samples are represented by J-dimensional vectors X_i. The goal could now be to cluster genes or samples. A cluster is a group of similar elements. Each cluster can be represented by a profile, either a summary measure such as a cluster mean or one of the elements itself, which is called a medoid or centroid.

13.2   Methods

13.2.1   Overview of clustering algorithms

For the sake of presenting a unified view of available clustering algorithms, we generalize the output of a clustering algorithm as a sequence of clustering results indexed by the number of clusters $k = 2, 3, \dots$ and options such as the choice of dissimilarity metric. The algorithm is a mapping from the empirical distribution of $X_1, \dots, X_I$ to this sequence of k-specific clustering results. For instance, this mapping could be the construction of an agglomerative hierarchical tree of gene clusters using 1 minus correlation as dissimilarity and single linkage as distance between clusters. Given a clustering algorithm, consider the output if the algorithm were applied to the data generating distribution P (i.e., infinite sample size). We call this output a *clustering parameter*, where we stress that any variation in the algorithm results in a different parameter. An example is the J-dimensional vector of gene cluster labels produced by applying a particular partitioning method (e.g., k-means using Euclidean distance) with a particular number of clusters (e.g., $k = 5$) to P. We might think of these as the true cluster labels, in contrast to the observed labels from a sample of size I. Another parameter is the k-dimensional vector of cluster sizes produced by the same algorithm.

We will focus on non-parametric clustering algorithms, in which one makes no assumptions about the data generating distribution P. Model based clustering algorithms are based on assuming that the vectors X_i are *i.i.d.* from a mixture of distributions (e.g., a multivariate normal mixture). The clustering result is typically a summary measure, such as the conditional probabilities of cluster membership (given the data), of the maximum likelihood estimator of the data generating distribution (Fraley and Raftery, 1998, 2000). Of course, if one only views this mixture model as a working model to define a clustering result, then these approaches fall in the category of non-parametric clustering algorithms. In this case, however, statistical inference cannot be based on the working model, and, contrary to the case in which one assumes this mixture model to contain the true data generating distribution, there does not exist a true number of clusters.

13.2.2   Ingredients of a clustering algorithm

We review here the choices one needs to consider before performing a cluster analysis.

Dissimilarity matrix: All clustering algorithms are (either implicitly or explicitly) indexed by a choice of dissimilarity measure, which quantifies the distinctness of each pair of elements (see Chapter 12). For clustering genes, this is a $J \times J$ symmetric matrix. Typical choices of dissimilarity include Euclidean distance, Manhattan distance, 1 minus correlation, 1 minus absolute correlation and 1 minus cosine-angle (i.e., : 1 minus uncentered correlation). The R function `dist` allows one to compute a variety of dissimilarities. Other distance functions are available in the function `daisy` from the cluster package or from the bioDist package. In hopach we have written `distancematrix` and implemented specialized versions of many of the standard distances. Data transformations, such as standardization of rows or columns, are some times performed before computing the dissimilarity matrix.

Number of clusters: One must specify the number of clusters or an algorithm for determining this number. In Section 13.2.7, we discuss and compare methods for selecting the number of clusters, including various data-adaptive approaches.

Criterion: Clustering algorithms are deterministic mappings that aim to optimize some criterion. This is often a real-valued function of the cluster labels that measures how similar elements are within clusters or how different elements are between clusters. The choice of criterion can have a dramatic effect on the clustering result. We recommend a careful study of a proposed criterion so that the user fully understands its strengths and weaknesses (i.e., its scoring strategy) in evaluating a clustering result. Simulations are a useful tool for comparing different criteria.

Searching strategy: One sensible goal is to find the clustering result that globally maximizes the selected criterion. Because of computational issues, heuristic search strategies (that guarantee convergence to a local maximum) are often needed. If the user prefers a tree structure linking all clusters, then forward or backward selection strategies are often used, and they do not correspond with local maxima of the criterion.

13.2.3   Building sequences of clustering results

We can classify clustering algorithms by their searching strategies. Figure 13.1 compares the clustering results from a partitioning (`pam`) and a hierarchical (`diana`) algorithm.

Partitioning: Partitioning methods, such as self-organizing maps (SOM) (Törönen et al., 1999), partitioning around medoids (PAM) (Kaufman and Rousseeuw, 1990), and k-means, map a collection of elements (e.g., genes) into $k \geq 2$ disjoint clusters by aiming to maximize a particular criterion. In this case, a clustering result for $k = 2$ is not used in computing the clustering result for $k = 3$.

Hierarchical: Hierarchical methods involve constructing a tree of clusters in which the root is a single cluster containing all the elements and the leaves each contain only one element. These trees are typically binary; that is, each node has exactly two children. The final level of the tree can be viewed as an ordered list of the elements, though most algorithms produce an ordering that is very dependent on the initial ordering of the data, and is thus not necessarily distance based.

A hierarchical tree can be divisive (i.e., built from the top down by re-cursively partitioning the elements) or agglomerative (i.e., built from the bottom up by recursively combining the elements) . The R function `diana` [R package cluster, Kaufman and Rousseeuw (1990)] is an example of a divisive hierarchical algorithm, while `agnes` (R package cluster, Kaufman and Rousseeuw (1990)) and Cluster (Eisen et al., 1998) are examples of agglomerative hierarchical algorithms. Agglomerative methods can be employed with different types of linkage, which refers to the distance between groups of elements and is typically a function of the dissimilarities between pairs of elements. In average linkage methods, the distance between two clusters is the average of the dissimilarities between the elements in one cluster and the elements in the other cluster. In single linkage methods (nearest neighbor methods), the dissimilarity between two clusters is the smallest dissimilarity between an element in the first cluster and an element in the second cluster.

Hybrid: The hierarchical ordered partitioning and collapsing hybrid (HOPACH) algorithm (van der Laan and Pollard, 2003) builds a tree of clusters, where the clusters in each level are ordered based on the pairwise dissimilarities between cluster medoids. This algorithm starts at the root node and aims to find the right number of children for each node by alternating partitioning (divisive) steps with collapsing (agglomerative) steps. The resulting tree is non-binary with a deterministically ordered final level.

Several R packages contain clustering algorithms. Table 13.2.3 provides a non-exhaustive list. We use the `agriculture` data set from the package cluster to demonstrate code and output of some standard clustering methods.

```
> library("cluster")
> data(agriculture)

> part <- pam(agriculture, k = 2)
> round(part$clusinfo, 2)
```

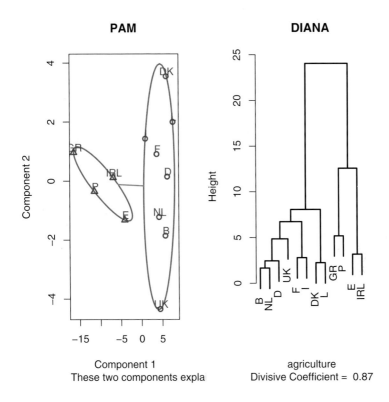

Figure 13.1. Partitioning versus hierarchical clustering. The **agriculture** data set from the package **cluster** contains two variables (Gross National Product per capita and percentage of the population working in agriculture) for each country belonging to the European Union in 1993. The countries were clustered by two algorithms from the package: (i) pam with two clusters and (ii) **diana**. The results are visualized as a **clusplot** for pam and a dendrogram for **diana**.

```
     size max_diss av_diss diameter separation
[1,]    8     5.42    2.89     8.05       5.73
[2,]    4     7.43    4.30    12.57       5.73

> hier <- diana(agriculture)

> par(mfrow = c(1, 2))
> plot(part, which.plots = 1, labels = 3, col.clus = 3,
+      lwd = 2, main = "PAM")
> plot(hier, which.plots = 2, lwd = 2, main = "DIANA")
```

Package	Functions	Description
cclust		Convex clustering methods
class	SOM	Self-organizing maps
cluster	agnes	AGglomerative NESting
	clara	Clustering LARge Applications
	diana	DIvisive ANAlysis
	fanny	Fuzzy Analysis
	mona	MONothetic Analysis
	pam	Partitioning Around Medoids
e1071	bclust	Bagged clustering
	cmeans	Fuzzy C-means clustering
flexmix		Flexible mixture modeling
fpc		Fixed point clusters, clusterwise regression and discriminant plots
hopach	hopach, boothopach	Hierarchical Ordered Partitioning and Collapsing Hybrid
mclust		Model-based cluster analysis
stats	hclust, cophenetic	Hierarchical clustering
	heatmap	Heatmaps with row and column dendrograms
	kmeans	k-means

Table 13.1. R functions and packages for cluster analysis (CRAN, Bioconductor).

13.2.4   Visualizing clustering results

Chapter 10 describes a variety of useful methods for visualizing gene expression data. The function heatmap, for example, implements the plot employed by Eisen et al. (1998) to visualize the $J \times I$ data matrix with rows and columns ordered by separate applications of their Cluster algorithm to both genes and arrays. Figure 13.2 shows an example of such a heat map. Heat maps can also be made of dissimilarity matrices (Figure 13.6 and Chapter 10), which are particularly useful when clustering patterns might not be easily visible in the data matrix, as with absolute correlation distance (van der Laan and Pollard, 2003).

As we see in Figures 13.2 and 13.1, there appears to be two clusters, one with four countries and another with eight. All visualizations make that reasonably obvious, although in different ways.

```
> heatmap(as.matrix(t(agriculture)), Rowv = NA,
+     labRow = c("GNP", "% in Agriculture"), cexRow = 1,
+     xlab = "Country")
```

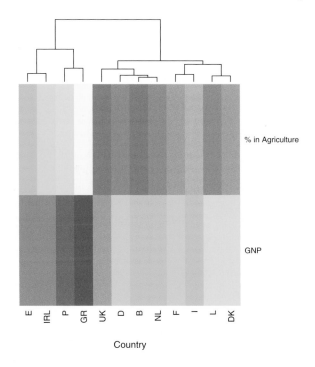

Figure 13.2. Heatmap for hierarchical clustering of the countries in the **agricul-ture** data set. The function **hclust** produces a dendrogram that is equivalent to that produced by **diana** with left and right children swapped at several nodes. Note that the ordering of countries in the **diana** tree depends a great deal on their order in the input data set, so that permuting the rows before running the algorithm will produce a different tree. The **hopach** (and to a lesser degree the **hclust** tree) is not sensitive to the initial order of the data.

13.2.5   Statistical issues in clustering

Exploratory techniques are capable of identifying interesting patterns in data, but they do not inherently lend themselves to statistical inference. The ability to assess reliability in an experiment is particularly crucial with the high dimensional data structures and relatively small samples presented by genomic experiments (Getz G., 2000; Hughes et al., 2000; Lockhart and Winzeler, 2000). Both jackknife (K.Y. et al., 2001) and bootstrap (Kerr and Churchill, 2001; van der Laan and Bryan, 2001) approaches have been used to perform statistical inference with gene expression data. van der Laan and Bryan (2001) present a statistical framework for clustering genes, where the clustering parameter θ is defined as a deterministic function $S(P)$ applied to the data generating distribution P. The parameter $\theta = S(P)$ is estimated by the observed sample subset $S(P_I)$, where the empirical distri-bution P_I is substituted for P. Most currently employed clustering methods fit into this framework, as they need only be deterministic functions of the empirical distribution. The authors also establish consistency of the clustering result under the assumption that $I/\log[J(I)] \to \infty$ (for a sample of I J-dimensional vectors), and asymptotic validity of the bootstrap in this context.

An interesting approach to clustering samples is to first cluster the genes and then cluster the samples using only the gene cluster profiles, such as medoids or means (Pollard and van der Laan, 2002b). In this way, the dimension of the data is reduced to the number of gene clusters so that the multiplicity problem for comparing subpopulations of samples is much less. Gene cluster profiles (particularly medoids) are very stable and hence the comparison of samples will not be affected by a few outlier genes [see also Nevins et al. (2003)]. Pollard and van der Laan (2002b) generalize the statistical framework proposed in van der Laan and Bryan (2001) to any clustering parameter $S(P)$, including algorithms that involve clustering both genes and samples.

13.2.6   Bootstrapping a cluster analysis

Though the clustering parameter $\theta = S(P)$ might represent an interesting clustering pattern in the true data generating distribution/population, when applied to empirical data P_I, it is likely to find patterns due to noise. To deal with this issue, one needs methods for assessing the variability of $\theta_I = S(P_I)$. One also needs to be able to test if certain components of θ_I are significantly different from the value of these components in a specified null experiment. Note that θ_I and P_I depend on the sample size I.

To assess the variability of the estimator θ_I, we propose to use the bootstrap. The idea of the bootstrap method is to estimate the distribution of θ_I with the distribution of $\theta_I^* = S(P_I^*)$, where P_I^* is the empirical distribution based on an $i.i.d.$ bootstrap sample [i.e., a sample of I $i.i.d.$ observations X_i^* $(i = 1, \ldots, I)$ from P_I]. The distribution of θ_I^* is obtained by applying the rule S to P_I^*, from each of B bootstrap samples, keeping track of parameters of interest. The distribution of a parameter is approximated by its empirical distribution over the B samples. There are several methods for generating bootstrap samples.

- **Non-parametric:** Resample I arrays with replacement.

- **Smoothed non-parametric:** Modify non-parametric bootstrap sampling with one of a variety of methods (e.g., Bayesian bootstrap or convex pseudo-data) for producing a smoother distribution.

- **Parametric:** Fit a model (e.g., multivariate normal, mixture of multivariate normals) and generate observations from the fitted distribution.

The non-parametric bootstrap avoids distributional assumptions about the parameter of interest. However, if the model assumptions are appropriate, or have little effect on the estimated distribution of θ_I, the parametric bootstrap might perform better.

13.2.7   Number of clusters

Consider a series of proposed clustering results. With a partitioning algorithm, these may consist of applying the clustering routine with $k = 2, 3, \ldots, K$ clusters, where K is a user-specified upper bound on the number of clusters. With a hierarchical algorithm the series may correspond to levels of the tree. With both types of methods, identifying cluster labels requires choosing the number of clusters. From a formal point of view, the question "How many clusters are there?" is essentially equivalent to asking "Which parameter is correct?" as each k defines a new parameter of the data generating distribution in the non-parametric model for P. Thus, selecting the correct number of clusters requires user input and typically there is no single right answer. Having said this, one is free to come up with a criterion for selecting the number of clusters, just as one might have an argument to prefer a mean above a median as location parameter. This criterion need not be the same as the criterion used to identify the clusters in the algorithm.

Overview of methods for selecting the number of clusters. Currently available methods for selecting the number of significant clusters include direct methods and testing methods. Direct methods consist of optimizing a criterion, such as functions of the within and between cluster sums of squares (Milligan and Cooper, 1985), occurrences of phase transitions in simulated annealing (Rose et al., 1990), likelihood ratios (Scott and Simmons, 1971), or average silhouette (Kaufman and Rousseeuw, 1990). The method of maximizing average silhouette is advantageous because it can be used with any clustering routine and any dissimilarity metric. A disadvantage of average silhouette is that, like many criteria for selecting the number of clusters, it measures the global clustering structure only. Testing methods take a different approach, assessing evidence against a specific null hypothesis. Examples of testing methods that have been used with gene expression data are the gap statistic (Tibshirani et al., 2000), the weighted average discrepant pairs (WADP) method (Bittner et al., 2000), a variety of permutation methods (Bittner et al., 2000; Hughes et al., 2000), and Clest (Fridlyand and Dudoit, 2001). Because they typically involve resampling, testing methods are computationally much more expensive than direct methods.

Median Split Silhouette. Median split silhouette (MSS) is a new direct method for selecting the number of clusters with either partitioning or hierarchical clustering algorithms (Pollard and van der Laan, 2002a). This method was motivated by the problem of finding relatively small, possibly nested clusters in the presence of larger clusters (Figure 13.3). It is frequently this finer structure that is of interest biologically, but most methods find only the global structure. The key idea is to evaluate how well the elements in a cluster belong together by applying a chosen clustering algorithm to the elements in that cluster alone (ignoring the other clusters) and then evaluating average silhouette after the split to determine the homogeneity of the parent cluster. We first define silhouettes and then describe how to use them in the MSS criterion.

Suppose we are clustering genes. The silhouette for a given gene is calculated as follows. For each gene j, calculate the average dissimilarity a_j of gene j with other genes in its cluster. For each gene j and each cluster l to which it does not belong, calculate the average dissimilarity b_{jl} of gene j with the members of cluster l. Let $b_j = \min_l b_{jl}$. The silhouette of gene j is defined by the formula: $S_j = (b_j - a_j)/\max(a_j, b_j)$. Heuristically, the silhouette measures how well matched an object is to the other objects in its own cluster versus how well matched it would be if it were moved to the next closest cluster. Note that the largest possible silhouette is 1, which occurs only if there is no dissimilarity within gene j's cluster (i.e., $a_j = 0$). A silhouette near 0 indicates that a gene lies between two clusters, and a silhouette near -1 means that the gene is very similar to elements in the neighboring cluster and hence is probably in the wrong cluster.

For a clustering result with k clusters, split each cluster into two or more clusters (the number of which can be determined, for example, by maximizing average silhouette). Each gene has a new silhouette after the split, which is computed relative to only those genes with which it shares a parent. We call the median of these for each parent cluster the split silhouette, SS_i, for $i = 1, 2, \ldots, k$, which is low if the cluster was homogeneous and should not have been split. $MSS(k) = median(SS_1, \ldots, SS_k)$ is a measure of the overall homogeneity of the clusters in the clustering result with k clusters. We advocate choosing the number of clusters which minimizes MSS. Note that all uses of median can be replaced with mean for a more sensitive, less robust criterion.

The following example of a data set with nested clusters demonstrates that MSS and average silhouette can identify different numbers of clusters. The data are generated by simulating a $J = 240$ dimensional vector consisting of eight groups of thirty normally distributed variables with the following means: $\mu \in (1, 2, 5, 6, 14, 15, 18, 19)$. The variables are uncorrelated with common standard deviation $\sigma = 0.5$. A sample of $I = 25$ is generated and the Euclidean distance computed.

```
> mu <- c(1, 2, 5, 6, 14, 15, 18, 19)
> X <- matrix(rnorm(240 * 25, 0, 0.5), nrow = 240,
+       ncol = 25)
> step <- 240/length(mu)
> for (m in 1:length(mu)) X[((m - 1) * step + 1):(m *
+       step), ] <- X[((m - 1) * step + 1):(m * step),
```

```
+      ] + mu[m]
> D <- dist(X, method = "euclidean")
```

Next, we check the number of clusters k identified by average silhouette with the function `silcheck` and by MSS with the function `msscheck`, both provided in the package hopach. These return a vector with the number of clusters optimizing the corresponding criterion in the first entry and the value of the criterion in the second.

```
> library("hopach")
> k.sil <- silcheck(X)[1]
> k.mss <- msscheck(as.matrix(D))[1]
> pam.sil <- pam(X, k.sil)
> pam.mss <- pam(X, k.mss)
```

We plot the distance matrix with the $J = 240$ variables ordered according to their `pam` cluster labels with each choice of k. We mark the two sets of cluster boundaries on each axis.

```
> image(1:240, 1:240, as.matrix(D)[order(pam.sil$clust),
+       order(pam.mss$clust)], col = topo.colors(80),
+       xlab = paste("Silhouette (k=", k.sil, ")", sep = ""),
+       ylab = paste("MSS (k=", k.mss, ")", sep = ""),
+       main = "PAM Clusters: Comparison of Two Criteria",
+       sub = "Ordered Euclidean Distance Matrix")
> abline(v = cumsum(pam.sil$clusinfo[, 1]), lty = 2, lwd = 2)
> abline(h = cumsum(pam.mss$clusinfo[, 1]), lty = 3, lwd = 2)
```

We have previously reported simulation results for MSS on a variety of data sets and relative to other direct methods (Pollard and van der Laan, 2002a). We refer the reader to the figures in that manuscript for further illustration of the MSS methodology.

HOPACH algorithm. The R package hopach implements the Hierarchical Ordered Partitioning and Collapsing Hybrid (HOPACH) algorithm for building a hierarchical tree of clusters (Figure 13.4). At each node, a cluster is split into two or more smaller clusters with an enforced ordering of the clusters. Collapsing steps uniting the two closest clusters into one cluster are used to correct for errors made in the partitioning steps. The hopach function uses the median split silhouette criterion to automatically choose (i) the number of children at each node, (ii) which clusters to collapse, and (iii) the main clusters (pruning the tree to produce a partition of homogeneous clusters). We describe the method as applied to clustering genes in an expression data set X, but the algorithm can be used much more generally. We will use the notation $PAM(X, k, d)$ for the PAM algorithm applied to the data X with k clusters ($k < 10$ for computational convenience) and dissimilarity d.

Initial level: Begin with all elements at the root node.

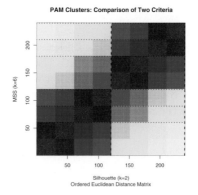

Figure 13.3. Median split silhouette (MSS) versus average silhouette. The Euclidean distance matrix from a data set with nested clusters is plotted here with the variables ordered according to their cluster labels. Blue corresponds to small and peach to large dissimilarity. The nested structure of the data is visible. Lines mark the boundaries of the PAM clusters, with the number of clusters k determined either by minimizing MSS or maximizing average silhouette. Average silhouette is more robust and therefore typically identifies fewer clusters.

1. *Partition*: Compute $PAM(X, k, d)$ and $MSS(k)$ for $k = 2, \ldots, 9$. Accept the minimizer $k1$ of $MSS(k)$ and corresponding partition $PAM(x, k1, d)$ as the first level of the tree. Also compute $MSS(1)$. If $MSS(1) < MSS(k1)$, print a warning message about the homogeneity of the data.
2. *Order*: Define the distance between a pair of clusters (i.e., linkage) as the dissimilarity between the corresponding medoids. If $k1 = 2$, then the ordering does not matter. If $k1 > 2$, then order the clusters by (a) building a hierarchical tree from the $k1$ medoids or (b) maximizing the empirical correlation between distance $j - i$ in the list and the corresponding dissimilarity $d(i, j)$ across all pairs (i, j) with $i < j$ with the function `correlationordering`.
3. *Collapse*: There is no collapsing at the first level of the tree.

Next level: For each cluster in the previous level of the tree, carry out the following procedure.
1. *Partition*: Apply PAM with $k = 1, \ldots, 9$ as in level 1, and select the minimizer of $MSS(k)$ and corresponding PAM partitioning.
2. *Order*: Order the child clusters by their dissimilarity with the medoid of the cluster next to the parent cluster in the previous level.
3. *Collapse*: Beginning with the closest pair of medoids (which may be on different branches of the tree), collapse the two clusters if doing so improves MSS. Continue collapsing until a collapse is rejected (or until all pairs of medoids are considered). The medoid of the new cluster can be chosen in a

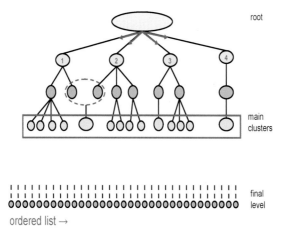

root

main
clusters

final
level

ordered list →

Figure 13.4. The HOPACH hierarchical tree unfolding through the steps of the clustering algorithm. First, the root node is partitioned and the children in the next level are ordered deterministically using the same dissimilarity matrix that is used for clustering. Next, each of these nodes is partitioned and its children are ordered. Before the next partitioning step, collapsing steps merge any similar clusters. The process is iterated until the main clusters are identified. Below the main clusters, the algorithm is run down without collapsing to produce a final ordered list.

variety of ways, including the nearest neighbor of the average of the two corresponding medoids.

Iterate: Repeat until each node contains no more than 2 genes or a maximum number of levels is reached (for computational reasons the limit is 16 levels in the current implementation).

Main clusters: The value of MSS at each level of the tree can be used to identify the level below which cluster homogeneity improves no further. The partition defined by the pruned tree at the selected level is identified as the main clusters.

The path that each gene follows through the HOPACH tree is encoded in a label with one digit for each level in the tree. Because we restrict the number of child clusters at each node to be less than ten, only a single digit is needed for each level. Zero denotes a cluster that is not split. A typical label of a gene at level 3 in the tree looks like 152, meaning that the gene is in the second child cluster of the fifth child cluster of the first cluster from level 1. In order to look at the cluster structure for level l of the tree, simply truncate the final cluster labels to l digits. Chapter 20 provides some relevant concepts and notation regarding paths and path labelling in graphs.

We refer the reader to van der Laan and Pollard (2003) for a comparison of HOPACH with other clustering algorithms. In simulations and real data analyses, we show that hopach is better able to identify small clusters and to produce a sensible final ordering of the elements than other algorithms discussed here.

13.3   Application: renal cell cancer

The renal cell cancer data package kidpack contains expression measures for 4224 genes and 74 patients. The tumor samples (labeled green) are compared to a common reference sample (labeled red). Log ratios measure expression in the control relative to each tumor.

13.3.1   Gene selection

To load the kidpack data set:

```
> library("kidpack")
> data(eset, package = "kidpack")
> data(cloneanno, package = "kidpack")
```

First, select a subset of interesting genes. Such a subset can be chosen in many ways, for example with the functions in the genefilter and multtest packages. For this analysis, we will simply take all genes (416 total) with log ratios greater than 3-fold in at least half of the arrays. This means that we are focusing on genes that are *suppressed* in the kidney tumor samples relative to the control sample. One would typically use a less arbitrary subset rule. We use the IMAGE ID (Lennon et al., 1996) as the gene name, adding the character "B" to the name of the second copy of any IMAGE ID.

```
> library("genefilter")
> ff <- pOverA(0.5, log10(3))
> subset <- genefilter(abs(exprs(eset)), filterfun(ff))
> kidney <- exprs(eset)[subset, ]
> dim(kidney)
> gene.names <- cloneanno[subset, "imageid"]
> is.dup <- duplicated(gene.names)
> gene.names[is.dup] <- paste(gene.names[is.dup],
+     "B", sep = "")
> rownames(kidney) <- gene.names
```

```
> colnames(kidney) <- paste("Sample", 1:ncol(kidney),
+      sep = "")
```

13.3.2 HOPACH clustering of genes

It is useful to compute the dissimilarity matrix before running `hopach`, because the dissimilarity matrix may be needed later in the analysis. The cosine-angle dissimilarity defined in Chapter 12 is often a good choice for clustering genes.

```
> gene.dist <- distancematrix(kidney, d = "cosangle")
> dim(gene.dist)
```

```
[1] 416 416
```

Now, run `hopach` to cluster the genes. The algorithm will take some time to run.

```
> gene.hobj <- hopach(kidney, dmat = gene.dist)
```

```
> gene.hobj$clust$k
```

```
[1] 84
```

```
> table(gene.hobj$clust$sizes)
```

```
 1   2   3   4   5   7   9  18  24  42  80 112
52   8  13   3   1   1   1   1   1   1   1   1
```

```
> gene.hobj$clust$labels[1:5]
```

```
[1] 22200 22200 21300 23200 43000
```

The `hopach` algorithm identifies 84 gene clusters. Many of the clusters are 1 to 4 genes, though some are much larger. The cluster labels show the relationships between the clusters and how they evolved in the first few levels of the tree. Next, we examine how close clones that represent the same gene (i.e., genes with a "B" in their name) are to one another in the HOPACH final ordering.

```
> gn.ord <- gene.names[gene.hobj$fin$ord]
> Bs <- grep("B", gn.ord)
> spaces <- NULL
> for (b in Bs) {
+      name <- unlist(strsplit(gene.names[gene.hobj$fin$ord][b],
+           "B"))
+      spaces <- c(spaces, diff(grep(name, gn.ord)))
+ }
> table(spaces)
```

```
spaces
 1   4   6  14  17  35  53  54  72  90 129
 5   1   1   1   1   1   1   1   1   1   1
```

Five of the fifteen pairs of replicate clones appear next to each other, and all of them appear closer to one another than expected for a random pair of clones.

13.3.3 Comparison with PAM

The hopach clustering results can be compared to simply applying PAM with the choice of k that maximizes average silhouette (using the function silcheck).

```
> bestk <- silcheck(dissvector(gene.dist), diss = TRUE)[1]
> pamobj <- pam(dissvector(gene.dist), k = bestk,
+      diss = TRUE)
> round(pamobj$clusinfo, 2)
```

	size	max_diss	av_diss	diameter	separation
[1,]	68	0.96	0.64	1.10	0.39
[2,]	348	0.94	0.45	1.21	0.39

While hopach identifies 84 clusters of median size 1, pam identifies 2 larger clusters. This result is typical in the sense that hopach tends to be more aggressive at finding small clusters, whereas pam is more robust and therefore only identifies the global patterns (i.e., fewer, larger clusters).

13.3.4 Bootstrap resampling

For each gene and each hopach cluster we can compute the proportion of bootstrap data sets where the gene is in the cluster. These are estimates of the membership of the gene in each cluster and can be considered as a form of fuzzy clustering.

```
> bobj <- boothopach(kidney, gene.hobj, B = 100)
```

The argument B controls the number of bootstrap resampled data sets used. The default value is B= 1000, which represents a balance between precision and speed. For this example, we use B= 100 since larger values have much longer run times. The bootplot function makes a barplot of the bootstrap reappearance proportions (see Figure 13.5).

```
> bootplot(bobj, gene.hobj, ord = "bootp", main = "Renal Cell Cancer",
+      showclusters = FALSE)
```

13.3.5 HOPACH clustering of arrays

The HOPACH algorithm can also be applied to cluster samples (i.e., arrays), based on their expression profiles across genes. This analysis method differs from classification, which uses knowledge of class labels associated with each sample (i.e., array). Euclidean distance may be a good choice for

Renal Cell Cancer Data
Barplot of Bootstrap Reappearance Proportions

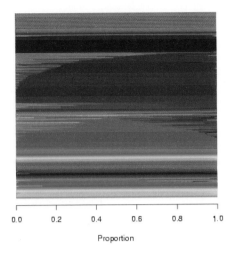

Proportion

Figure 13.5. The `bootplot` function makes a barplot of the bootstrap reappearance proportions for each gene and each cluster. These proportions can be viewed as fuzzy cluster memberships. Every cluster is represented by a different color. The genes are ordered by `hopach` cluster, and then by bootstrap estimated membership within cluster and plotted on the vertical axis. Each gene is represented by a very narrow horizontal bar. The length of this bar that is each color is proportional to the percentage of bootstrap samples in which that gene appeared in the cluster represented by that color. If the bar is all or mostly one color, then the gene is estimated to belong strongly to that cluster. If the bar is many colors, the gene has fuzzy membership in all these clusters. The continuity of colors across the genes indicates that nearby clusters are more likely to "swap" genes than more distant clusters.

clustering arrays, because it measures differences in magnitude, which is often what we are interested in detecting when comparing the expression profiles for different samples. A comparison of magnitude is valid, because we expect the data from different arrays to be on the same scale after normalization has been performed.

```
> array.hobj <- hopach(t(kidney), d = "euclid")
```

```
> array.hobj$clust$k
```

```
[1] 51
```

51 array clusters are identified. The function `dplot` can be used to visualize the ordered dissimilarity matrix corresponding with the HOPACH tree's final level. Clusters of similar arrays will appear as blocks on the diagonal of the matrix (Figure 13.6). We can label the arrays from patients with

Renal Cell Cancer: Array Clustering
Ordered Distance Matrix

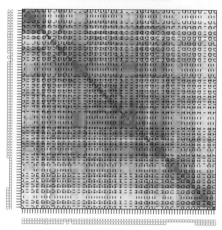

Figure 13.6. HOPACH clustering of patients with Euclidean distance. Patients are ordered according to the final level of the tree. Red corresponds to small distance and white to large distance. Dotted lines indicate the clusters boundaries in the level of the tree with minimum MSS. Many patients cluster alone, but there are several small groups of very similar patients. The ordering of patients by hopach coincides well with tumor type. cc: clear cell, p: papillary, ch: chromophobe.

different tumor types (clear cell, papillary, and chromophobe) and examine how these labels correspond with the clusters.

```
> tumortype <- unlist(strsplit(phenoData(eset)$type, "RCC"))
> dplot(distancematrix(t(kidney), d = "euclid"), array.hobj,
+     labels = tumortype, main = "Renal Cell Cancer: Array Clustering")
```

13.3.6   Output files

Gene clustering and bootstrap results table. The makeoutput function is used to write a tab delimited text file that can be opened in a spreadsheet application or text editor. The file will contain the hopach clustering results, plus possibly the corresponding bootstrap results, if these are provided. The argument gene.names can be used to insert additional gene annotation, in this case accession numbers.

```
> gene.acc <- cloneanno[subset, "AccNumber"]
> makeoutput(kidney, gene.hobj, bobj, file = "kidney.out",
+     gene.names = gene.acc)
```

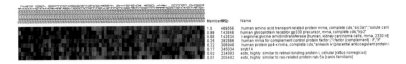

	Members	ID	Name
	1.0	489656	human amino acid transport-related protein mrna, complete cds,"slc3a1","solute carri
	0.99	143846	human glycoprotein receptor gp330 precursor, mrna, complete cds,"lrp2"
	0.88	142634	l-arginine:glycine amidinotransferase [human, kidney carcinoma cells, mrna, 2330 nt]
	0.25	292866	human mrna for complement control protein factor i,"i factor (complement)","if","if
	0.22	306948	human protein pp4-x mrna, complete cds,"annexin iv (placental anticoagulant protein i
	0.17	345034	scyb14
	0.02	234993	ests, highly similar to retinol-binding protein i, cellular [rattus norvegicus]
	0.01	205492	ests, highly similar to ras-related protein rab-5a [canis familiaris]

Figure 13.7. MapleTree zoom view of a single cluster in the kidney data. Genes are ordered according to their bootstrap membership. Red represents overexpression in control relative to tumor samples, and green is the opposite.

Bootstrap fuzzy clustering in MapleTree. MapleTree (Lisa Simirenko) is an open source, cross-platform, visualization tool for graphical browsing of results of cluster analyses. The software can be found at SourceForge. The boot2fuzzy function takes the gene expression data, plus corresponding hopach clustering output and bootstrap resampling output, and writes the (.cdt, .fct, and .mb) files needed to view these fuzzy clustering results in MapleTree.

```
> gene.desc <- cloneanno[subset, "description"]
> boot2fuzzy(kidney, bobj, gene.hobj, array.hobj,
+     file = "kidneyFzy", gene.names = gene.desc)
```

The three generated files can be opened in MapleTree by going to the Load menu and then Fuzzy Clustering Data. The heat map contains only the medoid genes (cluster profiles). Double clicking on a medoid opens a zoom window for that cluster, with a heat map of all genes ordered by their bootstrap estimated memberships in that cluster, with the highest membership first. Figure 13.7 contains the zoom window for gene cluster 15. The medoid and two other genes have high bootstrap reappearance probabilities.

HOPACH hierarchical clustering in MapleTree. The MapleTree software can also be used to view HOPACH hierarchical clustering results. The hopach2tree function takes the gene expression data, plus corresponding hopach clustering output for genes or arrays, and writes the (.cdt, .gtr, and optionally .atr) files needed to view these hierarchical clustering results in MapleTree. These files can also be opened in other viewers such as TreeView (Michael Eisen), jtreeview (Alok Saldanha), and GeneXPress (Eran Segal).

```
> hopach2tree(kidney, file = "kidneyTree", hopach.genes = gene.hobj,
+     hopach.arrays = array.hobj, dist.genes = gene.dist,
+     gene.names = gene.desc)
```

The hopach2tree function writes up to three text files. A .cdt file is always produced. When hopach.genesis not NULL, a .gtr is produced, and gene clustering results can be viewed, including ordering the genes in the heat map according to the final level of the hopach tree and drawing the dendrogram for hierarchical gene clustering. Similarly, when hopach.arrays is not NULL, an .atr file is produced, and array clustering results can

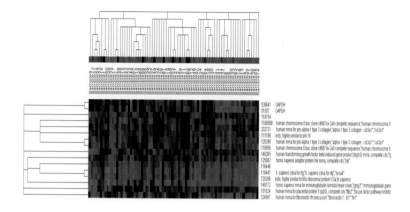

Figure 13.8. MapleTree HOPACH hierarchical view of a section of the gene tree and all of the array tree. Red represents overexpression in control relative to tumor samples, and green is the opposite. Two copies of the clone with I.M.A.G.E. ID 469566 appear near each other in the tree.

be viewed. These files can be opened in MapleTree by going to the Load menu and then HOPACH Clustering Data. By clicking on branches of the tree, a zoom window with gene names for that part of the tree is opened. Figure 13.8 illustrates this view for a section of the the kidney data in MapleTree.

13.4   Conclusion

This chapter has provided an overview of clustering methods in R, including the new hopach package. The variety of available dissimilarity measures, algorithms and criteria for choosing the number of clusters give the data analyst the ability to choose a clustering parameter that meets particular scientific goals. Viewing the output of a clustering algorithm as an estimate of this clustering parameter allows one to assess the reliability and repeatability of the observed clustering results. This kind of statistical inference is particularly important in the context of analyzing high-dimensional genomic data sets. Visualization tools, including data and distance matrix heatmaps, help summarize clustering results.

14

Analysis of Differential Gene Expression Studies

D. Scholtens and A. von Heydebreck

Abstract

In this chapter, we focus on the analysis of differential gene expression studies. Many microarray studies are designed to detect genes associated with different phenotypes, for example, the comparison of cancer tumors and normal cells. In some multifactor experiments, genetic networks are perturbed with various treatments to understand the effects of those treatments and their interactions with each other in the dynamic cellular network. For even the simplest experiments, investigators must consider several issues for appropriate gene selection. We discuss strategies for gene-at-a-time analyses, nonspecific and meta-data driven prefiltering techniques, and commonly used test statistics for detecting differential expression. We show how these strategies and statistical tools are implemented and used in Bioconductor. We also demonstrate the use of factorial models for probing complex biological systems and highlight the importance of carefully coordinating known cellular behavior with statistical modeling to make biologically relevant inference from microarray studies.

14.1   Introduction

Microarray technology is used in a wide variety of settings for detecting differential gene expression. Classic statistical issues such as appropriate test statistics, sample size, replicate structure, statistical significance, and outlier detection enter into the design and analysis of gene expression studies. Adding to the complexity is the fact that the number of samples I in a microarray experiment is inevitably much less than the number of genes J under investigation and that J is often on the scale of tens of thousands, thus creating a tremendous multiple testing burden (see Chapter 15 for further discussion). Investigators must ensure that the experimental design gives access to unambiguous tests of the key substantive hypotheses. This is a challenging task in the complex, dynamic cellular network. We begin our discussion in Section 14.2 by examining general issues in differential expression analysis relevant to most microarray experiments, illustrating these principles with case studies of the **ALL** and **kidpack** data in Sections 14.2.1 and 14.2.2. We then examine multifactor models in Section 14.3 with a case study of the **estrogen** data in Section 14.3.1.

14.2   Differential expression analysis

Fundamental to the task of analyzing gene expression data is the need to identify genes whose patterns of expression differ according to phenotype or experimental condition. Gene expression is a well coordinated system, and hence measurements on different genes are in general not independent. Given more complete knowledge of the specific interactions and transcriptional controls, it is conceivable that meaningful comparisons between samples can be made by considering the joint distribution of specific sets of genes. However, the high dimension of gene expression space prohibits a comprehensive exploration, while the fact that our understanding of biological systems is only in its infancy means that in many cases we do not know which relationships are important and should be studied. In current practice, differential expression analysis will therefore at least start with a gene-by-gene approach, ignoring the dependencies between genes.

A simple approach is to select genes using a fold-change criterion. This may be the only possibility in cases where no, or very few replicates, are available. An analysis solely based on fold change however does not allow the assessment of significance of expression differences in the presence of biological and experimental variation, which may differ from gene to gene. This is the main reason for using statistical tests to assess differential expression. Generally, one might look at various properties of the distributions of a gene's expression levels under different conditions, though most often location parameters of these distributions, such as the mean or the median, are considered. One may distinguish between parametric tests, such as the t-test, and non-parametric tests, such as the Mann-Whitney test or permutation tests. Parametric tests usually have a higher power if the underlying model assumptions, such as normality in the case of the t-test, are at least approximately fulfilled. Non-parametric tests do have the advantage of making less stringent assumptions on the data-generating distribution. In many microarray studies however, a small sample size leads to insufficient power for non-parametric tests. A pragmatic approach in these situations is to employ parametric tests, but to use the resulting p-values cautiously to rank genes by their evidence for differential expression.

When performing statistical analysis of microarray data, an important question is determining on which scale to analyze the data. Often the logarithmic scale is used in order to make the distribution of replicated measurements per gene roughly symmetric and close to normal. A variance-stabilizing transformation derived from an *error model* for microarray measurements (see Chapter 1) may be employed to make the variance of the measured intensities independent of their expected value (Huber et al., 2002). This can be advantageous for gene-wise statistical tests that rely on variance homogeneity, because it will diminish differences in variance between experimental conditions that are due to differences in the intensity level – however of course differences in variance between conditions may also have gene-specific biological reasons, and these will remain untouched.

One or two group t-test comparisons, multiple group ANOVA, and more general trend tests are all instances of linear models that are frequently used for assessing differential gene expression. As a parametric method, linear modeling is subject to the caveats discussed above, but the convenient interpretability of the model parameters often makes it the method of choice for microarray analysis. Due to the aforementioned lack of information regarding coregulation of genes, linear models are generally computed for each gene separately. When the lists of genes of interest are identified, investigators can hopefully begin to study their coordinated regulation for more sophisticated modeling of their joint behavior.

The approach of conducting a statistical test for each gene is popular, largely because it is relatively straightforward and a standard repertoire of methods can be applied. However, the approach has a number of drawbacks: most important is the fact that a large number of hypothesis tests is carried out, potentially leading to a large number of falsely significant results. *Multiple testing* procedures allow one to assess the overall significance of the results of a family of hypothesis tests. They focus on specificity by controlling type I (false positive) error rates such as the *family-wise error rate* or the *false discovery rate* (Dudoit et al., 2003). This topic is covered in detail in Chapter 15. Still, multiple hypothesis testing remains a problem, because an increase in specificity, as provided by p-value adjustment methods, is coupled with a loss of sensitivity, that is, a reduced chance of detecting true positives. Furthermore, the genes with the most drastic changes in expression are not necessarily the "key players" in the relevant biological processes. This problem can only be addressed by incorporating prior biological knowledge into the analysis of microarray data, which may lead to focusing the analysis on a specific set of genes. Also if such a biologically motivated preselection is not feasible, the number of hypotheses to be tested can often be reasonably reduced by non-specific filtering procedures, discarding, e.g., genes with consistently low intensity values or low variance across the samples. This is especially relevant in the case of genome-wide arrays, as often only a minority of all genes will be expressed at all in the cell type under consideration.

Many microarray experiments involve only few replicates per condition, which makes it difficult to estimate the gene-specific variances that are used, e.g., in the t-test. Different methods have been developed to exploit the variance information provided by the data of all genes (Baldi and Long, 2001; Tusher et al., 2001; Lönnstedt and Speed, 2002; Kendziorski et al., 2003). In the limma package, an Empirical Bayes approach is implemented that employs a global variance estimator s_0^2 computed on the basis of all genes' variances. The resulting test statistic is a moderated t-statistic, where instead of the single-gene estimated variances s_g^2, a weighted average of s_g^2 and s_0^2 is used. Under certain distributional assumptions, this test statistic can be shown to follow a t-distribution under the null hypothesis with the degrees of freedom depending on the data (Smyth, 2004).

In the following examples, we demonstrate the use of Bioconductor packages, especially multtest and limma, to identify differentially expressed genes.

14.2.1   Example: ALL data

In this example, we consider a subset of the ALL data representing 79 samples from patients with B-cell acute lymphoblastic leukemia that were investigated using HG-U95Av2 Affymetrix GeneChip arrays (Chiaretti et al., 2004). The probe-level data were preprocessed using RMA (Irizarry et al., 2003b), described in Chapter 2, to produce log (base 2) expression measurements. Of particular interest is the comparison of samples with the BCR/ABL fusion gene resulting from a translocation of the chromosomes 9 and 22 with samples that are cytogenetically normal. In the following code chunk, we load the data and define the subset of samples we are interested in – 37 BCR/ABL samples and 42 normal samples (labeled NEG). The *exprSet* object eset contains the relevant data.

```
> library("ALL")
> data(ALL)
> pdat <- pData(ALL)
> subset <- intersect(grep("^B", as.character(pdat$BT)),
+     which(pdat$mol %in% c("BCR/ABL", "NEG")))
> eset <- ALL[, subset]
```

Many of the genes represented by the 12625 probesets on the array are not expressed in B-cell lymphocytes (either in their normal condition or in any of the disease states being considered), which are the cells that were measured in this experiment. Hence the probesets for these genes can, and should, be removed from the analysis. Furthermore, we want to discard probesets with a low variability across all samples. In the next code chunk, we require expression measurements to be above 100 fluorescence units in at least 25% of the samples, and the interquartile range (IQR) across the samples on the log base 2 scale to be at least 0.5. This non-specific filtering is accomplished with functions from the package genefilter.

```
> library("genefilter")
> f1 <- pOverA(0.25, log2(100))
> f2 <- function(x) (IQR(x) > 0.5)
> ff <- filterfun(f1, f2)
> selected <- genefilter(eset, ff)
> sum(selected)
```

```
[1] 2391
```

```
> esetSub <- eset[selected, ]
```

We are left with 2391 probesets for further analysis. Using the multtest package, we perform a permutation test for equality of the mean expression levels in the two groups for each of these probesets. By default, the function mt.maxT computes Welch t-statistics, which allow for unequal variances in the two groups. The number of permutations B determines the granularity of the permutation p-values. Depending on the multiple testing procedure to be applied, the user may have to choose a value of B that is considerably larger than the number of tests being performed.

```
> cl <- as.numeric(esetSub$mol == "BCR/ABL")
> resT <- mt.maxT(exprs(esetSub), classlabel = cl,
+       B = 10000)
> ord <- order(resT$index)
> rawp <- resT$rawp[ord]
> names(rawp) <- geneNames(esetSub)
```

Figure 14.1 shows the histogram of unadjusted permutation p-values, as given by the vector rawp. The high proportion of small p-values suggests that a substantial fraction of the genes are differentially expressed between the two groups. In order to control the family-wise error rate (FWER), that is, the probability of at least one false positive in the set of significant genes, we have used the permutation-based maxT-procedure of Westfall and Young (Westfall and Young, 1993), as implemented in the function mt.maxT. We obtain 18 genes with an adjusted p-value below 0.05:

```
> sum(resT$adjp < 0.05)
```

```
[1] 18
```

A comparison of this number to the height of the leftmost bar in the histogram suggests that we may be missing a large number of differentially expressed genes. The FWER is a very stringent criterion, and in some microarray studies, only few genes may be significant in this sense, even if many more are truly differentially expressed. A more liberal criterion is provided by the false discovery rate (FDR), that is, the expected proportion of false positives among the genes that are called significant. We use the

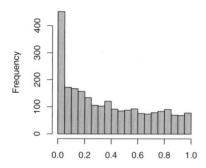

Figure 14.1. Histogram of *p*-values for the gene-by-gene comparison between BCR/ABL positive and cytogenetically normal leukemias.

procedure of Benjamini and Hochberg (1995) as implemented in **multtest** to control the FDR at a level of 0.05, which leaves us with 102 significant genes (note however that this procedure makes certain assumptions on the dependence structure between genes):

```
> res <- mt.rawp2adjp(rawp, proc = "BH")
> sum(res$adjp[, "BH"] < 0.05)
```

```
[1] 102
```

Effects of non-specific filtering

As indicated above, the aim of non-specific filtering is to remove genes that, e.g., due to their low overall intensity or variability, are unlikely to carry information about the phenotypes under investigation. The researcher will be interested in keeping the number of tests as low as possible while keeping the interesting genes in the selected subset.

If the truly differentially expressed genes are overrepresented among those selected in the filtering step, the FDR associated with a certain threshold of the test statistic will be lowered due to the filtering. This appears plausible for two commonly used global filtering criteria: *Intensity-based filtering* aims to remove genes that are not expressed at all in the samples studied, and therefore cannot be differentially expressed. Also concerning the *variability across samples*, a higher overall variance of the differentially expressed genes may be expected, because their between-class variance adds to their within-class variance.

To investigate these presumed effects, we compare the scores for intensity and variability that we used in the beginning for gene selection with the absolute values of the t-statistic, which we now compute for all 12625 probesets.

```
> IQRs <- esApply(eset, 1, IQR)
> intensityscore <- esApply(eset, 1, function(x) quantile(x,
+     0.75))
> abs.t <- abs(mt.teststat(exprs(eset), classlabel = cl))
```

The result is shown in Figure 14.2. Gene selection by the interquartile range (IQR) indeed seems to lead to a higher concentration of differentially expressed genes, whereas for the intensity-based criterion, the effect is less pronounced.

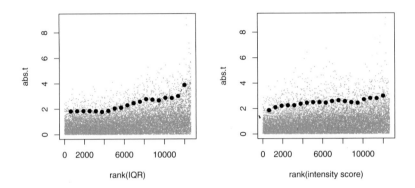

Figure 14.2. Plots of the absolute values of the t-statistic (y-axis) against the ranks of the values of the two filtering criteria: left, interquartile range (IQR), right, overall intensity score. The larger dark dots indicate the 95%-quantiles of the absolute value of the t-statistic computed for moving windows along the x-axis.

Using Gene Ontology data

A source of valuable biological data that is easily accessible through Bioconductor software is the Gene Ontology (GO). It is known that many of the effects due to the BCR/ABL translocation are mediated by tyrosine kinase activity. It will therefore be of interest to examine genes that are known to have tyrosine kinase activity. The term GO:0004713 from the *molecular function* portion of the GO hierarchy refers to protein-tyrosine kinase activity. We can obtain all Affymetrix probesets that are annotated at that node, either directly or by inheritance, using the following command.

```
> tykin <- unique(lookUp("GO:0004713", "hgu95av2",
+     "GO2ALLPROBES"))
> length(tykin)
```

```
[1] 352
```

We see that 352 probesets are annotated at this particular term, 48 of which were selected by our non-specific filtering step. We focus our attention on these 48 probesets and repeat the permutation t-test analysis. In the analysis of the GO-filtered data, 6 probesets have FWER-adjusted p-values less than 0.05. They are printed below, together with the adjusted p-values from the first analysis that involved 2391 genes.

```
[1] "GO analysis"

40480_s_at    1635_at  1636_g_at    39730_at  2039_s_at
    0.0001     0.0001     0.0001      0.0001     0.0005
   36643_at
    0.0286

[1] "All Genes"

    1635_at  1636_g_at    39730_at 40480_s_at  2039_s_at
    0.0001     0.0001      0.0001     0.0015     0.0149
   36643_at
    0.4691
```

Due to the reduced number of tests in the analysis focused on tyrosine kinases, we are left with more significant genes after correcting for multiple testing. For instance, the probeset `36643_at`, which corresponds to the gene DDR1, was not significant in the unfocused analysis, but would be if instead the investigation was oriented toward studying tyrosine kinases.

14.2.2 Example: Kidney cancer data

The kidpack package contains gene expression data from 74 renal cell carcinoma (RCC) patient biopsy samples, which were measured on two-color cDNA arrays together with a common reference sample. The data set is described in detail in the Appendix A.1.2 and in Sültmann et al. (2005). The RCC samples belong to three different histological types, clear cell (ccRCC), papillary (pRCC) and chromophobe (chRCC):

```
> pdat <- pData(esetSpot)
> table(pdat$type)

ccRCC chRCC  pRCC
   52     9    13
```

In the following, we illustrate how the differences in gene expression between these types can be investigated using the limma package (see also Chapter 23 for a more detailed description of limma). We are going to fit a linear model to the expression levels of each gene. limma expects the model to be specified by the *design matrix*, which can either be defined directly or be constructed from a formula via the function `model.matrix`, which is what we do here:

```
> design <- model.matrix(~-1 + factor(pdat$type))
> colnames(design) <- c("ccRCC", "chRCC", "pRCC")
```

This simple design matrix corresponds to the following parametrization:

$$y_{ik} = \alpha_k + \epsilon_{ik} \quad i = 1, 2, \ldots, n_k; \quad k = 1, 2, 3,$$

where k indicates the tumor type and i the individual samples. Note that the model is parameterized without an intercept term, and the estimated coefficients $\hat{\alpha}_k$ from a least squares fit are the mean expression values for the three cancer types.

To exploit the information of replicate measurements of each cDNA clone, limma allows fitting linear models to the spot intensities taking the correlation between replicate spots into account (Smyth et al., 2005). First, the correlation between replicate spots is estimated for each gene separately with restricted maximum likelihood (REML) based on a mixed effects linear model. An overall estimate of the correlation between replicates is computed as a robust average of the individual correlations on the hyperbolic arc tangent scale (atanh), and this overall estimate is then used when fitting a linear model for each gene. The same procedure can be applied in the case of several hybridizations (technical replicates) per cell or tissue sample (biological replicate). In our case, we estimate the correlation between the two replicate spots per clone (argument ndups). The 4224 different clones are listed in separate row blocks in the expression data matrix, hence their spacing is 4224:

```
> dupcor <- duplicateCorrelation(exprs(esetSpot),
+       design = design, ndups = 2, spacing = 4224)
> fit <- lmFit(esetSpot, design = design, ndups = 2,
+       spacing = 4224, correlation = dupcor$cor)

> dupcor$cor

[1] 0.407
```

By default, lmFit fits a linear model by the least squares method, but it also allows robust regression. We are now interested in the expression differences between any two of the cancer types. For this purpose, we set up a *contrast matrix* whose columns represent the pairwise differences between the model coefficients. With the function contrast.fit, we can compute estimated coefficients and standard errors for these contrasts from our original model fit:

```
> contrast.matrix <- makeContrasts(ccRCC - chRCC,
+       ccRCC - pRCC, chRCC - pRCC, levels = design)
> contrast.matrix
```

	ccRCC - chRCC	ccRCC - pRCC	chRCC - pRCC
ccRCC	1	1	0
chRCC	-1	0	1
pRCC	0	-1	-1

```
> fit2 <- contrasts.fit(fit, contrast.matrix)
```

Moderated *t*-statistics for these contrasts, where the gene-specific variances are augmented with a global variance estimator computed from the data of all genes, are obtained with the function `eBayes`:

```
> fit3 <- eBayes(fit2)
```

The `topTable` function produces a table of the top ranking genes, sorted by default by their log-odds for differential expression (see below). Here we show the output of `topTable` for the third contrast, referring to the comparison of chRCC and pRCC.

```
> topTable(fit3, coef = 3, n = 8, adjust.method = "fdr")
          ID      M        A      t  P.Value     B
2600 321496   2.68  -0.1154   18.5 1.12e-37  82.7
2729 502969   1.88  -0.1703   13.6 6.45e-25  53.4
1804 133812   1.81  -0.5036   13.3 3.00e-24  51.5
2859 725766   1.92  -0.1276   12.9 1.69e-23  49.5
3734 306257  -1.53   0.1353  -12.4 3.84e-22  46.3
1879 357297   1.36  -0.3215   11.7 3.38e-20  41.7
1905 774064   1.74  -0.4917   11.4 1.15e-19  40.4
2750 738532   1.37   0.0461   11.3 3.47e-19  39.2
```

For the column `P.value`, different methods to adjust the *p*-values for multiple testing can be chosen, which allow to control the family-wise error rate or the false discovery rate. Here we have chosen the FDR-based p-value adjustment according to Benjamini and Hochberg (1995). Further columns produced by `topTable` contain for each gene an identifier `Name` (in our case the Image ID of the respective cDNA clone), the estimated contrast coefficient `M`, the average expression value across all samples `A`, the moderated *t*-statistic `t`, and the log-odds for differential expression `B`, corresponding to a Bayesian interpretation of the moderated *t*-statistic. The interpretation of the values of `M` and `A` depends on the nature of the data used as input for `lmFit`. In our case, the column `M` contains expression differences on a generalized natural log scale relative to a common reference sample, and the values of `A` do not refer to absolute intensities but are given by the average of a gene's generalized log-ratio values with respect to the reference sample across all chips.

When testing different contrasts per gene simultaneously, the issue of *multiple comparisons* arises, that is, it is of interest to evaluate the significance of each single contrast in the light of the whole set of contrasts. The limma function `decideTests` allows the identification of significant test results in the sense of *multiple testing* across genes, as well as in the sense of *multiple comparisons* across contrasts. For the latter, the following approach is pursued with the argument `method="nestedF"`: The moderated *t*-statistic for a particular contrast is called significant at a certain level α (resulting from multiple testing adjustment across genes) if the moderated

F-test for that gene is still significant at level α when setting all the larger t-statistics for that gene to the same absolute value as the t-statistic in question. The function `decideTests` yields a matrix, where for each gene each contrast is marked as non-significant (zero), significantly positive (one), or significantly negative (minus one). In our example, we want to know how many genes are differentially expressed when fixing the significance level α of the moderated F-test so that it corresponds to a FDR of 0.05:

```
> clas <- decideTests(fit3, method = "nestedF",
+      adjust.method = "fdr", p = 0.05)
> colSums(abs(clas))

ccRCC - chRCC  ccRCC - pRCC  chRCC - pRCC
         1243           981           931
```

To assess the effect of using the single spot measurements opposed to the commonly used averaging across duplicate spots, we compare the results to those of an analogous analysis based on a data matrix `datAvDup` where the expression values of duplicate spots have been averaged.

```
> nclones <- 4224
> datAvDup <- (exprs(esetSpot)[1:nclones, ] +
+      exprs(esetSpot)[nclones + 1:nclones, ])/2
> fitAvDup <- lmFit(datAvDup, design = design)
> fit2AvDup <- contrasts.fit(fitAvDup, contrast.matrix)
> fit3AvDup <- eBayes(fit2AvDup)
```

The comparison of the resulting p-values (again for the comparison of chRCC and pRCC) suggests that the spot-wise analysis yields higher power (Figure 14.3).

14.3 Multifactor experiments

Multifactor microarray experiments often involve the application of treatments in combination to model organisms such as genetically identical cell lines or mice. The equal reference point from which these experiments start theoretically limits naturally occurring interindividual variability, thus allowing differential gene expression to be attributed to the treatments or experimental conditions under investigation. Frequently, these experiments are designed to investigate the perturbation of genetic networks by various combinations of treatments, thus allowing the initial steps of genetic network reconstruction. In factorial designs, effects of the treatments and their interactions can be conveniently quantified in a linear model. As long as the contrasts of interest are specified with careful accounting for the transcription and translation mechanisms affected by the treatments, investigators can often assign very meaningful biological interpretations to their results.

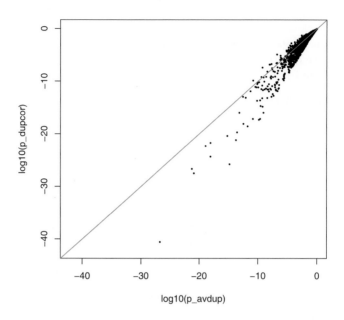

Figure 14.3. Comparison of base 10 logarithms of p-values for the comparison between chRCC and pRCC. x-axis: analysis based on average expression values across duplicate spots, y-axis: spot-wise analysis incorporating correlation between duplicate spots.

One significant difficulty with linear modeling in the microarray setting is model checking. Studentized residuals from the classic linear modeling paradigm are often inappropriate in designs with only a few replicates due to the large number of linear dependencies relative to the number of residuals. Specialized algorithms are often useful for very small designs; in Section 14.3.1, we discuss a technique for outlier detection in a factorial experiment with just two replicates. The development of multivariate permutation tests for the high-throughput setting would help alleviate this problem (Pesarin, 2001).

Multifactor linear models have been used for a variety of purposes in microarray studies. In addition to identifying differentially expressed genes due to treatments applied in combination, linear models have been very useful for data preprocessing of cDNA microarrays (see Chapter 4). In the estrogen example, we illustrate the interpretability of multifactor linear models for single channel arrays; the results extend naturally to two-color competitive hybridization platforms. We use the limma package for our analysis, but factDesign and daMA are also available for the analysis of factorial designed microarray experiments.

14.3.1 Example: Estrogen data

The package estrogen contains 8 Affymetrix HG-U95Av2 CEL files from an experiment involving breast cancer cells. We first perform quantile normalization and calculate expression estimates using RMA (Irizarry et al., 2003b).

```
> library("estrogen")
> library("limma")
> library("hgu95av2cdf")
> datadir <- system.file("extdata", package = "estrogen")
> targets <- readTargets("phenoData.txt", path = datadir,
+     sep = "")
> covdesc <- list("present or absent", "10 or 48 hours")
> names(covdesc) <- names(targets)[-1]
> pdata <- new("phenoData", pData = targets[, -1],
+     varLabels = covdesc)
> rownames(pData(pdata)) <- targets[, 1]
> gc()
> esAB <- ReadAffy(filenames = file.path(datadir,
+     targets$filename), phenoData = pdata)
> esEset <- rma(esAB)
```

This collection of eight arrays is a subset of 32 arrays from a 2^4 factorial experiment with two replicates for each treatment condition on an estrogen receptor positive (ER+) breast cancer cell line, the complete analysis of which is discussed in Scholtens et al. (2004). Upon binding to estrogen, the estrogen receptor (ER) acts as a transcription factor for specific genes, either stimulating or repressing their expression and causing a host of downstream effects. The investigators were interested in identifying primary and secondary targets of estrogen in these cells, and noting any changes in mRNA transcript behavior for the targets over time. After serum starvation of all eight samples, four samples were exposed to estrogen and then harvested for microarray analysis after 10 hours for two samples and 48 hours for the other two. The remaining four samples were left untreated and harvested after 10 hours for two samples, and 48 hours for the other two. An *exprSet* named esEset contains expression levels for 12,625 probesets for the 8 samples described above, as well as the corresponding *phenoData* that specify the 2^2 factorial design.

```
> esEset

Expression Set (exprSet) with
        12625 genes
        8 samples
```

```
                        phenoData object with 2 variables and 8 cases
              varLabels
                        estrogen: present or absent
                        time.h: 10 or 48 hours
> pData(esEset)

              estrogen time.h
low10-1.cel     absent     10
low10-2.cel     absent     10
high10-1.cel   present     10
high10-2.cel   present     10
low48-1.cel     absent     48
low48-2.cel     absent     48
high48-1.cel   present     48
high48-2.cel   present     48
```

Outlier detection. Before applying linear models to each gene, it may
be of interest to investigate the presence of outliers in the data. The single
outlier detection method available in factDesign focuses on differences between replicates, thus preserving the independence and normality assumed
for the original observations. First, replicate pairs with differences that are
significantly larger than expected are identified according to an adjusted
F-statistic using the outlierPair function. Next, a median absolute deviation filter is applied using madOutPair to ensure one of the observations
is indeed the single outlier. If no single outlier is detected, madOutPair will
return NA. For example, in Figure 14.4 728_at has a replicate pair with a
large difference, but neither observation appears to be outside the range
of the other data. On the other hand, 33379_at has one observation that
indeed appears to be a single outlier.

```
> library("factDesign")
> op1 <- outlierPair(exprs(esEset)["728_at", ],
+      INDEX = pData(esEset))
> op1

$test
[1] TRUE

$pval
[1] 0.0143

$whichPair
[1] 7 8

> madOutPair(exprs(esEset)["728_at", ], op1[[3]])

[1] NA

> op2 <- outlierPair(exprs(esEset)["33379_at", ],
+      INDEX = pData(esEset))
> madOutPair(exprs(esEset)["33379_at", ], op2[[3]])
```

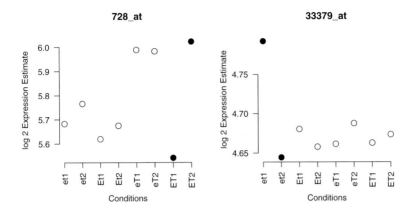

Figure 14.4. Both probesets contain a replicate pair with a larger difference than the other pairs for that probeset, however the single outlier is not obvious for 728_at.

[1] 1

The user must determine what to do with observations that appear to be single outliers, keeping in mind that removing single outliers assumes that the changes in expression across experimental conditions are small compared to the outlier effects. For probe 33379_at, it could be the second observation that is the outlier if true expression happens to be high at the earlier time in the absence of estrogen. In this application, we choose to leave the single outliers in the data set to preserve the balanced design.

Describing the Linear Model. The 2^2 factorial design of the estrogen experiment makes it a natural fit for linear model analysis. In Equation (14.1), y_{ji} is the observed expression level for gene j in sample i ($i = 1, ..., 8$) with $x_{ESi} = 1$ if estrogen is present and 0 otherwise and $x_{TIMEi} = 1$ if gene expression was measured at 48 hours and 0 otherwise. Using this parameterization, μ_j is the expression level of untreated gene j at 10 hours, β_{ESj} and β_{TIMEj} represent the effects of estrogen and time on the expression level of gene j, respectively, and the interaction term $\beta_{ES:TIMEj}$ quantifies any change in estrogen effect over time for gene j. The error term ϵ_{ji} is assumed to be normally distributed with mean 0 and variance σ_j^2.

$$y_{ji} = \mu_j + \beta_{ESj}x_{ESi} + \beta_{TIMEj}x_{TIMEi} + \beta_{ES:TIMEj}x_{ESi}x_{TIMEi} + \epsilon_{ji}$$
$$(14.1)$$

We use functions from the limma package to estimate the linear model parameters for every gene using least squares, and call the estimates $\hat{\mu}_j$, $\hat{\beta}_{ESj}$, $\hat{\beta}_{TIMEj}$, and $\hat{\beta}_{ES:TIMEj}$. For gene j, the samples that were not treated with estrogen and were measured at 10 hours will have estimated expression values of $\hat{\mu}_j$. The estrogen-treated, 10-hour samples will have estimates $\hat{\mu}_j + \hat{\beta}_{ESj}$. The untreated, 48-hour samples will have estimates $\hat{\mu}_j + \hat{\beta}_{TIMEj}$. The estrogen-treated, 48-hour samples will have estimates $\hat{\mu}_j + \hat{\beta}_{ESj} + \hat{\beta}_{TIMEj} + \hat{\beta}_{ES:TIMEj}$. In what follows, we drop the j subscripts for ease of notation, but the linear model parameters are understood to be gene-specific.

```
> pdat <- pData(esEset)
> design <- model.matrix(~factor(estrogen) * factor(time.h),
+       pdat)
> colnames(design) <- c("Intercept", "ES", "T48",
+       "ES:T48")
> fit <- lmFit(esEset, design)
> fit$coefficients[1:3, ]
```

	Intercept	ES	T48	ES:T48
1000_at	10.33	-0.3725	-0.122	0.2725
1001_at	5.80	0.1075	0.191	0.0350
1002_f_at	5.66	-0.0676	-0.215	0.1944

Suppose we are interested in identifying genes that demonstrate response to estrogen at 10 and/or 48 hours. Genes affected by estrogen at 10 hours will demonstrate a difference in their untreated 10-hour expression levels and their estrogen-treated 10-hour expression levels. Using the linear model parameterization, these genes can be identified as those for which the null hypothesis

$$H_{0,ES10} : \mu = \mu + \beta_{ES} \text{ or } H_{0,ES10} : \beta_{ES} = 0 \tag{14.2}$$

is rejected. Rejection of $H_{0,ES10}$ indicates a difference in the untreated 10-hour and estrogen-treated 10-hour experimental conditions. A similar null hypothesis can be constructed for genes affected by estrogen at 48 hours. We can compare the untreated, 48-hour expression levels to the estrogen-treated 48-hour expression levels by testing the null hypothesis

$$H_{0,ES48} : \mu + \beta_{TIME} = \mu + \beta_{TIME} + \beta_{ES} + \beta_{ES:TIME} \text{ or} \tag{14.3}$$
$$H_{0,ES48} : \beta_{ES} + \beta_{ES:TIME} = 0. \tag{14.4}$$

One way to select genes affected by estrogen at either or both time points is to simultaneously test both contrasts

$$H_{0,ES} : \begin{cases} \beta_{ES} = 0 \\ \beta_{ES} + \beta_{ES:TIME} = 0 \end{cases} \tag{14.5}$$

and then classify the genes according to whether they were affected by estrogen at 10 hours, 48 hours, or both.

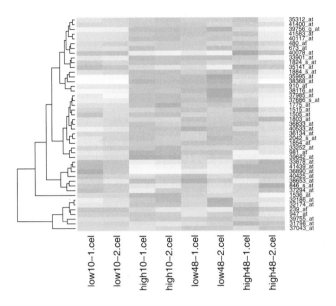

Figure 14.5. Heatmap of expression levels for genes identified as estrogen targets at both 10- and 48-hour time points.

```
> contM <- cbind(es10 = c(0, 1, 0, 0), es48 = c(0,
+      1, 0, 1))
> fitC <- contrasts.fit(fit, contM)
> fitC <- eBayes(fitC)
> esClas <- classifyTestsF(fitC, p = 1e-05)
> print(colSums(abs(esClas)))

es10 es48
  51   83
```

Heatmaps can be a helpful way to visualize the results of linear model analyses for factorial designed experiments. Here we examine three separate heatmaps for genes affected at 10 and 48, only 10, and only 48 hours.

The heatmap in Figure 14.5 helps identify collections of genes that show similar, consistent patterns of up- or down-regulation by estrogen. For these genes, we notice consistent effects for both time points. No genes in this example demonstrate up-regulation at one time point and down-regulation at the other, although such expression behavior could be detected by the contrasts we tested. Further experiments examining the joint behavior of these genes could clarify effects of estrogen on breast cancer cellular pathways that are consistent over time.

Estrogen target at 10 hours only

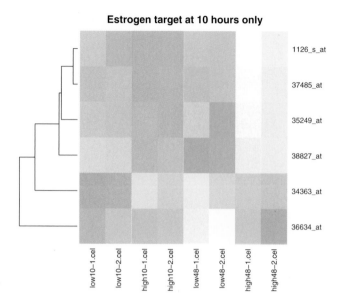

Figure 14.6. Heatmap of expression levels for genes identified as estrogen targets at 10 hours only.

For the 10-hour only target genes, the heatmap in Figure 14.6 identifies two similar clusters. Note that the genes that are up-regulated at 10 hours return to their original expression level at 48 hours, whereas the genes that are down-regulated at 10 hours stay down throughout the course of the experiment.

The heatmap in Figure 14.7 shows genes that were chosen as estrogen targets at 48 hours only and reveals that most of those genes had changes in expression earlier in the experiment. One might conclude that the genes affected at 10 hours comprise the direct targets of estrogen, that is, those genes that are directly stimulated or inhibited by the estrogen-bound ER. The 48-hour targets may be genes further downstream in estrogen-affected pathways. While that is appealing, the time sequence data alone are not strong enough to allow such conclusions.

Multifactor experiments, when designed very carefully with the appropriate biological context and estimable contrasts in mind, can lead to highly informative information regarding the genetic network. As stated previously, the estrogen data set consists of a subset of a larger 2^4 factorial experiment. In additional to estrogen and time, the investigators also exposed the breast cancer cells to cyclohexamide (CX), a translational inhibitor, as well as a drug, here called Z. Translational inhibition by CX

Estrogen target at 48 hours only

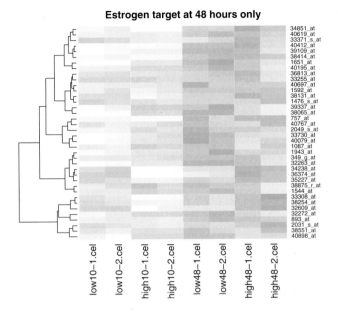

Figure 14.7. Heatmap of expression levels for genes identified as estrogen targets at 48 hours only.

presents a problem for normalization because the presumption that most genes are not differentially expressed is violated. Nevertheless, CX was crucial to the interpretable experimental design as explained in what follows. The full linear model for this factorial experiment consists of all four main effects for CX, ES, Z, and TIME, as well as all possible interactions.

Rather than rely on time sequence alone, CX was a key factor in this experiment for correctly identifying primary and secondary targets. For primary targets, estrogen can cause changes in mRNA levels regardless of the presence of CX. In the presence of CX, however, mRNA from the primary targets cannot be translated into protein, therefore preventing downstream transcriptional changes for the secondary targets. At the ten hour time point, primary estrogen targets were identified by testing

$$H_{0,primary} : \mu + \beta_{CX} = \mu + \beta_{CX} + \beta_{ES} + \beta_{CX:ES} = 0 \text{ or}$$
$$H_{0,primary} : \beta_{ES} + \beta_{CX:ES} = 0 \qquad (14.6)$$

as a low p-value for this test of contrast would indicate that the expression level of the gene when exposed only to CX was different than when exposed to both ES and CX. Estrogen targets for which $H_{0,primary}$ (14.6) was not rejected, but $H_{0,ES10}$ (14.2) was rejected, were identified as secondary targets because they were affected by estrogen, but not in the presence of CX. The fact that CX prevented expression level change due to ES indicated that translation of some other ES target gene's mRNA into protein was required for ES stimulation or repression of the secondary ES target. Similar tests of contrasts were also performed in this experiment to determine which genes were affected by the drug Z, and whether Z executed its action through transcriptional of translational control of the gene expression mechanism.

14.4   Conclusion

In summary, microarrays are used in a wide variety of experimental settings for the detection of differential gene expression. Although the goals and design concerns of these experiments vary, concepts including gene filtering, multiple comparisons adjustment, and gene selection according to the appropriate test statistic apply in general to these experiments. The Bioconductor packages help address these concerns, thereby providing insight into biological pathways and providing a platform for future hypothesis development.

15

Multiple Testing Procedures: the multtest Package and Applications to Genomics

K. S. Pollard, S. Dudoit, and M. J. van der Laan

Abstract

The Bioconductor R package multtest implements widely applicable resampling-based single-step and stepwise multiple testing procedures (MTP) for controlling a broad class of Type I error rates. The current version of multtest provides MTPs for tests concerning means, differences in means, and regression parameters in linear and Cox proportional hazards models. Typical testing scenarios are illustrated by applying various MTPs implemented in multtest to the Acute Lymphoblastic Leukemia (ALL) data set of Chiaretti et al. (2004), with the aim of identifying genes whose expression measures are associated with (possibly censored) biological and clinical outcomes.

15.1 Introduction

Current statistical inference problems in biomedical and genomic data analysis routinely involve the simultaneous test of thousands, or even millions, of null hypotheses. Examples include:

- identification of differentially expressed genes in microarray experiments, i.e., genes whose expression measures are associated with possibly censored responses or covariates;

- tests of association between gene expression measures and Gene Ontology (GO) annotation;

- identification of transcription factor binding sites in ChIP-Chip experiments (Keleş et al., 2004);

- genetic mapping of complex traits using single nucleotide polymorphisms (SNP).

The above testing problems share the following general characteristics: inference for high-dimensional multivariate distributions, with complex and unknown dependence structures among variables; a broad range of parameters of interest, e.g. regression coefficients and correlations; many null hypotheses, in the thousands or even millions; complex dependence structures among test statistics.

Motivated by these applications, we have developed resampling-based single-step and stepwise multiple testing procedures (MTP) for controlling a broad class of Type I error rates. The main steps in applying a MTP are listed in the flowchart of Table 15.1. The different components of our multiple testing methodology are treated in detail in a collection of related articles (Dudoit et al., 2004a,b; Pollard and van der Laan, 2004; van der Laan et al., 2004a,b) and a book in preparation (Dudoit and van der Laan, 2004). In order to make this general methodology accessible, we have implemented several MTPs in the Bioconductor R package **multtest**, which is the subject of the current chapter. An expanded version of this chapter is available on-line as a technical report (Pollard et al., 2004).

15.2   Multiple hypothesis testing methodology

15.2.1   Multiple hypothesis testing framework

Hypothesis testing is concerned with using observed data to test hypotheses, i.e., make decisions, regarding properties of the unknown data generating distribution. For example, microarray experiments might be conducted on a sample of patients in order to identify genes whose expression levels are associated with survival. Below, we discuss in turn the main ingredients of a multiple testing problem.

Data. Let $X_1, \ldots, X_n$ be a *random sample* of n independent and identically distributed *(i.i.d.)* random variables, $X \sim P \in \mathcal{M}$, where the *data generating distribution* P is an element of a particular *statistical model* $\mathcal{M}$ (i.e., a set of possibly non-parametric distributions). In a microarray experiment, for example, X is a vector of gene expression measurements, which we observe for each of n arrays.

Null and alternative hypotheses. Define M *null hypotheses* $H_0(m) \equiv$ I$[P \in \mathcal{M}(m)]$ in terms of a collection of *submodels*, $\mathcal{M}(m) \subseteq \mathcal{M}$, $m = 1, \ldots, M$, for the data generating distribution P. The corresponding *alter-*

Table 15.1. *Multiple hypothesis testing flowchart.*

Provide data set

MTP arguments: X, W, Y, Z, Z.incl, and Z.test

⇓

Define parameters of interest, $\psi(m)$

⇓

Define null and alternative hypotheses, $H_0(m)$ and $H_1(m)$

⇓

Specify test statistics, $T_n(m)$

MTP arguments: test, robust, standardize, alternative, and psi0

⇓

Estimate test statistics null distribution, Q_{0n}

MTP arguments: nulldist and B

⇓

Select Type I error rate, $\theta(F_{V_n, R_n})$

MTP arguments: typeone and alpha (and also k and q)

⇓

Apply MTP

MTP argument: method

FWER	$Pr(V_n > 0)$	Single-step maxT procedure (sec. 15.2.3)
		Single-step minP procedure (sec. 15.2.3)
		Step-down maxT procedure (sec. 15.2.4)
		Step-down minP procedure (sec. 15.2.4)
gFWER	$Pr(V_n > k)$	Single-step $T(k+1)$ procedure (sec. 15.2.3)
		Single-step $P(k+1)$ procedure (sec. 15.2.3)
		Augmentation procedure (sec. 15.2.5)
TPPFP	$Pr(V_n/R_n > q)$	Augmentation procedure (sec. 15.2.5)
General	$\theta(F_{V_n})$	Single-step common cutoff procedure (sec. 15.2.3)
		Single-step common quantile procedure (sec. 15.2.3)

⇓

Summarize results

adjusted *p*-values, rejection regions, and confidence regions

MTP arguments: get.adjp, get.cutoff, and get.cr

native hypotheses are $H_1(m) \equiv \mathrm{I}[P \notin \mathcal{M}(m)]$. In many testing problems, the submodels concern *parameters*, i.e., functions of the data generating distribution P, $\Psi(P) = \psi = (\psi(m) : m = 1, \ldots, M)$, such as means, differences in means, correlation coefficients, and regression parameters.

Test statistics. A testing procedure is a *data-driven* rule for deciding whether or not to *reject* each of the M null hypotheses $H_0(m)$ based on an M-vector of *test statistics*, $T_n = (T_n(m) : m = 1, \ldots, M)$, that are functions of the observed data. Denote the typically unknown (finite sample) *joint distribution* of the test statistics T_n by $Q_n = Q_n(P)$.

Single-parameter null hypotheses are commonly tested using *t-statistics*, i.e., standardized differences,

$$T_n(m) \equiv \frac{\text{Estimator} - \text{Null value}}{\text{Standard error}} = \sqrt{n}\,\frac{\psi_n(m) - \psi_0(m)}{\sigma_n(m)}. \tag{15.1}$$

For tests of means, $T_n(m)$ is the usual one-sample or two-sample t-statistic, where $\psi_n(m)$ and $\sigma_n(m)$ are based on empirical means and variances, respectively. In some settings, it may be appropriate to use (unstandardized) *difference statistics*, $T_n(m) \equiv \sqrt{n}[\psi_n(m) - \psi_0(m)]$ (Pollard and van der Laan, 2004). Test statistics for other types of null hypotheses include F-statistics, χ^2-statistics, and likelihood ratio statistics.

Multiple testing procedure. A *multiple testing procedure* (MTP) provides *rejection regions*, $C_n(m)$, i.e., sets of values for each test statistic $T_n(m)$ that lead to the decision to reject the null hypothesis $H_0(m)$. In other words, a MTP produces a random (i.e., data-dependent) subset $\mathcal{R}_n$ of rejected hypotheses that estimates the set of true positives,

$$\mathcal{R}_n = \mathcal{R}(T_n, Q_{0n}, \alpha) \equiv \{m : H_0(m) \text{ is rejected}\} = \{m : T_n(m) \in C_n(m)\}, \tag{15.2}$$

where the long notation $\mathcal{R}(T_n, Q_{0n}, \alpha)$ emphasizes that the MTP depends on: (i) the *data* through the *test statistics* T_n; (ii) a (estimated) test statistics *null distribution*, Q_{0n}, for deriving rejection regions; and (iii) the *nominal level* α, i.e., the desired upper bound for a suitably defined Type I error rate. Unless specified otherwise, it is assumed that large values of the test statistic, $T_n(m)$, provide evidence against the corresponding null hypothesis $H_0(m)$.

Example. Suppose that, as in the analysis of the ALL data set of Chiaretti et al. (2004) (Section 15.4), one is interested in identifying genes that are differentially expressed in two populations of ALL cancer patients, those with the B-cell subtype and those with the T-cell subtype. The data consist of random vectors X of microarray expression measures on M genes and an indicator Y for the ALL subtype (1 for B-cell, 0 for T-cell). Then, the parameter of interest is an M-vector of differences in mean expression mea-

Table 15.2. *Type I and Type II errors in multiple hypothesis testing.* $\mathcal{H}_0$ is the set of true null hypotheses, $\mathcal{H}_1$ is the set of false null hypotheses (i.e., true positives), and $\mathcal{R}_n$ is the set of rejected null hypotheses.

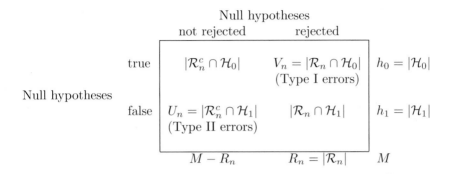

		Null hypotheses		
		not rejected	rejected	
Null hypotheses	true	$\|\mathcal{R}_n^c \cap \mathcal{H}_0\|$	$V_n = \|\mathcal{R}_n \cap \mathcal{H}_0\|$ (Type I errors)	$h_0 = \|\mathcal{H}_0\|$
	false	$U_n = \|\mathcal{R}_n^c \cap \mathcal{H}_1\|$ (Type II errors)	$\|\mathcal{R}_n \cap \mathcal{H}_1\|$	$h_1 = \|\mathcal{H}_1\|$
		$M - R_n$	$R_n = \|\mathcal{R}_n\|$	M

sures in the two populations, $\psi(m) = E[X(m)|Y = 1] - E[X(m)|Y = 0]$, $m = 1, \ldots, M$. To identify genes with higher mean expression measures in the B-cell compared to T-cell ALL subjects, one can test the one-sided null hypotheses $H_0(m) = I[\psi(m) \leq 0]$ vs. the alternative hypotheses $H_1(m) = I[\psi(m) > 0]$, using two-sample Welch t-statistics

$$T_n(m) \equiv \frac{\bar{X}_{1,n_1}(m) - \bar{X}_{0,n_0}(m)}{\sqrt{n_0^{-1}(m)\sigma_{0,n_0}^2(m) + n_1^{-1}(m)\sigma_{1,n_1}^2(m)}}, \qquad (15.3)$$

where $n_k(m)$, $\bar{X}_{k,n_k}(m)$, and $\sigma_{k,n_k}^2(m)$ denote, respectively, the sample sizes, sample means, and sample variances, for patients with tumor sub-type k, $k = 0, 1$.

Type I and Type II errors. In any testing situation, two types of errors can be committed: a *false positive*, or *Type I error*, is committed by rejecting a true null hypothesis, and a *false negative*, or *Type II error*, is committed when the test procedure fails to reject a false null hypothesis. The situation can be summarized by Table 15.2.

Type I error rates. When testing multiple hypotheses, there are many possible definitions for the Type I error rate and power of a test procedure. Accordingly, we define Type I error rates as *parameters*, $\theta_n = \theta(F_{V_n, R_n})$, of the joint distribution F_{V_n, R_n} of the numbers of Type I errors V_n and rejected hypotheses R_n (Dudoit et al., 2004b; Dudoit and van der Laan, 2004). Such a general representation covers the following commonly-used Type I error rates.

Generalized family-wise error rate (gFWER), or probability of at least $(k + 1)$ Type I errors,

$$gFWER(k) \equiv Pr(V_n > k). \tag{15.4}$$

When $k = 0$, the gFWER is the usual *family-wise error rate* (FWER), or probability of at least one Type I error, $FWER \equiv Pr(V_n > 0)$.

Tail probabilities for the proportion of false positives (TPPFP) among the rejected hypotheses,

$$TPPFP(q) \equiv Pr(V_n/R_n > q), \qquad q \in (0, 1). \tag{15.5}$$

False discovery rate (FDR), or expected value of the proportion of false positives among the rejected hypotheses (Benjamini and Hochberg, 1995),

$$FDR \equiv E[V_n/R_n]. \tag{15.6}$$

The convention that $V_n/R_n \equiv 0$ if $R_n = 0$ is used. Error rates based on the *proportion* of false positives (e.g., TPPFP and FDR) are especially appealing for large-scale testing problems such as those encountered in genomics, compared to error rates based on the *number* of false positives (e.g., gFWER), as they do not increase exponentially with the number of tested hypotheses.

Adjusted *p*-values. The notion of *p*-value extends directly to multiple testing problems, as follows. Given a MTP $\mathcal{R}_n(\alpha) = \mathcal{R}(T_n, Q_{0n}, \alpha)$, the *adjusted p-value* $\widetilde{P}_{0n}(m) = \widetilde{P}(T_n, Q_{0n})(m)$, for null hypothesis $H_0(m)$, is defined as the smallest Type I error level α at which one would reject $H_0(m)$, that is,

$$
\begin{aligned}
\widetilde{P}_{0n}(m) &\equiv \inf\{\alpha \in [0, 1] : m \in \mathcal{R}_n(\alpha)\} \tag{15.7} \\
&= \inf\{\alpha \in [0, 1] : T_n(m) \in \mathcal{C}_n(m)\}, \qquad m = 1, \dots, M.
\end{aligned}
$$

As in single hypothesis tests, the smaller the adjusted *p*-value, the stronger the evidence against the corresponding null hypothesis. Reporting the results of a MTP in terms of adjusted *p*-values, as opposed to the binary decisions to reject or not the hypotheses, provides *flexible summaries* that can be used to compare different MTPs and do not require specifying the level α ahead of time.

Confidence regions. For the test of single-parameter null hypotheses and for any Type I error rate of the form $\theta(F_{V_n})$, Pollard and van der Laan (2004) and Dudoit and van der Laan (2004) provide results on the correspondence between single-step MTPs and θ-specific *confidence regions*.

15.2.2   Test statistics null distribution

The choice of null distribution Q_0 is crucial, in order to ensure that (finite sample or asymptotic) control of the Type I error rate under the *assumed* null distribution Q_0 does indeed provide the required control under the *true* distribution $Q_n(P)$. For error rates $\theta(F_{V_n})$ (e.g., gFWER), defined as arbitrary parameters of the distribution of the number of Type I errors V_n, we propose as null distribution the asymptotic distribution $Q_0 = Q_0(P)$ of the M-vector Z_n of null value shifted and scaled test statistics (Dudoit and van der Laan, 2004; Dudoit et al., 2004b; Pollard and van der Laan, 2004; van der Laan et al., 2004b),

$$Z_n(m) \equiv \sqrt{\min\left\{1, \frac{\tau_0(m)}{Var[T_n(m)]}\right\}}\left\{T_n(m) + \lambda_0(m) - E[T_n(m)]\right\}. \quad (15.8)$$

For the test of single-parameter null hypotheses using t-statistics, the null values are $\lambda_0(m) = 0$ and $\tau_0(m) = 1$. For testing the equality of K population means using F-statistics, the null values are $\lambda_0(m) = 1$ and $\tau_0(m) = 2/(K - 1)$, under the assumption of equal variances in the different populations. By shifting the test statistics $T_n(m)$ as in Equation (15.8), the number of Type I errors V_0 under the null distribution Q_0, is asymptotically stochastically greater than the number of Type I errors V_n under the true distribution $Q_n = Q_n(P)$.

Note that we are only concerned with Type I error control under the *true data generating distribution* P. The notions of weak and strong control (and associated subset pivotality, Westfall and Young (Westfall and Young, 1993), p. 42-43) are therefore irrelevant to our approach. In addition, we propose a *null distribution for the test statistics*, $T_n \sim Q_0$, and not a data generating null distribution, $X \sim P_0 \in \cap_{m=1}^{M}\mathcal{M}(m)$. The latter practice does not necessarily provide proper Type I error control, as the test statistics' *assumed* null distribution $Q_n(P_0)$ and their *true* distribution $Q_n(P)$ may have different dependence structures, in the limit, for the true null hypotheses.

Resampling procedures, such as the bootstrap procedure of section 15.2.2, may be used to conveniently obtain consistent estimators Q_{0n} of the null distribution Q_0 and of the corresponding test statistic cutoffs and adjusted p-values (Dudoit and van der Laan, 2004; Dudoit et al., 2004b; Pollard and van der Laan, 2004; van der Laan et al., 2004b). This bootstrap procedure is implemented in the internal function `boot.resample` and may be specified via the arguments `nulldist` and `B` of the main user-level function `MTP`.

Having selected a suitable test statistics null distribution, there remains the main task of specifying rejection regions for each null hypothesis, i.e., cutoffs for each test statistic, such that the Type I error rate is controlled at a desired level α. Next, we summarize the approaches to this task that

Bootstrap estimation of the null distribution Q_0

1. Let $P_n^\star$ denote an estimator of the data generating distribution P.

2. Generate B bootstrap samples, each consisting of n *i.i.d.* realizations of a random variable $X^\# \sim P_n^\star$. For the *non-parametric bootstrap*, samples of size n are drawn at random, with replacement from the observed data.

3. For the bth bootstrap sample, $b = 1, \ldots, B$, compute an M-vector of test statistics, and arrange these in an $M \times B$ matrix, $\mathbf{T}_n^\# = \left[T_n^\#(m, b) \right]$, with rows corresponding to the M null hypotheses and columns to the B bootstrap samples.

4. Compute row means, $E[T_n^\#(m, \cdot)]$, and row variances, $Var[T_n^\#(m, \cdot)]$, of the matrix $\mathbf{T}_n^\#$, to yield estimates of the true means $E[T_n(m)]$ and variances $Var[T_n(m)]$ of the test statistics, respectively.

5. Obtain an $M \times B$ matrix, $\mathbf{Z}_n^\# = \left[Z_n^\#(m, b) \right]$, of null value shifted and scaled bootstrap statistics $Z_n^\#(m, b)$, by row-shifting and scaling the matrix $\mathbf{T}_n^\#$ as in Equation (15.8) using the bootstrap estimates of $E[T_n(m)]$ and $Var[T_n(m)]$ and the user-supplied null values $\lambda_0(m)$ and $\tau_0(m)$.

6. The bootstrap estimate Q_{0n} of the null distribution Q_0 is the empirical distribution of the B columns $Z_n^\#(\cdot, b)$ of matrix $\mathbf{Z}_n^\#$.

have been implemented in the multtest package. The chosen procedure is specified using the method argument to the function MTP.

15.2.3   Single-step procedures for controlling general Type I error rates $\theta(F_{V_n})$

Control of a Type I error rate $\theta(F_{V_n})$ can be obtained by substituting the *known, null distribution* F_{R_0} of the number of rejected hypotheses for the *unknown, true distribution* F_{V_n} of the number of Type I errors. We propose the following single-step common cutoff and common quantile procedures (Dudoit et al., 2004b; Pollard and van der Laan, 2004).

General θ-controlling single-step common cutoff procedure

The set of rejected hypotheses is of the form $\mathcal{R}_n(\alpha) \equiv \{m : T_n(m) > c_0\}$, where the common cutoff c_0 is the *smallest* (i.e., least conservative) value for which $\theta(F_{R_0}) \le \alpha$. For $gFWER(k)$ control, the procedure is based on the $(k+1)st$ *ordered test statistic*. The adjusted p-values for the *single-step*

$T(k+1)$ *procedure* are given by

$$\widetilde{p}_{0n}(m) = Pr_{Q_0}\left(Z^\circ(k+1) \geq t_n(m)\right), \qquad m = 1, \ldots, M, \qquad (15.9)$$

where $Z^\circ(m)$ denotes the mth ordered component of $Z = (Z(m) : m = 1, \ldots, M) \sim Q_0$, so that $Z^\circ(1) \geq \ldots \geq Z^\circ(M)$. For FWER control $(k = 0)$, one recovers the *single-step maxT procedure*.

General θ-controlling single-step common quantile procedure

The set of rejected hypotheses is of the form $\mathcal{R}_n(\alpha) \equiv \{m : T_n(m) > c_0(m)\}$, where $c_0(m) = Q_{0,m}^{-1}(\delta_0)$ is the δ_0-quantile of the marginal null distribution $Q_{0,m}$ of the test statistic for the mth null hypothesis, i.e., the smallest value c such that $Q_{0,m}(c) = Pr_{Q_0}(Z(m) \leq c) \geq \delta_0$ for $Z \sim Q_0$. Here, δ_0 is chosen as the *smallest* (i.e., least conservative) value for which $\theta(F_{R_0}) \leq \alpha$.

For $gFWER(k)$ control, the procedure is based on the $(k+1)st$ *ordered unadjusted p-value*. Specifically, let $\bar{Q}_{0,m} \equiv 1 - Q_{0,m}$ denote the survivor functions for the marginal null distributions $Q_{0,m}$ and define unadjusted p-values $P_0(m) \equiv \bar{Q}_{0,m}[Z(m)]$ and $P_{0n}(m) \equiv \bar{Q}_{0,m}[T_n(m)]$, for $Z \sim Q_0$ and $T_n \sim Q_n$, respectively. The adjusted p-values for the *single-step $P(k+1)$ procedure* are given by

$$\widetilde{p}_{0n}(m) = Pr_{Q_0}\left[P_0^\circ(k+1) \leq p_{0n}(m)\right], \qquad m = 1, \ldots, M, \qquad (15.10)$$

where $P_0^\circ(m)$ denotes the mth ordered component of the M-vector of unadjusted p-values $P_0 = [P_0(m) : m = 1, \ldots, M]$, so that $P_0^\circ(1) \leq \ldots \leq P_0^\circ(M)$. For FWER control $(k = 0)$, one recovers the *single-step minP procedure*.

15.2.4   Step-down procedures for controlling the family-wise error rate

Step-down MTPs consider hypotheses successively, from most significant to least significant, with further tests depending on the outcome of earlier ones. van der Laan et al. (2004b) propose step-down common cutoff (maxT) and common quantile (minP) procedures for controlling the family-wise error rate, FWER.

FWER-controlling step-down common cutoff (maxT) procedure

Let $O_n(m)$ denote the indices for the ordered test statistics $T_n(m)$, so that $T_n(O_n(1)) \geq \ldots \geq T_n(O_n(M))$. Consider the distributions of maxima of test statistics over the nested subsets of ordered null hypotheses $\overline{\mathcal{O}}_n(h) \equiv \{O_n(h), \ldots, O_n(M)\}$. The adjusted p-values are given by

$$\widetilde{p}_{0n}[o_n(m)] = \max_{h=1,\ldots,m} Pr_{Q_0}\left\{\max_{l \in \bar{l}_n(h)} Z(l) \geq t_n[o_n(h)]\right\}, \qquad (15.11)$$

where $Z = [Z(m) : m = 1, \ldots, M] \sim Q_0$.

FWER-controlling step-down common quantile (minP) procedure.

Let $O_n(m)$ denote the indices for the ordered unadjusted p-values $P_{0n}(m)$, so that $P_{0n}[O_n(1)] \leq \ldots \leq P_{0n}[O_n(M)]$. Consider the distributions of minima of unadjusted p-values over the nested subsets of ordered null hypotheses $\overline{\mathcal{O}}_n(h) \equiv \{O_n(h), \ldots, O_n(M)\}$. The adjusted p-values are given by

$$\widetilde{p}_{0n}(o_n(m)) = \max_{h=1,\ldots,m} Pr_{Q_0} \left\{ \min_{l \in \overline{l}_n(h)} P_0(l) \leq p_{0n}[o_n(h)] \right\}, \quad (15.12)$$

where $P_0(m) \equiv \bar{Q}_{0,m}[Z(m)]$ and $P_{0n}(m) \equiv \bar{Q}_{0,m}[T_n(m)]$, for $Z \sim Q_0$ and $T_n \sim Q_n$, respectively.

15.2.5 Augmentation multiple testing procedures for controlling tail probability error rates

van der Laan et al. (2004a), and subsequently Dudoit et al. (2004a) and Dudoit and van der Laan (2004), propose *augmentation multiple testing procedures* (AMTP), obtained by adding suitably chosen null hypotheses to the set of null hypotheses already rejected by an initial gFWER-controlling MTP. Adjusted p-values for the AMTP are shown to be simply shifted versions of the adjusted p-values of the original MTP. Denote the adjusted p-values for the initial FWER-controlling procedure $\mathcal{R}_n(\alpha)$ by $\widetilde{P}_{0n}(m)$. Order the M null hypotheses according to these p-values, from smallest to largest, that is, define indices $O_n(m)$, so that $\widetilde{P}_{0n}[O_n(1)] \leq \ldots \leq \widetilde{P}_{0n}[O_n(M)]$.

gFWER-controlling augmentation multiple testing procedure

For control of $gFWER(k)$ at level α, given an initial FWER-controlling procedure $\mathcal{R}_n(\alpha)$, reject the $R_n(\alpha) = |\mathcal{R}_n(\alpha)|$ null hypotheses specified by this MTP, as well as the next $A_n(\alpha)$ most significant hypotheses,

$$A_n(\alpha) = \min\{k, M - R_n(\alpha)\}. \quad (15.13)$$

The adjusted p-values $\widetilde{P}_{0n}^+[O_n(m)]$ for the new gFWER-controlling AMTP are simply k-shifted versions of the adjusted p-values of the initial FWER-controlling MTP, with the first k adjusted p-values set to zero. That is,

$$\widetilde{P}_{0n}^+[O_n(m)] = \begin{cases} 0, & \text{if } m \leq k \\ \widetilde{P}_{0n}[O_n(m-k)], & \text{if } m > k \end{cases}. \quad (15.14)$$

The AMTP thus guarantees at least k rejected hypotheses.

TPPFP-controlling augmentation multiple testing procedure

For control of $TPPFP(q)$ at level α, given an initial FWER-controlling procedure $\mathcal{R}_n(\alpha)$, reject the $R_n(\alpha) = |\mathcal{R}_n(\alpha)|$ null hypotheses specified by this MTP, as well as the next $A_n(\alpha)$ most significant hypotheses,

$$
\begin{aligned}
A_n(\alpha) &= \max\left\{ m \in \{0,\dots,M-R_n(\alpha)\} : \frac{m}{m+R_n(\alpha)} \le q \right\} \quad (15.15) \\
&= \min\left\{ \left\lfloor \frac{qR_n(\alpha)}{1-q} \right\rfloor , M - R_n(\alpha) \right\},
\end{aligned}
$$

where the *floor* $\lfloor x \rfloor$ denotes the greatest integer less than or equal to x, i.e., $\lfloor x \rfloor \le x < \lfloor x \rfloor + 1$. That is, keep rejecting null hypotheses until the ratio of additional rejections to the total number of rejections reaches the allowed proportion q of false positives. The adjusted p-values $\widetilde{P}_{0n}^{+}[O_n(m)]$ for the new TPPFP-controlling AMTP are simply mq-shifted versions of the adjusted p-values of the initial FWER-controlling MTP. That is,

$$
\widetilde{P}_{0n}^{+}(O_n(m)) = \widetilde{P}_{0n}(O_n(\lceil (1-q)m \rceil)), \qquad m = 1,\dots,M, \qquad (15.16)
$$

where the *ceiling* $\lceil x \rceil$ denotes the least integer greater than or equal to x.

FDR-controlling procedures

Given any TPPFP-controlling procedure, van der Laan et al. (2004a) derive two simple (conservative) FDR-controlling procedures. The more general and conservative procedure controls the FDR at nominal level α, by controlling $TPPFP(\alpha/2)$ at level $\alpha/2$. The less conservative procedure controls the FDR at nominal level α, by controlling $TPPFP(1 - \sqrt{1-\alpha})$ at level $1 - \sqrt{1-\alpha}$. The reader is referred to the original article for details and proofs of FDR control (Section 2.4, Theorem 3). In what follows, we refer to these two MTPs as *conservative* and *restricted*, respectively.

15.3 Software implementation: R multtest package

The MTPs proposed in Sections 15.2.3 - 15.2.5 are implemented in the latest version of the Bioconductor R package multtest (Version 1.5.4). We stress that *all* the bootstrap-based MTPs implemented in multtest can be performed using the main user-level function MTP. Note that the multtest package also provides several simple, marginal FWER-controlling MTPs, available through the mt.rawp2adjp function, which takes a vector of unadjusted p-values as input and returns the corresponding adjusted p-values. For greater detail on multtest functions, the reader is referred to the package documentation, in the form of help files, e.g., ?MTP, and vignettes, e.g., openVignette("multtest").

15.3.1 Resampling-based multiple testing procedures: MTP function

The main user-level function for resampling-based multiple testing is MTP.

```
> args(MTP)

function (X, W = NULL, Y = NULL, Z = NULL, Z.incl = NULL,
    Z.test = NULL, na.rm = TRUE, test = "t.twosamp.unequalvar",
    robust = FALSE, standardize = TRUE, alternative = "two.sided",
    psi0 = 0, typeone = "fwer", k = 0, q = 0.1,
    fdr.method = "conservative", alpha = 0.05, nulldist = "boot",
    B = 1000, method = "ss.maxT", get.cr = FALSE, get.cutoff = FALSE,
    get.adjp = TRUE, keep.nulldist = FALSE, seed = NULL)
```

INPUT.

Data. The data, X, consist of a J-dimensional random vector, observed on each of n sampling units (patients, cell lines, mice, etc.). Other data components include weights W, a possibly censored continuous or polychotomous outcome Y, and additional covariates Z, whose use is specified with the arguments Z.incl and Z.test. The argument na.rm controls the treatment of missing values (NA). It is TRUE by default, so that an observation with a missing value in any of the data objects' jth component $(j = 1, \ldots, J)$ is excluded from the computation of any test statistic based on this jth variable.

Test statistics. In the current implementation of multtest, the following test statistics are available through the argument test: one-sample t-statistics for tests of means; equal and unequal variance two-sample t-statistics for tests of differences in means; paired t-statistics; multi-sample F-statistics for tests of differences in means in one-way and two-way designs; t-statistics for tests of regression coefficients in linear models and Cox proportional hazards survival models. *Robust, rank-based* versions of the above test statistics can be specified by setting the argument robust to TRUE (the default value is FALSE).

Type I error rate. The MTP function controls by default the FWER (argument typeone="fwer"). Augmentation procedures (Section 15.2.5), controlling other Type I error rates such as the gFWER, TPPFP, and FDR, can be specified through the argument typeone. Details regarding the related arguments k, q, and fdr.method are available in the package documentation. The nominal level of the test is determined by the argument alpha, by default 0.05.

Test statistics null distribution. The test statistics null distribution is estimated by default using the non-parametric version of the bootstrap procedure of section 15.2.2 (argument `nulldist="boot"`). Permutation null distributions are also available via `nulldist="perm"`. The number of resampling steps is specified by the argument B, by default 1,000.

Multiple testing procedures. The MTP function implements the single-step and step-down (common cutoff) maxT and (common quantile) minP MTPs for FWER control, described in Sections 15.2.3 and 15.2.4, and specified through the argument `method`. In addition, augmentation procedures (AMTPs) are implemented in the functions `fwer2gfwer`, `fwer2tppfp`, and `fwer2fdr`, which take FWER adjusted p-values as input and return augmentation adjusted p-values for control of the gFWER, TPPFP, and FDR, respectively. These AMTPs can also be applied directly via the `typeone` argument of the main function MTP.

Output control. Additional arguments allow the user to specify which combination of MTP results should be returned.

OUTPUT.

The S4 class/method object-oriented programming approach was adopted to summarize the results of a MTP. The output of the MTP function is an instance of the *class MTP*, with the following *slots*,

```
> slotNames("MTP")

[1] "statistic" "estimate"  "sampsize"  "rawp"
[5] "adjp"      "conf.reg"  "cutoff"    "reject"
[9] "nulldist"  "call"      "seed"
```

MTP results. An instance of the *MTP* class contains slots for the following MTP results: `statistic`, an M-vector of test statistics; `estimate`, an M-vector of estimated parameters; `rawp`, an M-vector of unadjusted p-values; `adjp`, an M-vector of adjusted p-values; `conf.reg`, lower and upper simultaneous confidence limits for the parameter vector; `cutoff`, cutoffs for the test statistics; `reject`, rejection indicators (TRUE for a rejected null hypothesis).

Null distribution. The `nulldist` slot contains the $M \times B$ *matrix* for the estimated test statistics null distribution.

Reproducibility. The slot `call` contains the call to the function MTP, and `seed` is an integer specifying the state of the random number generator used to create the resampled data sets.

15.3.2   *Numerical and graphical summaries*

The following *methods* were defined to operate on *MTP* instances and summarize the results of a MTP. The `print` method returns a description of an object of class *MTP*. The `summary` method returns a list with the the following components: `rejections`, number(s) of rejected hypotheses; `index`, indices for ordering the hypotheses according to significance; `summaries`, six number summaries of the distributions of the adjusted *p*-values, unadjusted *p*-values, test statistics, and parameter estimates. The `plot` method produces graphical summaries of the results of a MTP. The type of display may be specified via the `which` argument. Methods are also provided for subsetting ([) and conversion (`as.list`).

15.4   Applications: ALL microarray data set

15.4.1   *ALL data package and initial gene filtering*

We illustrate some of the functionality of the multtest package using the Acute Lymphoblastic Leukemia (ALL) microarray data set of Chiaretti et al. (2004), available in the data package ALL. The main object in this package is ALL, an instance of the class *exprSet*. The genes-by-subjects matrix of 12,625 Affymetrix *expression measures* (chip series HG-U95Av2) for each of 128 ALL patients is provided in the `exprs` slot of ALL. The `phenoData` slot contains 21 *phenotypes* (i.e., patient level responses and covariates) for each patient. Note that the expression measures have been obtained using the three-step robust multichip average (RMA) preprocessing method, implemented in the package affy. In particular, the expression measures have been subject to a base 2 logarithmic transformation. For greater detail, please consult the ALL package documentation and Appendix A.1.1.

```
> library("ALL")
> library("hgu95av2")
> data(ALL)
```

Our goal is to identify genes whose expression measures are associated with (possibly censored) biological and clinical outcomes such as: tumor cellular subtype (B-cell vs. T-cell), tumor molecular subtype (BCR/ABL, NEG, ALL1/AF4), and time to relapse. Alternative analyses of this data set are discussed in Chapters 10, 12, 16, 17, and 23. Before applying the MTPs, we perform initial gene filtering as in Chiaretti et al. (2004) and retain only those genes for which: (i) at least 20% of the subjects have a measured intensity of at least 100 and (ii) the coefficient of variation (i.e., the ratio of the standard deviation to the mean) of the intensities across samples is between 0.7 and 10. These two filtering criteria can be readily applied using functions from the genefilter package.

```
> ffun <- filterfun(pOverA(p = 0.2, A = 100), cv(a = 0.7,
+       b = 10))
> filt <- genefilter(2^exprs(ALL), ffun)
> filtALL <- ALL[filt, ]

> filtX <- exprs(filtALL)
> pheno <- pData(filtALL)
```

The new filtered data set, `filtALL`, contains expression measures on 431 genes, for 128 patients.

15.4.2 Association of expression measures and tumor cellular subtype: Two-sample t-statistics

In this example we examine use of FWER-controlling step-down minP MTP with two-sample Welch t-statistics and bootstrap null distribution.

Different tissues are involved in ALL tumors of the B-cell and T-cell subtypes. The phenotypic data include a variable, `BT`, which encodes the tissue type and stage of differentiation. In order to identify genes with higher mean expression measures in B-cell ALL patients compared to T-cell ALL patients, we create an indicator variable, `Bcell` (1 for B-cell, 0 for T-cell), and compute, for each gene, a two-sample Welch (unequal variance) t-statistic. We choose to control the FWER using the bootstrap-based step-down minP procedure with $B = 100$ bootstrap iterations, although more bootstrap iterations are recommended in practice.

```
> table(pData(ALL)$BT)

 B B1 B2 B3 B4  T T1 T2 T3 T4
 5 19 36 23 12  5  1 15 10  2

> Bcell <- rep(0, length(pData(ALL)$BT))
> Bcell[grep("B", as.character(pData(ALL)$BT))] <- 1

> seed <- 99
> BT.boot <- cache("BT.boot", MTP(X = filtX, Y = Bcell,
+      alternative = "greater", B = 100, method = "sd.minP",
+      seed = seed))

running bootstrap...
iteration = 100
```

Let us examine the results of the MTP stored in the object `BT.boot`.

```
> summary(BT.boot)

MTP:  sd.minP
Type I error rate:  fwer

   Level Rejections
1  0.05         273
```

	Min.	1st Qu.	Median	Mean	3rd Qu.	Max.
adjp	0.00	0.000	0.000	0.364	1.000	1.00
rawp	0.00	0.000	0.000	0.354	1.000	1.00
statistic	-34.40	-1.570	2.010	2.060	5.380	22.30
estimate	-4.66	-0.317	0.381	0.326	0.995	4.25

The summary method prints the name of the MTP (here, sd.minP, for step-down minP), the Type I error rate (here, fwer), the number of rejections at each Type I error rate level specified in alpha (here, 273 at level $\alpha = 0.05$), and six number summaries (mean and quantiles) of the adjusted p-values, unadjusted p-values, test statistics, and parameter estimates (here, difference in means).

The following commands may be used to obtain a list of genes that are differentially expressed in B-cell vs. T-cell ALL patients at nominal FWER level $\alpha = 0.05$, i.e., genes with adjusted p-values less than or equal to 0.05. Functions from the annotate and annaffy packages may then be used to obtain annotation information on these genes (e.g., gene names, PubMed abstracts, GO terms) and to generate HTML tables of the results (see Chapters 7 and 9). Here, we list the names of the first two genes only.

```
> BT.diff <- BT.boot@adjp <= 0.05
> BT.AffyID <- geneNames(filtALL)[BT.diff]
> mget(BT.AffyID[1:2], env = hgu95av2GENENAME)

$"1005_at"
[1] "dual specificity phosphatase 1"

$"1065_at"
[1] "fms-related tyrosine kinase 3"
```

Various graphical summaries of the results may be obtained using the plot method, by selecting appropriate values of the argument which. Figure 15.1 displays four such plots. We see (top left) that the number of rejections increases slightly when nominal FWER is greater than 0.6, and then increases quickly as FWER approaches 1. Similarly, the adjusted p-values for many genes are close to either 0 or 1 (top right) and the test statistics for genes with small p-values do not overlap with those for genes with p-values close to 1 (bottom left). Together these results indicate that there is a clear separation between the rejected and accepted hypotheses, i.e., between genes that are declared differentially expressed and those that are not.

```
> par(mfrow = c(2, 2))
> plot(BT.boot)
```

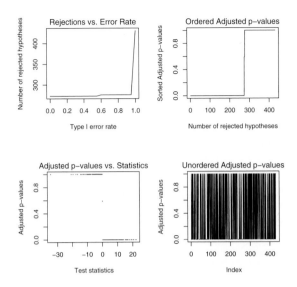

Figure 15.1. B-cell vs. T-cell ALL – FWER-controlling step-down minP MTP. By default, four graphical summaries are produced by the `plot` method for instances of the class *MTP*.

15.4.3 Augmentation procedures

In the context of microarray gene expression data analysis or other high-dimensional inference problems, one is often willing to tolerate some false positives, provided their number is small in comparison to the number of rejected hypotheses. In this case, the FWER is not a suitable choice of Type I error rate, and one should consider other rates that lead to larger sets of rejected hypotheses. The augmentation procedures of Section 15.2.5, implemented in the function `MTP`, allow one to reject additional hypotheses, while controlling an error rate such as the generalized family-wise error rate (gFWER), the tail probability for the proportion of false positives (TPPFP), or the false discovery rate (FDR). We illustrate the use of the `fwer2tppfp` and `fwer2fdr` functions, but note that the gFWER, TPPFP, and FDR can also be controlled directly using the main `MTP` function, with appropriate choices of arguments `typeone`, `k`, `q`, and `fdr.method`.

TPPFP control.

```
> q <- c(0.05, 0.1, 0.25)
> BT.tppfp <- fwer2tppfp(adjp = BT.boot@adjp, q = q)
> comp.tppfp <- cbind(BT.boot@adjp, BT.tppfp)
> mtps <- c("FWER", paste("TPPFP(", q, ")", sep = ""))
> mt.plot(adjp = comp.tppfp, teststat = BT.boot@statistic,
+     proc = mtps, leg = c(0.1, 430), col = 1:4,
```

```
+      lty = 1:4, lwd = 3)
> title("Comparison of TPPFP(q)-controlling AMTPs\n based on SD minP MTP")
```

Figure 15.2 (left) shows that, as expected, the number of rejections increases with the allowed proportion q of false positives when controlling $TPPFP(q)$ at a given level α.

FDR control. Given any TPPFP-controlling MTP, van der Laan et al. (2004a) derive two simple (conservative) FDR-controlling MTPs. Here, we compare these two FDR-controlling approaches, based on a TPPFP-controlling augmentation of the step-down minP procedure, to the marginal Benjamini and Hochberg (Benjamini and Hochberg, 1995) and Benjamini and Yekutieli (Benjamini and Yekutieli, 2001) procedures, implemented in the function mt.rawp2adjp. The following code chunk first computes adjusted p-values for the augmentation procedures, then for the marginal procedures, and finally makes a plot of the numbers of rejections vs. the nominal FDR for the four MTPs.

```
> BT.fdr <- fwer2fdr(adjp = BT.boot@adjp, method = "both")$adjp
> BT.marg.fdr <- mt.rawp2adjp(rawp = BT.boot@rawp,
+      proc = c("BY", "BH"))
> comp.fdr <- cbind(BT.fdr, BT.marg.fdr$adjp[
+      order(BT.marg.fdr$index), -1])
> mtps <- c("AMTP Cons", "AMTP Rest", "BY", "BH")
> mt.plot(adjp = comp.fdr, teststat = BT.boot@statistic,
+      proc = mtps, leg = c(0.1, 430), col = c(2,
+          2, 3, 3), lty = rep(1:2, 2), lwd = 3)
> title("Comparison of FDR-controlling MTPs")
```

Figure 15.2 (right) shows that the AMTPs based on conservative bounds for the FDR ("AMTP Cons" and "AMTP Rest") are more conservative than the Benjamini and Hochberg ("BH") MTP for nominal FDR less than 0.4, but less conservative than "BH" for larger FDR. The Benjamini and Yekutieli ("BY") MTP, a conservative version of the Benjamini and Hochberg MTP (with $\sim \log M$ penalty on the p-values), leads to the fewest rejections.

15.4.4   Association of expression measures and tumor molecular subtype: Multi-sample F-statistics

The phenotype data include a variable, mol.bio, which records chromosomal abnormalities, such as the BCR/ABL gene rearrangement; these abnormalities concern primarily patients with B-cell ALL and may be related to prognosis. To identify genes with differences in mean expression measures between different tumor molecular subtypes (BCR/ABL, NEG, ALL1/AF4, E2A/PBX1, p15/p16), within B-cell ALL subjects, one can perform a family of F-tests. Tumor subtypes with fewer than 10 subjects are removed from the analysis. Adjusted p-values and test statistic cutoffs

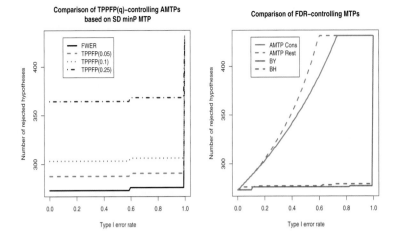

Figure 15.2. *B-cell vs. T-cell ALL – TPPFP and FDR-controlling AMTPs.* Plots of number of rejected hypotheses vs. nominal Type I error rate. *Left:* Comparison of TPPFP-controlling AMTPs, based on the FWER-controlling bootstrap-based step-down minP procedure, for different allowed proportions q of false positives. *Right:* Comparison of four FDR-controlling MTPs.

(for nominal levels α of 0.01 and 0.10) are computed as follows for the FWER-controlling bootstrap-based single-step maxT procedure.

```
> BX <- filtX[, Bcell == 1]
> Bpheno <- pheno[Bcell == 1, ]
> mb <- as.character(Bpheno$mol.biol)
> table(mb)

mb
ALL1/AF4  BCR/ABL E2A/PBX1     NEG  p15/p16
      10       37       5       42        1
> other <- c("E2A/PBX1", "p15/p16")
> mb.boot <- cache("mb.boot", MTP(X = BX[, !(mb %in%
+     other)], Y = mb[!(mb %in% other)], test = "f",
+     alpha = c(0.01, 0.1), B = 100, get.cutoff = TRUE,
+     seed = seed))

running bootstrap...
iteration = 100

> mb.rej <- summary(mb.boot)$rejections

> mb.rej

  Level Rejections
1  0.01        416
2  0.10        418
```

For control of the FWER at nominal level $\alpha = 0.01$, the bootstrap-based single-step maxT procedure with F-statistics identifies 416 genes as having significant differences in mean expression measures between tumor molecular subtypes.

15.4.5   Association of expression measures and time to relapse: Cox t-statistics

The bootstrap-based MTPs implemented in the main MTP function (nulldist="boot") allow the test of hypotheses concerning regression parameters in models for which the subset pivotality condition may not hold (e.g., logistic and Cox proportional hazards models). The phenotype information in the ALL package includes the original remission status of the ALL patients (remission variable in the *data.frame* pData(ALL)). There are 66 B-cell ALL subjects who experienced original complete remission (remission="CR") and who were followed up for remission status at a later date. We apply the single-step maxT procedure to test for a significant association between expression measures and time to relapse amongst these 66 subjects, adjusting for sex. Note that most of the code below is concerned with extracting the (censored) time to relapse outcome and covariates from slots of the *exprSet* instance ALL.

```
> cr.ind <- (Bpheno$remission == "CR")
> cr.pheno <- Bpheno[cr.ind, ]
> times <- strptime(cr.pheno$"date last seen", "%m/%d/%Y") -
+     strptime(cr.pheno$date.cr, "%m/%d/%Y")
> time.ind <- !is.na(times)
> times <- times[time.ind]
> cens <- ((1:length(times)) %in% grep("CR", cr.pheno[time.ind,
+     "f.u"]))
> rel.times <- Surv(times, !cens)
> patients <- (1:ncol(BX))[cr.ind][time.ind]
> relX <- BX[, patients]
> relZ <- Bpheno[patients, ]

> cox.boot <- cache("cox.boot", MTP(X = relX, Y = rel.times,
+     Z = relZ, Z.incl = "sex", Z.test = NULL, test = "coxph.YvsXZ",
+     B = 100, get.cr = TRUE, seed = seed))
```

For control of the FWER at nominal level $\alpha = 0.05$, the bootstrap-based single-step maxT procedure identifies 22 genes whose expression measures are significantly associated with time to relapse. Using the function mget, we examine the names of these genes.

```
> cox.diff <- cox.boot@adjp <= 0.05
> sum(cox.diff)
```

[1] 22

```
> cox.AffyID <- geneNames(filtALL)[cox.diff]
> mget(cox.AffyID, env = hgu95av2GENENAME)
```

$"106_at"
[1] "runt-related transcription factor 3"

$"1403_s_at"
[1] "chemokine (C-C motif) ligand 5"

$"182_at"
[1] "inositol 1,4,5-triphosphate receptor, type 3"

$"286_at"
[1] "histone 2, H2aa"

$"296_at"
[1] "tubulin, beta 2"

$"33232_at"
[1] "cysteine-rich protein 1 (intestinal)"

$"34308_at"
[1] "histone 1, H2ac"

$"35127_at"
[1] "histone 1, H2ae"

$"36638_at"
[1] "connective tissue growth factor"

$"37027_at"
[1] "AHNAK nucleoprotein (desmoyokin)"

$"37218_at"
[1] "BTG family, member 3"

$"37343_at"
[1] "inositol 1,4,5-triphosphate receptor, type 3"

$"38124_at"
[1] "midkine (neurite growth-promoting factor 2)"

$"39182_at"
[1] "epithelial membrane protein 3"

$"39317_at"
[1] "cytidine monophosphate-N-acetylneuraminic acid
    hydroxylase (CMP-N-acetylneuraminate monooxygenase)"

```
$"39331_at"
[1] "tubulin, beta 2"

$"39338_at"
[1] "S100 calcium binding protein A10 (annexin II ligand,
    calpactin I, light polypeptide (p11))"

$"40147_at"
[1] "vesicle amine transport protein 1 homolog (T californica)"

$"40567_at"
[1] "tubulin, alpha 3"

$"40729_s_at"
[1] "allograft inflammatory factor 1"

$"41071_at"
[1] "serine protease inhibitor, Kazal type 2 (acrosin-
    trypsin inhibitor)"

$"41164_at"
[1] "immunoglobulin heavy constant mu"
```

Figure 15.3 is a plot of the Cox regression coefficient estimates (circles) and corresponding confidence regions (text indicating the level) for the five genes with the smallest adjusted p-values. The plot illustrates that the level $\alpha = 0.05$ confidence regions corresponding to the significant gene does not include the null value $\psi_0 = 0$ for the Cox regression parameters (red line). The confidence regions for the next four genes, do include 0.

```
> plot(cox.boot, which = 5, top = 5, sub.caption = NULL)
> abline(h = 0, col = "red")
```

15.5   Discussion

The multtest package implements resampling-based multiple testing procedures that can be applied to a broad range of testing problems in biomedical and genomic data analysis. Ongoing efforts involve expanding the class of MTPs implemented in multtest, enhancing software design and the user interface, and increasing computational efficiency. Specifically, regarding the offering of MTPs, we envisage the following new developments.

- Expanding the class of available tests, by adding test statistic closures for tests of correlations, quantiles, and parameters in generalized linear models (e.g., logistic regression).

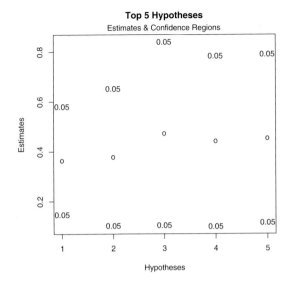

Figure 15.3. Time to relapse – FWER-controlling single-step maxT MTP. Plot of Cox regression coefficient estimates and corresponding confidence intervals for the fifteen genes with the smallest adjusted p-values, based on the FWER-controlling bootstrap-based single-step maxT procedure (`plot` method, `which=5`).

- Expanding the class of resampling-based estimators for the test statistics null distribution (e.g., parametric bootstrap, Bayesian bootstrap), possibly using a function closure approach.

- Providing parameter confidence regions and test statistic cutoffs for other Type I error rates than the FWER.

- Implementing the new augmentation multiple testing procedures proposed in Dudoit et al. (2004a) and Dudoit and van der Laan (2004), for controlling tail probabilities $Pr(g(V_n, R_n) > q)$ for an arbitrary function $g(V_n, R_n)$ of the numbers of false positives V_n and rejected hypotheses R_n.

Efforts regarding software design and the user interface include the following.

- Providing a formula interface for a symbolic description of the tests to be performed (cf. model specification in `lm`).

- Providing an `update` method for objects of class *MTP*, to facilitate the reuse of available estimates of the null distribution when implementing new MTPs.

- Extending the *MTP* class to keep track of results for several MTPs.

16

Machine Learning Concepts and Tools for Statistical Genomics

V. J. Carey

Abstract

In this chapter, supervised machine learning methods are described in the context of microarray applications. The most widely used families of machine learning methods are described, along with various approaches to learner assessment. The Bioconductor interfaces to machine learning tools are described and illustrated. Key problems of model selection and interpretation are reviewed in examples.

16.1 Introduction

Machine learning refers to computational and statistical inference processes employed to create, on the basis of observational data, reusable algorithms for prediction. The term *machine* is introduced to reflect the view that the creation of the predictive algorithm should occur with minimal human intervention, and the predictions for future observations should occur with no human intervention.

This chapter is focused on the subdiscipline of machine learning known as *supervised learning*, in which some *a priori* knowledge about the phenomena under investigation is available to guide the learning process. This is in contrast to unsupervised learning methods such as cluster analysis, where structure or labelings are imposed solely on the basis of the data configuration, with no *a priori* classification or labeling available.

Primary concerns of supervised learning methods are

- establishing acceptable quantitative representations of features and distance concepts supporting quantitative comparison of feature sets (see Chapter 12),

- creating algorithmic classifiers that reach decisions without human intervention,

- evaluating the performance of algorithmic classifiers,

- interpreting features of the performance of an algorithmic classifier to engender new substantive knowledge in the domain to which it is applied, and

- devising general principles of classifier construction.

Each one of these concerns leads to difficult research questions in a variety of domains. Accounts used in the preparation of this chapter include texts by Ripley (1996a), Duda et al. (2001), Hastie et al. (2001), Schölkopf and Smola (2001), and Vapnik (1998). The *Journal of Machine Learning Research* is published electronically at `www.jmlr.org/papers`. A special issue of the journal *Machine Learning*, devoted to applications in functional genomics, was published in 2003 (v. 52).

An abstract statement of a *pattern recognition* problem is as follows. We have I objects, each bearing J-vector of features x_i, $i = 1, \ldots, J$, and a class label y. The solution to the pattern classification problem consists of a function $C(x)$ that computes the class label y of the object bearing feature vector x. Machine learning involves estimating the function C on the basis of "training samples," and evaluating the performance of an estimated function $\hat{C}$ using "test samples." Thus we have y, the true class label associated with feature vector x, $C(\cdot)$, an element of a set $\mathfrak{C}$ of computable functions from elements of the feature space to elements of the set of class labels, and $\hat{C}(\cdot)$, an element of $\mathfrak{C}$ selected on the basis of a training set of feature vectors.

Statistical considerations are central to the theory of machine learning. We will consider applications in genomics, where commitments to parametric statistical models for features and class-feature relations are hard to justify. *Sample splitting*, *cross-validation* and the *bootstrap* are fundamental tools for non-parametric evaluation of machine learning applications.

The next Section introduces applied machine learning with a very restricted example involving two probesets in the ALL data. Subsequent sections review families of learning methods, principles of model assessment, and detailed examples with the ALL data package.

16.2   Illustration: Two continuous features; decision regions

To set the stage for review of machine learning concepts and implementation, Figure 16.1 depicts decision regions produced by four different machine

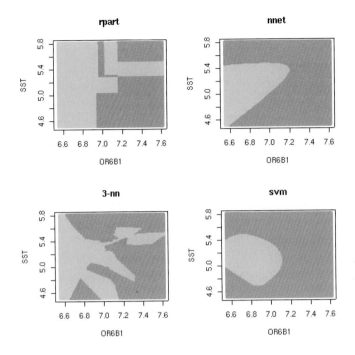

Figure 16.1. Decision regions for the distinction between two subtypes of acute lymphatic leukemia (ALL). The subtypes are those tumors with fusion of the BCR and ABL genes and without. As features, two randomly selected genes are used: olfactory receptor 6 B1 (OR6B1, probeset 31424_at) and somatostatin (SST, probeset 37782_at). Four different machine learning procedures are used to derive the decision regions. If the color of (x, y) is peach, the subject bearing gene expression levels (x, y) is predicted to have status NEG; if the color is turquoise, the subject is predicted to have BCR/ABL positive status.

learning procedures to a single set of ALL data. We will explicitly define the different procedures in subsequent sections.

Each of the panels in Figure 16.1 constitutes a solution to the pattern recognition problem as described above. Given values of expression on HG-U95Av2 probesets 31424_at (OR6B1) and 37782_at (SST) measured on B-cells derived from an ALL patient, the class of the patient's malignancy is predicted according to the color of the point (OR6B1, SST). The display exposes a number of issues in the deployment of machine learning tools that must be addressed in applications.

- The decision regions are confined by the ranges of feature values observed up to the time of model construction. Applications of machine learning tools need to explicitly define how decisions are reached when feature values are presented that are outside the range of those used to construct the model. Reasonable procedures for *extrapolation* may be available in some contexts; in others, a declaration of *outlier* may be necessary.

- The use of sharp decision boundaries is often artificial. Feature combinations that are on or are very close to a boundary may more realistically lead to decisions of *doubt*. A declaration of doubt may be of interest in its own right, or may be an indication that further studies are in order for the subject bearing the hard-to-classify features.

- It is clear that the different procedures do not agree in detail. In fact, each of the procedures depends upon a configuration of tuning parameters. Changes to the values of the tuning parameters can substantially alter the appearance of the decision regions. Effective choice of tuning parameter configuration is a central problem in applications of machine learning methods.

- Selection of machine learning procedure families or tuning parameter configurations should be governed by accuracy and practical utility of the resulting decision procedure. Sample reuse methods are widely employed for accuracy assessment, but there is no widely applicable optimality framework for structuring the accuracy assessment or matching the learning procedure to the problem at hand.

- The decision regions have very different geometries; the CART procedure forms compound dichotomies that are simple to express but may lead to unintuitive decision procedures corresponding to islands or narrow channels in the feature space. Such complex feature configurations may represent statistical artifacts, or may be indications of important feature interactions.

In summary, machine learning applications must address extrapolation, acknowledgment of doubt, tuning parameter selection, uncertain appraisal of performance, and plausibility of the detailed decision procedure created by the learner. Software tools for machine learning, in conjunction with the comprehensive data analysis environment R, can help the analyst tackle these problems. There are many methods for machine learning implemented and available as packages in R and Bioconductor.

16.3   Methodological issues

16.3.1   Families of learning methods

Most machine learning methods construct classifiers through optimization. A family of classifiers $\hat{C}(\cdot)$ is indexed by a parameter $\psi \in \Psi$. The predictions

$\hat{C}^{\psi}(x)$ are compared to the true classes y using some loss measure $l(\hat{C}, y)$, and the $\hat{C}^{\psi}$ that minimizes this loss among $\psi \in \Psi$ is selected for future use. Classes of machine learning methods can be defined by the structure of l or by features of the decision regions that can be formed.

The mathematical details of methodologies to be catalogued here can be obtained in the references cited in Section 16.1. We focus on conceptual definitions and illustrations to allow room for software demonstration.

Linear methods. Chapter 5 of Duda et al. (2001) defines linear discriminant functions for classification

$$g(x) = w^t x + w_0 \tag{16.1}$$

where $x \in R^p$ is a p-dimensional feature vector, w is a p-dimensional weight vector, and w_0 is called a "threshold weight." For a two-category problem, classification proceeds by determining the sign of $g(x)$. For $K > 2$ categories, category-specific weight vectors and threshold weights are defined leading to the system

$$g_i(x) = w_i^t x + w_{0i}, \quad i = 1, \dots, K$$

and classification proceeds by determining the value of i for which g_i is maximized.

It is common to include in the family of linear methods those that do not employ only linear combinations of x, the raw features, but also allow various transformation of x to enter the linear form of Equation (16.1).

The learning process determines values of w and w_0 from observed data. There are many approaches reviewed in the references noted above. Algorithms for obtaining linear discriminant functions are neatly schematized in Table 5.1 of Duda et al. (2001).

Nonlinear methods.

Nonlinear models are familiar in applied statistics (Bates and Watts, 1988). The basic statistical setup for a random continuous response Y with predictor x is

$$Y = f(x; \theta) + \epsilon, \quad f \in \mathfrak{F}_\Theta, \theta \in \Theta$$

where $\mathfrak{F}_\Theta$ is a family of functions indexed by a parameter $\theta \in \Theta$, and ϵ is a random error term. Key challenges in the application of nonlinear statistical models include specifying models for errors, motivating restriction to function families $\mathfrak{F}_\Theta$, and optimizing the resulting objective functions.

A widely used nonlinear model for classification is the neural network. There is a complex taxonomy of these models. Figure 16.2 schematizes a "generic feed-forward network" after Figure 5.1 of Ripley (1996a). Let y_k denote the kth output element, then the model corresponding to Figure 16.2 is

$$y_k = f_k \left[\alpha_k + \sum_{j \to k} w_{jk} f_j \left(\alpha_j + \sum_{i \to j} w_{ij} x_i \right) \right] \tag{16.2}$$

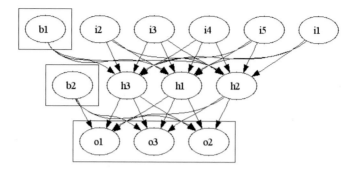

Figure 16.2. Generic feed-forward neural network with one hidden layer of size three, five inputs and three outputs.

where the $f.$ have linear or nonlinear forms [the logistic function $f(x) = (1 + e^{-x})^{-1}$ is commonly used], x_i is the ith input feature, $\sum_{j \to k}$ denotes summation restricted to connected units, and $\alpha.$ and $w..$ are parameters to be estimated. See Ripley (1996a) for generalization to multiple hidden layers with skip-layer connections. A review of Bayesian inference methods with neural network models is given in Titterington (2004).

Regularized methods. Regularized methods are those for which the loss function has the form

$$l(\hat{C}, c) = l_0(\hat{C}, c) + \lambda K, \tag{16.3}$$

where K is a non-negative measure of classifier complexity, and λ is a tuning parameter. Small values of λ diminish the penalty associated with high values of K, allowing more complex models to be competitive in the optimization search space. Smoothing splines (Wahba, 1990) are a widely used implementation of this framework for continuous responses, and these methods are basic to the generalized additive models of Hastie and Tibshirani (1990), the logspline models of Stone et al. (1997), and to certain formulations of support vector machines [see Section 12.3.2 of Hastie et al. (2001)].

The measurement of classifier complexity is a basic concern of computational learning theory. We examine some of the relevant concepts in Section 16.3.2 below.

Local methods. The classifier definition of expression (16.1) leads to decision boundaries that are hyperplanes in the (possibly transformed) feature space. Such boundaries may be too rigid in certain applications. Local methods of estimation and classification provide greater flexibility.

A basic tool for localization is the concept of a *kernel function*. A kernel K is a bounded function on the feature space which integrates to unity [section 6.1 of Ripley (1996a)]. The function should peak around zero, and for two points x and y in feature space, $K(x-y)$ is regarded as a measure of their proximity. The multivariate Gaussian density is a frequently encountered kernel. Kernel estimation of probability densities is widely practiced; the *Parzen estimator* of a density on the basis of N points has the form

$$\hat{f}(x) = (N\lambda)^{-1} \sum_i K_\lambda(x, x_i)$$

where λ plays the role of a bandwidth, e.g., $K_\lambda(x, y) = \phi(|x-y|/\lambda)$, where ϕ is the standard Gaussian density. For classification, the class-specific density estimates can be used to compute Bayes rule:

$$\hat{Pr}(C = j | x = x_0) = \frac{\hat{\pi}_j \hat{f}_j(x_0)}{\sum_k \hat{\pi}_k \hat{f}_j(x_0)}.$$

An alternative to the use of a proper kernel (with integral 1) defines K to be constant over k nearest neighbors, and zero elsewhere. The proportions of classes in this k-neighborhood give a local estimate of the posterior class distribution, and the class prediction for the neighborhood is the most common class in the neighborhood. When $k = 1$, the feature space is tiled by the Dirichlet tessellation, and all points on each tile are classified to the class of the data point contained in the tile (Section 6.2 of Ripley (1996a)).

Tree-structured models. Tree-structured models for classification have a long history. Breiman et al. (1984) defined the *classification and regression tree (CART)* procedures implemented in R through packages rpart and tree. The basic output is a sequence of predicates in x that define the nodes (splits) and leaves (terminal groupings) of a binary tree.

For concreteness, Figure 16.3 displays the tree corresponding to the CART panel of Figure 16.1. The splitting sequence begins with a predicate that divides the entire data set into two subsets. In Figure 16.3, the predicate is "OR6B1 < 7.167". This leads to two nodes, one predominantly occupied by samples of class BCR/ABL, the other predominantly NEG. Tree construction proceeds recursively, with the objective of creating nodes that are purest with respect to the distribution of the response. Options are available for selecting the measure of node purity, and for defining when the splitting procedure terminates.

A generalization of the CART procedure is the *random forests* methodology of Breiman (2001). Whereas CART uses all variables and all relevant cases when creating nodes in a single tree that represents the outcome of the learning process, random forests creates a large number of trees developed on random samples of the input cases. The input data for each tree is based on a bootstrap sample from the original data. The variables used for constructing splits are a random subsample of the complete set of variables. All trees are grown fully, with no pruning. The classification for a given feature vector is given by the majority vote over all trees on its class.

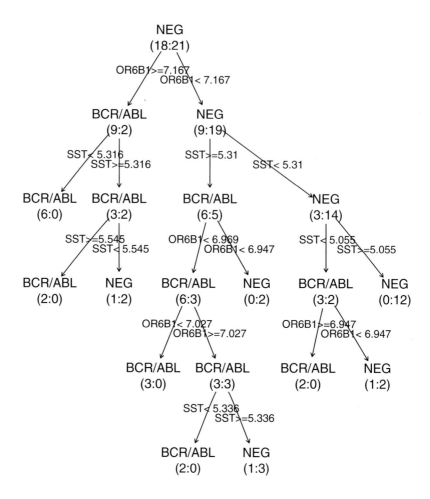

Figure 16.3. Tree-structured model for prediction of BCR/ABL vs NEG status in ALL data. Only two probesets were employed; the **rpart minsplit** parameter was set to 4. Notation: each node is labeled with the name of the majority class, and the composition of the node is given as (n:m), where n is the number of BCR/ABL outcomes, and m is the number of NEG outcomes.

Boosting. *Boosting* refers to an iterative approach to classifier construction. Briefly, one begins with a training set and a "weak learner" (procedure with generalization error just better than chance). In the first stage, a weak classifier is built on the basis of all the training data. In subsequent stages, observations that are difficult to classify are given greater weights than those that are classified correctly, and weak learners are constructed with reweighted data at each stage. After a prespecified number of iterations, a weighted vote is used to classify each training instance. See Chapter 17 for full details.

16.3.2 Model assessment

PAC learning theory. "Probably almost correct" (PAC) learners have measurable generalization properties. Let F denote a probability model. The data are realizations of $(X, c(X))$, where $X \overset{iid}{\sim} F$, and the class labels $c(X) \in \{1, \ldots, K\}$. F itself is of no interest. Instead, we seek $\hat{C}(\cdot) \in \mathfrak{C}$ close to $c(\cdot)$, where $\mathfrak{C}$ is a set of classifiers that may be of finite or infinite cardinality. PAC learning theory defines the following framework: Let $T(n)$ denote a population of training sets of size n, then the objective is $\hat{C}$ satisfying

$$\Pr_{T(n)} \left[\Pr_F \{\hat{C}(X) \neq c(X)\} < \epsilon \right] > 1 - \delta$$

where ϵ and δ are small positive numbers. In words, the proportion of training data sets of size n for which the classification error of $\hat{C}$ is less than ϵ (that is, $\hat{C}$ is almost correct), is no greater than $1 - \delta$ (that is, $\hat{C}$ is highly probably almost correct). Section 2.8 of Ripley (1996a) and Section 5.2 of Schölkopf and Smola (2001) review probability inequalities that permit probabilistic bounding of generalization error using estimates of training set error.

An important concept in PAC learning theory is the complexity of the class $\mathfrak{C}$, as measured by the *Vapnik-Chervonenkis (VC) dimension*. Consider a set $\mathfrak{F}$ of functions f_j, $j = 1, \ldots$, with subsets of the feature space $\mathfrak{X}$ as domains and the set of class labels as common range. (This set $\mathfrak{F}$ is unrelated to the parametric function set introduced in Section 16.3.1.) $\mathfrak{F}$ *shatters* a set of points $X \subset \mathfrak{X}$ if for any class labeling of points in X, some $f_j \in \mathfrak{F}$ may be found that computes the labeling. The VC dimension of $\mathfrak{F}$ is the cardinality of the largest subset X of $\mathfrak{X}$ for which some $f \in \mathfrak{F}$ shatters X. For example, if the class labels are binary, $\mathfrak{X}$ is R^2, and $\mathfrak{F}$ is the set of linear discriminators, it can be seen that the VC dimension of $\mathfrak{F}$ is three.

Proposition 2.6 of Ripley (1996a) is a characteristic application of VC dimension calculations, relating the discrepancy between true and observed classification error rates to the number of training samples n. Let $err(g)$ [$\widehat{err}(g)$] denote the true [observed] classification error rate for classifier g. If d is the (finite) VC dimension of $\mathfrak{F}$ and

$$n \geq \epsilon^{-2} 16[\log(4/\delta) + d \log 32e/\epsilon^2], \tag{16.4}$$

then

$$Pr \left\{ \sup_{g \in \mathfrak{F}} |\widehat{err}(g) - err(g)| > \epsilon \right\} < \delta.$$

For use in practice, one must know the VC dimension for the family of classifiers in use. Typically only upper bounds on the VC dimension are available. Ripley (1996b, section 5.7) reports illustrative results for a class of neural networks. If $\mathfrak{C}$ is the class models of form (16.2) where the total number of input, output, hidden, and bias units is M, the total number of weights is W, and the f. are all threshold functions [that is, of the form $f(x) = I(x > c)$, where $I(\cdot)$ is an indicator function], then the VC dimension d of $\mathfrak{C}$ satisfies

$$d \leq 2W \log_2 eM.$$

This may be plugged in to the sample size relation (16.4) to design a training experiment leading to a classifier with specified probability that the discrepancy between true generalization error and estimated classification error is greater than ϵ. In practice, the theoretical bounds may be quite loose.

Sample splitting. *Data splitting* is the practice of dividing data into training and test sets, using the test set to assess predictive ability of the trained model. *Cross-validation* is the practice of systematically partitioning the data, using each partition as a test set for the model built on its complement. The "leave-one-out" species of cross-validation for a data set of size n leads to n model fits, each using $n - 1$ records.

A basic concern in data splitting in both the training vs. test and cross-validation paradigms is the choice of split, which should be justified on the grounds of maximizing both predictive ability of the final model and the accuracy of the assessment of the predictive ability. Picard and Berk (1990) provide useful formalism. In the framework of the linear model $y = X\beta + e$, where y is an n-vector of responses, X is $n \times p$ full rank, and e is an n-sample from $N(0, \sigma^2)$, the problem is to partition X and y as $X^t = (E^t \| V^t)$, $y^t = (y_E^t \| y_V^t)$, where E connotes estimation (training) resource, V connotes validation (testing) resource and superscript t denotes the matrix transpose operation. There are n_E estimation units and n_V validation units, $n_E + n_V = n$. Picard and Berk derive a goodness of split criterion by summing (squared) integrated mean squared error estimated on the training set with residual variance estimated on the validation set. They demonstrate that this criterion is approximately minimized for least squares problems when $E^t E / n_E = V^t V / n_V = X^t X / n$. These authors recommend reserving 25% to 50% of the available data for validation, using an approximately matched split, to correct for optimism arising from model selection and other data analytic practices, and to confront extrapolation requirements that may emerge when new predictions are required.

Unfortunately, computing tools for conveniently identifying optimal splits do not seem to exist. Methods for the construction of optimal designs (Johnson and Nachtsheim, 1983) and for the development of matched subsets of observational study cohorts (Rosenbaum, 1995) will require adaptation to support application in data splitting. A very recent discussion paper by Efron (2004) reviews cross-validation and model-based alternatives for accurately assessing prediction error.

Bootstrapping and bagging. The sample splitting concept is intuitive but simple implementations are generally inefficient. Resampling and ensemble methods are an important advance for both learning and appraisal of learner performance. See Chapter 17 for details.

16.3.3 Metatheorems on learner and feature selection

The *No Free Lunch Theorem* [Section 9.2.1 of Duda et al. (2001)] states that all learning algorithms have the same expected generalization error, when the expectation is taken over all possible classification functions $c(\cdot)$. The interpretation given by Duda et al. is that

> [i]f the goal is to obtain good generalization performance, there are no context-independent or usage-independent reasons to favor one learning or classification method over another. If one algorithm seems to outperform another in a particular situation, it is a consequence of its fit to the particular pattern recognition problem, not the general superiority of the algorithm (p.454 of Duda et al. (2001))

The *ugly duckling theorem* [Section 9.2.2 of Duda et al. (2001)] states that there is no problem- or purpose-independent selection of features that may be used to define similarity among objects for classification. Here similarity is measured by counting the number of predicates (drawn from a finite stock) shared by the two feature vectors being compared. The theorem establishes that the number of predicates shared by any pair of patterns is a fixed constant, independent of the choice of patterns. Thus domain-specific knowledge will play an essential role in the identification of genuinely informative feature sets.

Finally, there is interest in understanding the role of classifier complexity as a determinant of classifier generalization ability. Domingos (1999) distinguishes two forms of *Occam's razor, nunquam ponenda est pluralitas sin necesitate*, that have been espoused in the literature of machine learning methods. The first form states that the simpler of two classifiers with common generalization error should be preferred, because simplicity is intrinisically desirable. The second form states that the simpler of two classifiers with common training set error should be preferred, because simplicity is associated with lower generalization error. Domingos provides extensive argumentation against the second form. A basic theme is that the first form is valid and plays a significant role in science in general, but that apparent simplicity of machine learning methods is often secured through hidden violations of the first form of Occam's razor (Domingos, 1999).

16.3.4   Computing interfaces

S-PLUS™ and R have been distinguished among statistical computing environments for their support of a wide variety of machine learning methods with relatively straightforward interfaces. In R, the package bundle VR has long provided functions for discriminant analysis, classification and regression trees, nearest neighbor classification, and fitting of feed-forward neural networks. The e1071 package provides tools for fitting support vector machine models and for systematically tuning a number of machine learning methods. Other packages such as locfit, logspline, gpls, gam and mgcv implement local and regularization-based procedures for density estimation and regression, which can serve as components of classification methods (Hastie et al., 2001, Section 6.6). A number of other packages related to machine learning can be found on CRAN.

All of the fitting functions provided in the packages noted here have a standard formula interface. Given a data frame or environment d in which bindings of formula variables may be determined, an R command with form

```
ans <- f(y~x, data=d, p1=P1, ...)
```

will return an object describing the model specified by function f on responses y and predictors x in data frame d with values of additional parameters such as p1 set optionally. In most cases the resulting object ans will respond to the method call predict() (with optional newdata parameter) with a vector of predicted of the same type as y. However, there is considerable diversity in the structures and behaviors of the objects returned by machine learning routines in various R packages. This diversity makes it difficult to compare procedures, or to create generic downstream processing methods that work on the basis of the learning results or use the trained learner to do additional predictions.

Bioconductor provides a new package, MLInterfaces, that aims to simplify the use of machine learning tools and to simplify and make more uniform the use of their outputs. Presently, MLInterfaces caters for input data structures derivable from *exprSet* instances. The general pattern is

```
ans <- fB( eset, respname, inds, opt1=O1, ... )
```

where f is the name of an established fitting function (examples are nnet, knn, rpart, so that the corresponding methods to be called are nnetB, knnB, rpartB.) The first argument is always an *exprSet* instance, the second argument is a character value naming a *phenoData* variable in the *exprSet* instance, and the third argument is an integer vector identifying the proper subset of the samples in the *exprSet* instance that are to be used as a training set. The object returned by an MLInterfaces method is of class *MLOutput*, which is extended by classes *classifOutput* and *clustOutput*. See the manual pages for details.

	Package	Functions covered
1	class	`knn1, knn.cv, lvq1, lvq2, lvq3, olvq1, som` `SOM`
2	cluster	`agnes, clara, diana, fanny, silhouette`
3	e1071	`bclust, cmeans, cshell, hclust, lca` `naiveBayes, svm`
4	gbm	`gbm`
5	ipred	`bagging, ipredknn, lda, slda`
6	MASS	`isoMDS, qda`
7	nnet	`nnet`
8	pamr	`cv, knn, pam, pamr`
9	randomForest	`randomForest`
10	rpart	`rpart`
11	stats	`kmeans`

Table 16.1. Packages and functions covered by MLInterfaces.

Table 16.1 gives the names of the functions for which interfaces have been constructed. S4 generic methods are identified by appending "B" to the native function name.

16.4 Applications

16.4.1 Exploring and comparing classifiers with the ALL data

In order to allow illustrative computations to complete rapidly on cheap hardware, we filter the ALL data (original size 12625 genes × 128 arrays) to those samples that have molecular biology classification NEG or BCR/ABL, and we use limma to identify the top 500 differentially expressed genes.

First we filter on tumor status selecting only B-cell samples, and we add the two-level indicator factor bcrabl to the *exprSet* all2:

```
> bio <- which(ALL$mol.biol %in% c("BCR/ABL", "NEG"))
> isb <- grep("^B", as.character(ALL$BT))
> kp <- intersect(bio, isb)
> all2 <- ALL[, kp]
> tmp <- all2$mol.biol == "BCR/ABL"
> tmp <- ifelse(tmp, "BCR/ABL", "NEG")
> pData(all2)$bcrabl <- factor(tmp)
```

Now we work with limma to identify the Ndiff = 500 most differentially expressed genes, using the moderated *t*-statistic to measure differential expression:

```
> library("limma")
> des <- model.matrix(~all2$bcrabl)
> fit <- lmFit(all2, des)
> fit2 <- eBayes(fit)
> Tdiff <- topTable(fit2, coef = 2, Ndiff)
> all2 <- all2[as.numeric(rownames(Tdiff)), ]
```

We begin with an example of linear discriminant analysis. The full set of parameters to method ldaB is found by

```
> args(ldaB)
```

```
function (exprObj, classifLab, trainInd, prior, tol = 1e-04,
    method, CV = FALSE, nu, metric = "euclidean", ...)
NULL
```

To obtain the interpretations of the parameters listed here, see the manual page for class::lda. We will use the first 40 observations as the training set and accept all default parameter settings.

```
> l1 <- ldaB(all2, "bcrabl", 1:40)
```

The report on the classification includes basic identifying information, tabulation of predictions on the test set, and summaries of the distribution of estimated class membership probabilities computed for each test record. The confusion matrix for the test set (all observations beyond the 40th) is:

```
> confuMat(l1)
```

```
          predicted
given      BCR/ABL NEG
  BCR/ABL       13   3
  NEG            3  20
```

The rows are the true class labels and the columns are the predicted class labels. The test set misclassification rate is 6/39.

k-Nearest neighbor (k-NN, here with $k = 1$) is computed by calling a different function with identical arguments.

```
> k1 <- knnB(all2, "bcrabl", 1:40)
```

```
> confuMat(k1)
```

```
          predicted
given      BCR/ABL NEG
  BCR/ABL       15   1
  NEG            2  21
```

Here the test set misclassification rate is 3/39.

16.4.2   Neural net initialization, convergence, and tuning

Feed-forward neural nets can be used quite conveniently. The large number of inputs (500 in this case) necessitates manual setting of the `MaxNWts` parameter. We begin with a 5 hidden unit model:

```
> n1 <- nnetB(all2, "bcrabl", 1:40, size = 5, MaxNWts = 10000)

> confuMat(n1)
```

```
          predicted
given      BCR/ABL NEG
  BCR/ABL       16   0
  NEG           10  13
```

Note that if we perform the same call again, the behavior can be quite different.

```
> n1b <- nnetB(all2, "bcrabl", 1:40, size = 5, MaxNWts = 10000)

> confuMat(n1b)
```

```
          predicted
given      BCR/ABL
  BCR/ABL       16
  NEG           23
```

This is because the `nnet` procedure employs random initializations. One can obtain reproducible `nnet` fits by calling `set.seed` prior to invoking the procedure. Setting a fixed seed prior to invoking any randomized algorithm is a good idea in case data-dependent debugging is required. Handling seeds in simulation contexts requires care.

The behavior of a neural net model in `nnet` depends on the number of units (size), functional form of of the f. of Equation (16.2), and the setting of a decay parameter corresponding to λ of Equation (16.3).

```
> n2 <- nnetB(all2, "bcrabl", 1:40, size = 6, decay = 0.05,
+       MaxNWts = 10000)

> confuMat(n2)
```

```
          predicted
given      BCR/ABL NEG
  BCR/ABL       14   2
  NEG            2  21
```

The e1071 package includes software for automatically searching through the tuning parameter space for `nnet` models.

16.4.3   Other methods

MLInterfaces makes it convenient to experiment with a variety of learning tools. The following code can be used for the gradient boosting machine:

```
> g1 <- gbmB(all2, "bcrabl", 1:40, n.minobsinnode = 3,
+       n.trees = 1000)
> confuMat(g1)
```

```
          predicted
given     BCR/ABL NEG
  BCR/ABL      14   2
  NEG           6  17
```

For random forests:

```
> rf1 <- randomForestB(all2, "bcrabl", 1:40, importance = TRUE)
> confuMat(rf1)
```

```
          predicted
given     BCR/ABL NEG
  BCR/ABL      13   3
  NEG           4  19
```

Finally, a default support vector machine:

```
> s1 <- svmB(all2, "bcrabl", 1:40)
> confuMat(s1)
```

```
          predicted
given     BCR/ABL NEG
  BCR/ABL      14   2
  NEG           3  20
```

16.4.4   Structured cross-validation support

Cross-validation refers to a family of methods whereby estimates or predictions are developed through a sequence of partitions or subsamplings of the full data set. The MLInterfaces package provides a generic method xval to support controlled application of cross-validation methods to data in exprSet objects. The arguments are:

```
> args(xval)
```

```
function (data, classLab, proc, xvalMethod, group, indFun, niter,
    ...)
NULL
```

The data parameter should be an instance of the exprSet class. classLab is a string identifying the phenoData element to be used for classification. proc is the name of a machine learning generic method provided in MLInterfaces (e.g., knnB, nnetB). xvalMethod takes value "LOO" (leave one out), "LOG" (leave group out), or "FUN" (evaluate a subsampling function to derive the training and validation sets for each cross-validation iteration). We illustrate with the k-NN analysis of the ALL data:

```
> xvloo <- xval(all2, "bcrabl", knnB, "LOO")
```

```
> table(given = all2$bcrabl, predicted = xvloo)

          predicted
given    BCR/ABL NEG
  BCR/ABL     35   2
  NEG          5  37
```

We now have a different appraisal of performance of the k-NN procedure. In our initial example in Section 16.4.1, we explicitly partitioned the data and evaluated performance based on the test set. Here we have a sequence of partitions in which each data point is used as a test set of size one.

16.4.5 Assessing variable importance

Two of the learning tools supported by MLInterfaces provide very convenient measures of variable importance. The random forests algorithm has a bootstrap-based measure. For every tree grown in the forest, the out-of-bag cases are classified and the number of correct classifications is recorded. The values of variable m are randomly permuted across out-of-bag cases and the classification is performed again. The accuracy importance measure reported by the varImpPlot function is the average decrease in accuracy induced by permutation divided by an estimate of its standard error.

The gradient boosting machine also offers a measure of relative importance of variables. An abstract expression of this measure for the mth variable is (Friedman, 2000)

$$I_m = \left(E_X \left[\frac{\partial \hat{C}(x)}{\partial x_m} \right]^2 \cdot \mathrm{var}_X[x_m] \right)^{1/2},$$

but Friedman notes that only an approximation can be used. See his paper for the specific definition. The gbm package produces the approximate relative importance through the summary function. The MLInterfaces package includes generic methods for extracting and structuring variable importance measures from those procedures that compute them, and for plotting these measures; see Figure 16.4.

16.4.6 Expression density diagnostics

Machine learning methods can be applied to other aspects of data analysis. The edd package includes tools to permit the use of various machine learning techniques to the classification of gene-specific marginal densities of expression values across samples. Assessment of these marginal densities is important for decisions about transformation and choice of method for comparing expression distributions.

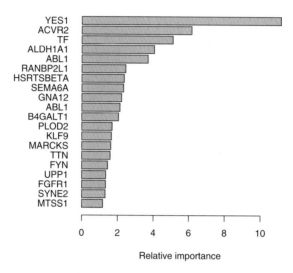

Figure 16.4. Relative variable importance measures from the gradient boosting machine algorithm.

To illustrate, let us consider a deliberately selected pair of genes from the ALL data. For the NEG and BCR/ABL phenotypes, we have the histograms shown in Figure 16.5.

We see suggestions of bimodality in several of these histograms. The edd method compares the empirical distribution of expression for each gene in an exprSet to a catalogue of parametrically or empirically defined distributions that may be specified by the user. The documentation and vignettes for the edd package should be consulted for details.

```
> neg <- edd(gg[, gg$mol.biol == "NEG"])
> as.character(neg)

[1] ".75N(0,1)+.25N(4,1)" "logN(0,1)"

> bcr <- edd(gg[, gg$mol.biol == "BCR/ABL"])
> as.character(bcr)

[1] "B(2,8)" "t(3)"
```

The procedure classifies the shapes of the NEG genes as Gaussian mixture and lognormal, and of the BCR/ABL genes as Gaussian and t_3 (a heavy-tailed distribution). This particular classification employed k-NN

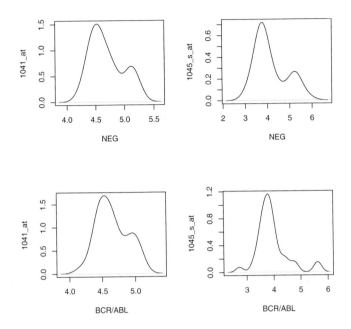

Figure 16.5. Gene- and phenotype-specific histograms for expression of two genes in the ALL data.

with $k = 1$ to predict the class of distributional shape of the four vectors of gene expression values (two genes by two strata) using a large collection of simulated samples from specified distributions as the training set. All vectors are scaled to unit MAD and zero median before the training and predictions are carried out, so that this procedure is focused on distributional shape. The edd function is a general interface to various forms of training-set construction and machine learning approaches to this distributional shape classification problem. For additional applications, including a display of diversity of distributional shapes in vectors of expression values derived from Golub's leukemia data (Golub et al., 1999), see the example for edd.

16.5   Conclusions

We have seen that application of machine learning procedures always requires informed involvement of human investigators. The metatheorems of Section 16.3.3 indicate that effective learning via feature selection and optimization cannot be purely mechanical. Thus there will always be a role for comparative experimental application of various machine learning procedures to a given problem. Bioconductor interfaces that simplify routine aspects of comparative machine learning in genomic contexts have been illustrated. Future developments will support more convenient integration of machine learning methods with proper accommodation of doubt and outlier decisions into the high-level analytic workflow.

17

Ensemble Methods of Computational Inference

T. Hothorn, M. Dettling, and P. Bühlmann

Abstract

Prognostic modeling of tumor classes, disease status, and survival time based on information obtained from gene expression profiling techniques is studied in this chapter. The basic principles of ensemble methods like bagging, random forests, and boosting are explained. The application of those methods to data from patients suffering acute lymphoblastic leukemia or renal cell cancer is illustrated. The problem of identifying the best method for a certain prediction task is addressed by means of benchmark experiments.

17.1   Introduction

Recent technological developments have opened new perspectives in biomedical research and clinical practice, and they also have changed the way in which statistics makes progress. The main challenge is to develop computational inference tools that can deal with very high dimensional data sets containing thousands of input variables for a few dozens of experiments only. In particular, such data structures emerge from the gene expression microarray technology and from protein mass spectroscopy analysis.

In this chapter, we focus on tumor prognosis with gene expression data. Given efficient data analysis tools, information from biotechnology may represent a promising supplement to tumor prediction based on traditional clinical factors. Let us assume that we are given previous experience from a cohort of I patients, which mathematically amounts to a learning sample

$$\mathcal{L} = \{(y_1, x_1), (y_2, x_2), \ldots, (y_I, x_I)\}.$$

The input variable $x_i \in R^J$ contains the vector-valued gene expression profile of J genes for the ith patient. Initially, we focus on a dichotomous response, $y_i \in \{0, 1\}$, coding for two different populations or classes, as for example different cancers or tumor subtypes. Multiclass prediction and the analysis of survival data are discussed in Sections 17.4 and 17.7

We systematically exploit the information in the learning sample, $\mathcal{L}$, to define a rule that predicts disease status, the response variable, from the gene expression profile x. This rule can then be applied to novel cancer patients for establishing an early and precise prognosis, which is often crucial for a treatment with few side-effects and good cure rates. From a mathematical viewpoint, this amounts to learning a function $g : R^J \to R$, which can be applied in order to predict the class membership of an observation with expression profile x via some fixed transformation f. For example, $g(x|\mathcal{L}) = \log\{\hat{p}(x)/[1 - \hat{p}(x)]\}$, where $\hat{p}(x) = \widehat{Pr}[Y = 1|X = x]$ is the probability of class 1 given the inputs x, and $f(g) = \chi(g > 0)$ is a threshold that yields a classification rule respecting equal misclassification costs, where χ denotes the indicator function. In the sequel, such procedures are referred to as classifiers. Although supervised classification is a well-known methodology in statistics, finding solutions to problems with many input variables or genes J but a very limited number of patients I is challenging. A promising approach that is suitable for high-dimensional prediction problems in bioinformatics is to use ensemble methods that aggregate many simple classifiers into a powerful committee.

In contrast to using just a single fit from a particular method, ensemble techniques aim at improving the predictive ability of a relatively simple statistical learning technique by constructing a weighted sum thereof. Technically, the ensemble is written as

$$g_{E(M)}(\cdot|\mathcal{L}) = \sum_{m=1}^{M} \alpha_m g_m(\cdot|\mathcal{L}_m),$$

where $\mathcal{L}_m$ is a reweighted version of the learning sample $\mathcal{L}$ and α_m are weights. Here, g_m is a base procedure, which we term the *weak learner*. For example, $g_m(x)$ may be a simple estimate of the log-odds ratio, $\log\{\hat{p}(x)/[1 - \hat{p}(x)]\}$. The choice of the reweighted sample, $\mathcal{L}_m$, the aggregation weights, α_m, as well as of the weak learner are the art in this methodology. Bagging, boosting and random forests are such methods, which will be discussed below. The final ensemble estimator depends on the choice of the weak learner. The most prominent choice in high-dimensional problems are recursive partitioning methods, "classification and regression trees", as they incorporate some form of feature selection and tend to work well when there are more input variables than there are observations.

We refer to Breiman et al. (1984) for a detailed description on recursive partitioning methods. The rpart package (Therneau and Atkinson, 1997) implements the methodology described by Breiman et al. (1984).

17.2   Bagging and random forests

Bagging (Breiman, 1996a) is a rather simple but effective ensemble method. Its name is an acronym for **bootstrap agg**eg**ating**, which suggests the principal idea of this procedure: predictions of weak learners fitted to bootstrap samples $\mathcal{L}_m$ of the original learning sample $\mathcal{L}$ are aggregated by a majority vote.

A bootstrap sample from the original learning sample $\mathcal{L}$ is an i.i.d. random sample of I observations $(y_i^*, x_i^*), i = 1, \ldots, I$, where the probability of selecting observation (y_i^*, x_i^*) from $\mathcal{L}$ is $1/I$. A bootstrap sample is thus a sample from the empirical distribution function of the learning sample. The algorithm is as follows:

1. Draw M bootstrap samples $\mathcal{L}_m, m = 1, \ldots M$, from the original learning sample $\mathcal{L}$.

2. Fit a weak learner for each of the bootstrap samples $g_m(\cdot|\mathcal{L}_m)$, and construct the classifiers, $f(g_m(\cdot|\mathcal{L}_m))$.

3. Aggregate the classifiers using weights $\alpha_m = 1/M$, yielding the ensemble

$$g_{E(M)}(\cdot|\mathcal{L}) = M^{-1} \sum_{m=1}^{M} f(g_m(\cdot|\mathcal{L}_m)).$$

The bagging ensemble votes for class 1 if the majority of the M weak learners votes for 1; and vice versa. More generally, the ensemble predicts class 1 when the fraction of weak learners predicting class 1 exceeds some number, ν.

It has been argued that for unstable estimation methods such as decision trees, bagging reduces the variance while the bias remains approximately the same (Bühlmann and Yu, 2002).

It is rather straightforward to implement the bagging procedure in high-level languages like S. The basic ingredient is a tree building algorithm used for fitting trees to bootstrap samples of the learning sample.

```
> simple_bagging <- function(x, lsample, M = 100,
+        nu = 0.5) {
+        I <- nrow(lsample)
+        bsample <- rmultinom(M, I, rep(1, I)/I)
+        pred <- rep(0, nrow(x))
+        rpc <- rpart.control(xval = 0, cp = 0.01)
+        for (m in 1:M) {
+            weaktree <- rpart(y ~ ., data = lsample,
+                weights = bsample[, m], control = rpc)
+            prtree <- predict(weaktree, newdata = x,
+                type = "class")
+            pred <- pred + (prtree == levels(lsample$y)[2])
+        }
```

```
+       factor(pred/M > nu, levels = levels(lsample$y))
+ }
```

First, M random index vectors representing M bootstrap samples are drawn from the multinomial distribution. A classification tree is fitted to the learning sample `lsample` using `weights = bsample[,m]` representing the mth bootstrap sample. Large trees, without applying any form of pruning whatsoever, are grown. The vector `pred`, of length I, which counts the number of trees predicting the second class is updated. Finally, the function returns a factor coding the ensemble predictions obtained for observations `x` from a simple majority vote.

A simple but extremely successful modification to this algorithm is the random forest approach (Breiman, 2001). The basic idea is to modify the tree growing algorithm leading to even weaker components of the ensemble. This is achieved by choosing a small random subset of inputs available for splitting at each stage of the recursive partitioning algorithm building the tree. Thus, the input variables actually used in each of the weak learners are, to a large extent, determined at random.

17.3   Boosting

Boosting was introduced to the machine learning literature by Freund and Schapire (1996) and has demonstrated empirical success on a wide variety of especially high-dimensional prediction problems. The basic notion was that in each boosting iteration, the cases that were misclassified in the previous round get their weights increased, whereas the weights are decreased for cases that were correctly classified. Thus, unlike bagging and random forests, both the aggregation weights α_m and the reweighted learning samples $\mathcal{L}_m$ depend on the previous function fits $g_1(\cdot|\mathcal{L}_1), \ldots, g_{m-1}(\cdot|\mathcal{L}_{m-1})$. However, boosting can be viewed as a forward stagewise strategy, working by iterative optimization of an empirical risk function

$$R[\mathcal{L}, p(x), L] = \frac{1}{I} \sum_{i=1}^{I} L[y_i, p(x_i)], \ p(x) = Pr[Y = 1|X = x],$$

from the learning set $\mathcal{L}$ via constrained (imposed by the weak learner) functional gradient descent, where $L(\cdot, \cdot)$ is a statistically motivated loss function. If we employ the binomial log-likelihood

$$L[y, p(x)] = y \cdot \log[p(x)] + (1 - y) \cdot [1 - p(x)],$$

a continuous surrogate for the 0/1-misclassification loss and a very established criterion for binary classification, it has been shown that the resulting logitboost algorithm (Friedman et al., 2000) yields an approximation to half of the log-odds ratio. That is,

$$g_{E(M)}(x|\mathcal{L}) = \sum_{m=1}^{M} \alpha_m g_m(x|\mathcal{L}_m) \approx \frac{1}{2} \log \left(\frac{p(x)}{1 - p(x)} \right).$$

Hence, logitboost is a linear expansion in terms of weak learners $g_m(\cdot|\mathcal{L}_m)$ on the logit scale, constructed by stagewise optimization of the binomial log-likelihood. Estimated conditional class probabilities are obtained by the simple transformation,

$$p(x) = \frac{1}{1 + \exp[-2g_{E(M)}(x)]},$$

which can be used for class prediction, using a threshold that depends on the misclassification costs. Logitboost has been demonstrated to be a competitive prediction algorithm for tumor classification with microarray data (Dettling and Bühlmann, 2003, 2004). The procedure works as follows.

1. Initialize $p(x_i) = 1/2, i = 1, \ldots, I; m = 1; g_{E(0)}(\cdot|\mathcal{L}) \equiv 0$.

2. Build the pseudo-response for each observation i

$$\begin{aligned} w_i &= p(x_i)[1 - p(x_i)] \\ u_i &= \frac{y_i - p(x_i)}{w_i}, \end{aligned}$$

   setup-up the new learning sample

$$\mathcal{L}_m = \{(u_i, x_i); i = 1, \ldots, I\}$$

   and fit the weak learner $g_m(\cdot|\mathcal{L}_m)$ with case weights w_i.

3. Update ensemble $g_{E(m)}(\cdot|\mathcal{L}) = g_{E(m-1)}(\cdot|\mathcal{L}) + \frac{1}{2}g_m(\cdot|\mathcal{L}_m)$ and $p(x_i) = 1/\{1 + \exp[-2g_{E(m)}(x_i|\mathcal{L})]\}$

4. Repeat until $m = M$.

Predictions are computed with $\alpha_m = 1/2$ and $f(z) = \chi[\exp(1 - 2z) > \frac{1}{2}]$.

The definition of the weights w_i in the logitboost algorithm is such that each weak learner is forced to focus on observations close to the decision boundary, i.e., data points where the boosting classifier is in doubt about the predicted class. The final number of boosting iterations M regulates the complexity of the prediction model, early stopping is a form of shrinkage.

In the context of microarray data, we recommend a default value of $M = 100$, which is a reasonable compromise between computing time, predictive accuracy and prevention of overfitting. This choice was shown to be empirically superior to approaches where M was estimated on the training data via cross validation (Dettling and Bühlmann, 2003). Provided that an interface to the weak learning algorithm is present, an implementation of boosting in high-level languages, like S, is straightforward.

17.4   Multiclass problems

A popular approach for dealing with multiclass problem is to split them in multiple binary ones. In the context of microarray data, we have collected some empirical evidence for the success of this strategy (Dettling and Bühlmann, 2003). The simplest solution is the *one-against-all* approach, which works by defining the response in the kth problem as $y^{(k)} = 1$ if $y = k$, and $y^{(k)} = 0$ else. Then, we boost K times on the modified data $\mathcal{L}^{(k)} = \{(y_1^{(k)}, x_1), \ldots, (y_I^{(k)}, x_I)\}$. The estimated conditional class probabilities are normalized and can in turn be used for maximum likelihood classification via,

$$\widehat{p}^{(k)}(x) = \frac{\widehat{Pr}_{\mathcal{L}^{(k)}}(y^{(k)} = 1|x)}{\sum\limits_{k=0}^{K-1} \widehat{Pr}_{\mathcal{L}^{(k)}}(y^{(k)} = 1|x)}, k = 1, \ldots, K$$

$$\widehat{y}(x) = \operatorname*{argmax}_{k \in \{0, \ldots, K-1\}} \widehat{p}^{(k)}(x).$$

Depending on the data, other schemes may be more accurate for splitting polytomous into multiple binary problems.

Note that no additional procedures are necessary in order to deal with multiclass responses for bagging trees or random forests: Estimated conditional class probabilities arise directly from the ensemble of trees.

17.5   Evaluation

The choice of an appropriate classifier for a prediction problem at hand is by no means obvious. The problem can be separated into the method selection and the error rate estimation tasks. The first task is concerned with choosing the best method available from a, possibly huge, set of statistical procedures capable of dealing with the problem. For error rate estimation, we try to come up with a realistic assessment of the prediction error of the selected procedure. This information is extremely important when we need to take the decision whether it is worth or even ethical to apply a certain classifier in realistic settings.

The prediction error can be measured by a scalar loss function $L(y, \widehat{y})$ assessing the goodness of the prediction $\widehat{y}$ for some response y. When the response is a categorical variable with classes $\{0, \ldots, K - 1\}$, the misclassification error $L(y, \widehat{y}) = \chi(y \neq \widehat{y})$ is an often used loss function. However, this choice is not necessarily the one we are interested in. When we are faced with a two class problem aiming at predicting whether a person suffers a rare but dangerous disease or not the loss of missing an affected person is much higher compared to the loss induced by a false positive. For such problems, the misclassification loss is of the form

$L(y,\hat{y}) = c_1\chi(y = 0, \hat{y} = 1) + c_2\chi(y = 1, \hat{y} = 0)$ for some misclassification costs c_1, c_2. We may be also interested in measuring the accuracy of the estimated probability $\hat{p}(x)$. Then, the negative log-likelihood can serve as a useful loss function.

A major problem is that the learning sample used for model fitting can not be directly used for the estimation of the prediction error of that model. Such an estimate would be optimistically biased because the same observations were used for model building and evaluation. Resampling methods like the bootstrap or cross-validation have been studied extensively, a practical introduction can be found in Hastie et al. (2001). Here, we will use the notion of out-of-bag estimation (Breiman, 1996b) for error rate estimation. When a bootstrap sample of the original learning sample $\mathcal{L}$ is drawn some observations are left out due to sampling *with* replacement. Those observations are an independent sample from the distribution of interest and can be used for the assessment of the prediction error. We can draw random samples from the distribution of the prediction error as follows.

1. Draw B bootstrap samples $\mathcal{L}_b, b = 1, \ldots B$, from the original learning sample $\mathcal{L}$.

2. Fit a model to each bootstrap sample $\mathcal{L}_b$, i.e. $g(\cdot|\mathcal{L}_b)$ and assess the error p_b by the loss averaged over the predictions of the observations in the out-of-bootstrap sample

$$p_b = |\mathcal{L} \setminus \mathcal{L}_b|^{-1} \sum_{(y,x)\in\mathcal{L}\setminus\mathcal{L}_b} L(f(g(x|\mathcal{L}_b)), y), \ b = 1, \ldots, B.$$

The B error rates can now be visualized, for example by boxplots, or can be described via estimates of parameters such as the mean or variance. When multiple candidate models are under consideration, we obtain a sample of B error rates for each of them. Those data distributions can be used to identify the best or a set of the best algorithms. The null-hypothesis of equality of the bootstrap-distributions of the prediction error of several algorithms can be tested, for example by means of the Friedman test. It should be noted that a rejection of this hypothesis should not be generalized to the population from which the data were drawn, as the inference is conditional on the given learning sample. All that we can conclude is whether a finite number B of bootstrap replicates is sufficient for detecting performance differences in the exact bootstrap distributions, with $B = \infty$ (or from the typically infeasible exact multinomial distribution induced by the bootstrap), of the different algorithms. Theoretical justification and illustration of this approach can be found elsewhere (Hothorn et al., 2004b).

17.6   Applications: tumor prediction

17.6.1   Acute lymphoblastic leukemia

In this first example, we apply ensemble methods to construct a model that regresses the stage of acute lymphoblastic leukemia (ALL) on the microarray expression levels. We are primarily interested in two questions. We would like to investigate whether there is any information about the stage of the disease covered by the microarray expression levels. That is, we will test the null-hypothesis that the stage of the disease is independent of the expression levels measured. If we are able to reject this global null-hypothesis, we then want to build a model that allows us to predict the stage of the disease of a patient based on the microarray expression levels only. Data from patients suffering from both T- and B-cell ALL are available from the study of Chiaretti et al. (2004), and we restrict ourselves to patients with B-cell leukemia.

The package **ALL** offers a data object **ALL**, an object of class *exprSet*, which contains the data of patients suffering from both T- and B-cell leukemia.

We are provided with expression levels of 12625 genes for 128 patients. For the analysis here, we restrict the data to patients suffering B-cell leukemia, with 4 subclasses, and to expression levels of genes with standard deviation on the log-scale between 0.08 and 0.18.

```
> cvv <- apply(exprs(ALL), 1, function(x) sd(log(x)))
> ok <- cvv > 0.08 & cvv < 0.18
> BStagelev <- paste("B", 1:4, sep = "")
> BALL <- ALL[ok, ALL$BT %in% BStagelev]
> pData(phenoData(BALL))$BStage <- factor(BALL$BT,
+     levels = BStagelev)
```

Although ordered, we treat the disease stages as nominal. The class distribution of the stages of the disease (BALL$BStage) can be inspected via

```
> table(BALL$BStage)
```

```
B1 B2 B3 B4
19 36 23 12
```

Although we found 3716 genes showing a reasonable variation among the patients, some mild form of univariate variable selection will help to circumvent computational difficulties and speedup the computations. Here, we measure the association between the rank of the expression levels of each gene and the stage of the disease, our response variable, by means of a linear statistic based on the genewise ranks of the expression levels and a matrix of dummy-codings for the response. The statistics are standardized by their conditional expectation and variance (Strasser and Weber, 1999) and the 100 genes associated with the largest standardized statistics are selected.

```
> response <- BALL$BStage
> I <- ncol(exprs(BALL))
> expressions <- t(apply(exprs(BALL), 1, rank))
> Iindx <- 1:I
> var_selection <- function(indx, expressions, response,
+     p = 100) {
+     y <- switch(class(response), factor = {
+         model.matrix(~response - 1)[indx, , drop = FALSE]
+     }, Surv = {
+         matrix(cscores(response[indx]), ncol = 1)
+     }, numeric = {
+         matrix(rank(response[indx]), ncol = 1)
+     })
+     x <- expressions[, indx, drop = FALSE]
+     n <- nrow(y)
+     linstat <- x %*% y
+     Ey <- matrix(colMeans(y), nrow = 1)
+     Vy <- matrix(rowMeans((t(y) - as.vector(Ey))^2),
+         nrow = 1)
+     rSx <- matrix(rowSums(x), ncol = 1)
+     rSx2 <- matrix(rowSums(x^2), ncol = 1)
+     E <- rSx %*% Ey
+     V <- n/(n - 1) * kronecker(Vy, rSx2)
+     V <- V - 1/(n - 1) * kronecker(Vy, rSx^2)
+     stats <- abs(linstat - E)/sqrt(V)
+     stats <- do.call("pmax", as.data.frame(stats))
+     return(which(stats > sort(stats)[length(stats) -
+         p]))
+ }
> selected <- var_selection(Iindx, expressions,
+     response)
```

The function var_selection takes an index vector of observations between 1 and I and returns a vector of length p indicating which genes have been selected. Now, we are able to fit a random forest model to the data using the package MLInterfaces, see Chapter 16, which provides an unified interface to machine learning procedures including the randomForest package (Liaw and Wiener, 2002). Here, we use the data of all observations except the Ith one as learning sample and obtain information from the model on the Ith observation.

```
> rf <- randomForestB(BALL[selected, ], "BStage",
+     Iindx[-1], sampsize = I - 1)
> print(rf)

MLOutput instance, method= randomForest
Call:
```

```
randomForestB(exprObj = BALL[selected, ], classifLab = "BStage",
    trainInd = Iindx[-1], sampsize = I - 1)
predicted class distribution:
B2
 1
```

The forest predicts the disease status of the first patient being B2, which is the correct decision in this case.

The framework for the evaluation of classifiers sketched in Section 17.5 can be implemented as follows. We loop over B bootstrap samples, perform variable selection for the current bootstrap sample and fit four models to the selected genes of this bootstrap sample. First, a random forest model where only mtry = 3 genes are evaluated in each node of the classification trees, bagging (which corresponds to random forest without random sampling of genes) and logitboost with $M = 100$ boosting iterations. In addition, we include a model that does not use any information about the expression levels. To be more specific, for each bootstrap sample we estimate (guess) the class with maximal prior probability for all observations. The misclassification errors computed for each model and bootstrap sample are stored into a dataframe performance.

```
> set.seed(290875)
> B <- 100
> performance <- as.data.frame(matrix(0, nrow = B,
+      ncol = 4))
> colnames(performance) <- c("RF", "Bagg", "LBoost",
+      "Guess")
> for (b in 1:B) {
+      bsample <- sample(Iindx, I, replace = TRUE)
+      selected <- var_selection(bsample, expressions,
+          response)
+      rf3 <- randomForestB(BALL[selected, ], "BStage",
+          bsample, mtry = 3, sampsize = I)
+      predicted3 <- factor(rf3@predLabels, levels = levels(response))
+      performance[b, 1] <- mean(response[-bsample] !=
+          predicted3)
+      rfBagg <- randomForestB(BALL[selected, ],
+          "BStage", bsample, mtry = length(selected),
+          sampsize = I)
+      predictedBagg <- factor(rfBagg@predLabels,
+          levels = levels(response))
+      performance[b, 2] <- mean(response[-bsample] !=
+          predictedBagg)
+      lb <- logitboostB(BALL[selected, ], "BStage",
+          bsample, 100)
+      predictedlb <- factor(lb@predLabels, levels = levels(response))
+      performance[b, 3] <- mean(response[-bsample] !=
+          predictedlb)
```

```
+      performance[b, 4] <- mean(response[-bsample] !=
+          levels(response)[which.max(tabulate(response[bsample]))])
+ }
```

The distributions of the misclassification errors of all four models can be analyzed by the appropriate procedures for paired observations. The global null-hypothesis of equality of all three models can be tested using

```
> friedman.test(as.matrix(performance))

        Friedman rank sum test

data:  as.matrix(performance)
Friedman chi-squared = 181, df = 3, p-value <
2.2e-16
```

which leads to a rejection indicating that there are global differences between the performances of the four candidate models. This result allows us to conclude that there is a relationship between expression levels and the stage of B-cell leukemia. A parallel coordinate plot and boxplots help us to investigate where the differences come from:

```
> matplot(1:ncol(performance), t(performance), xlab = "",
+     ylab = "Misclassification error", type = "l",
+     col = "#377EB8", lty = 1, axes = FALSE)
> axis(1, at = 1:ncol(performance), labels = colnames(performance))
> axis(2)
> box()

> boxplot(performance, col = "#4DAF4A")
```

The result is shown in Figure 17.1. Figure 17.1 gives impression that the three ensemble methods perform better than guessing. Random forests and bagging seem to perform equally well, the amount of the difference can be inspected by confidence intervals, for example via

```
> t.test(performance$RF, performance$Bagg, paired = TRUE,
+     conf.int = TRUE)$conf.int

[1] 0.000231 0.022961
attr(,"conf.level")
[1] 0.95
```

We emphasize again, see end of Section 17.5, that the Friedman test and the graphical illustrations indicate whether there are differences among the theoretical bootstrap distributions (with $B = \infty$), although we compute with $B = 100$ only.

17.6.2 Renal cell cancer

In this second application, we focus on the relationship between gene expression levels and response variables describing either clinical subtypes of

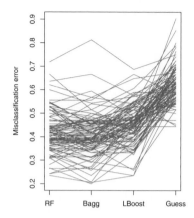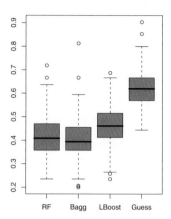

Figure 17.1. Parallel coordinates plot and boxplots of the misclassification errors of random forest (RF), bagging (Bagg), logitboost (LBoost) and guessing (Guess) for 100 bootstrap samples for the ALL data.

renal cell cancer or, most interesting in a clinical setting, the survival time of the patients. The package **kidpack** offers data of a study by Sültmann et al. (2005); for more background information we refer to the Appendix.

The analysis is very similar to the steps described for the ALL data in Section 17.6.1.

First, we load the package **kidpack** which offers the data in form of an *exprSet*.

```
> set.seed(290875)
> data(eset)
> pData(phenoData(eset))$type <- as.factor(eset$type)
```

The class distribution of clear cell renal cell cancer ccRCC, papillary renal cell cancer pRCC and chromophobe renal cell cancer chRCC is

```
> table(pData(phenoData(eset))$type)
```

```
ccRCC chRCC  pRCC
   52     9    13
```

Again, a standardized linear statistic for each gene is applied to perform variable selection:

```
> response <- eset$type
> expressions <- t(apply(exprs(eset), 1, rank))
> I <- ncol(exprs(eset))
> Iindx <- 1:I
```

```
> selected <- var_selection(Iindx, expressions,
+      response)
```

and a random forest model for the Ith patient can be fitted to 100 selected genes using

```
> rf <- randomForestB(eset[selected, ], "type",
+      Iindx[-1], sampsize = I - 1)
> rf

MLOutput instance, method= randomForest
Call:
 randomForestB(exprObj = eset[selected, ], classifLab = "type",
    trainInd = Iindx[-1], sampsize = I - 1)
predicted class distribution:
ccRCC
    1
```

The misclassification error for the four models (random forest, bagging, log-itboost, and guessing) applied to bootstrap samples of the data is computed along the following lines:

```
> B <- 100
> performance <- as.data.frame(matrix(0, nrow = B,
+      ncol = 4))
> colnames(performance) <- c("RF", "Bagg", "LBoost",
+      "Guess")
> for (b in 1:B) {
+      bsample <- sample(Iindx, I, replace = TRUE)
+      selected <- var_selection(bsample, expressions,
+          response)
+      rf3 <- randomForestB(eset[selected, ], "type",
+          bsample, mtry = 3, sampsize = I)
+      predicted3 <- factor(rf3@predLabels, levels = levels(response))
+      performance[b, 1] <- mean(response[-bsample] !=
+          predicted3)
+      rfBagg <- randomForestB(eset[selected, ],
+          "type", bsample, mtry = length(selected),
+          sampsize = I)
+      predictedBagg <- factor(rfBagg@predLabels,
+          levels = levels(response))
+      performance[b, 2] <- mean(response[-bsample] !=
+          predictedBagg)
+      lb <- logitboostB(eset[selected, ], "type",
+          bsample, 100)
+      predictedlb <- factor(lb@predLabels, levels = levels(response))
+      performance[b, 3] <- mean(response[-bsample] !=
+          predictedlb)
+      performance[b, 4] <- mean(response[-bsample] !=
+          levels(response)[which.max(tabulate(response[bsample]))])
+ }
```

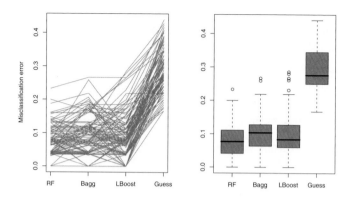

Figure 17.2. Parallel coordinates plot and boxplots of the misclassification errors of random forest (RF), bagging (Bagg), logitboost (LBoost), and guessing (Guess) for 100 bootstrap samples for the renal cell cancer data.

Again, the global null-hypothesis of equality of the performance of the candidate models can be rejected

```
> friedman.test(as.matrix(performance))

        Friedman rank sum test

data:  as.matrix(performance)
Friedman chi-squared = 202, df = 3, p-value <
2.2e-16
```

which, in the light of Figure 17.2, can be explained by the superior performance of the ensemble methods compared to the model where we guess the prediction from the prior distribution of the classes itself. Random forests seem to perform a little bit better compared to bagging and logitboost, as the (non-simultaneous) confidence intervals for the difference indicate:

```
> t.test(performance$RF, performance$Bagg, paired = TRUE,
+     conf.int = TRUE)$conf.int

[1] -0.02847 -0.00833
attr(,"conf.level")
[1] 0.95

> t.test(performance$RF, performance$LBoost, paired = TRUE,
+     conf.int = TRUE)$conf.int

[1] -0.02741 -0.00245
attr(,"conf.level")
[1] 0.95
```

17.7   Applications: Survival analysis

Mainly in clinical studies, the disease-free survival time or overall survival time is of major interest. For each observation, we are provided with the time for which each patient was at risk and whether an event, recurrence or death, occurred. Observations may have been stopped due to other reasons (a lethal accident) or end of follow-up. More formally, the response variable is now bivariate with $y_i \in R^+ \times \{0, 1\}$. We are interested in an estimate of the expected survival time for a patient with expression levels x, i.e., an estimate of the conditional probability that a patient will survive time t, given the patients gene expression levels.

   Basically, the ingredients of the bagging algorithm for the two-class problem can be applied to those problems as well. However, it is a challenge to aggregate the predictions of the multiple trees in order to come up with a ensemble prediction. A simple average of the predicted survival time did not prove to be of much use (Dannegger, 2000). As an alternative, Hothorn et al. (2004a) suggested to aggregate observations instead of predictions directly and compute one single prediction based on aggregated observations. For each of the M survival trees fitted to bootstrap samples $\mathcal{L}_m, m = 1, \ldots, M$, we extract the observations from the bootstrap sample that are elements of the same terminal node as the predictor value x of interest. Those observations, of course containing many tied values, are collected over all M bootstrap rounds, and then one single Kaplan-Meier curve is computed which serves as the ensemble's prediction.

```
> remove <- is.na(eset$survival.time)
> seset <- eset[, !remove]
> response <- Surv(seset$survival.time, seset$died)
> response[response[, 1] == 0] <- 1
> expressions <- t(apply(exprs(seset), 1, rank))
> exprDF <- as.data.frame(t(expressions))
> I <- nrow(exprDF)
> Iindx <- 1:I
```

The survival time of patients with renal cell cancer is now treated as the response variable of interest. First, we remove all observations where the survival time is missing (14 observations) from the data set and define the survival time of one patient, who died immediately, as one.

   The response variable is now an object of class *Surv*. A slightly different test statistic needs to be used for the variable selection. Here, we use logrank scores as implemented in the cscores method from package exactRankTests and the expression values for each gene and measure the association between the rank of the expression level of each gene and the logrank scores by a standardized statistic. Because of the small number of patients with information on the survival time being available, we only select 25 genes at a time.

```
> selected <- var_selection(Iindx, expressions,
+      response)
```

The ensemble of survival trees is fitted to the data using the `bagging` method of package ipred (Peters et al., 2002). The predictions for each observation in the learning sample are computed based on trees whose corresponding bootstrap samples did not contain the specific observation (out-of-bootstrap prediction) and are therefore not affected by overfitting problems. In addition, we compute a simple Kaplan-Meier curve for the survival times via `survfit`. The *model fit* of both procedures may be assessed by means of the Brier score for censored data (Graf et al., 1999) implemented in the function `sbrier`.

```
> bagg <- bagging(response ~ ., data = exprDF[,
+      selected], ntrees = 100)
> prKM <- predict(bagg)
> sbrier(response, prKM)

integrated Brier score
              0.170
attr(,"time")
[1]   1 65

> sbrier(response, survfit(response))

integrated Brier score
              0.207
attr(,"time")
[1]   1 65
```

The Brier scores for the Kaplan-Meier estimate and bagging seem to indicate that the survival ensemble fits the data better than a model without knowledge of the gene expression data and hence that there is information about the survival time contained in the expression levels of the genes. The predicted survival curves for each patient and the simple Kaplan-Meier estimate of all survival times are depicted in Figure 17.3 where we see a reasonable differentiation between the patients. The clinical subtypes of the patients are color encoded. Except for one patient suffering chromophobe renal cell cancer with poor prognosis, the estimated survival curves for patients suffering papillary and chromophobe subtypes indicate a longer survival compared to patients with clear cell renal cell cancer. To assess this finding, we need to perform a benchmark comparison with a model that predicts the survival time without knowledge of the gene expression values, i.e., a simple Kaplan-Meier estimate.

```
> set.seed(290875)
> B <- 100
> performance <- as.data.frame(matrix(0, nrow = B,
+      ncol = 2))
> colnames(performance) <- c("Bagging", "Kaplan-Meier")
```

```
> plot(survfit(response), lwd = 4, conf.int = FALSE,
+     xlab = "Survival time in month", ylab = "Probability")
> col <- c("lightgray", "darkblue", "red3")
> type <- factor(seset$type)
> table(type)

type
ccRCC chRCC  pRCC
   45     3    12

> for (i in 1:length(prKM)) lines(prKM[[i]], lty = 2,
+     col = col[as.numeric(type)[i]])
```

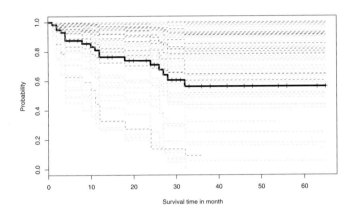

Figure 17.3. Out-of-bootstrap predicted survival curves for each patient (dashed lines) and overall Kaplan-Meier curve (thick solid line). Clinical subtypes of patients are color encoded: gray (ccRCC), blue (pRCC), and red (chRCC).

```
> for (b in 1:B) {
+     bsample <- sample(Iindx, I, replace = TRUE)
+     selected <- var_selection(bsample, expressions,
+         response)
+     bagg <- bagging(response ~ ., data = exprDF[,
+         selected], subset = bsample, ntrees = 100)
+     pr <- predict(bagg, newdata = exprDF[-bsample,
+         ])
+     KM <- survfit(response[bsample])
+     performance[b, 1] <- sbrier(response[-bsample],
+         pr)
+     performance[b, 2] <- sbrier(response[-bsample],
+         KM)
+ }
```

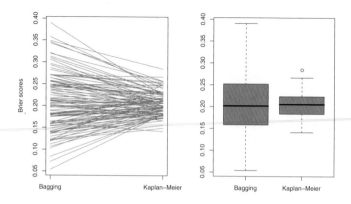

Figure 17.4. Parallel coordinates plot and boxplots of the Brier scores of 100 bootstrap samples for renal cell cancer survival.

The visualizations of the Brier scores in Figure 17.4 indicate differences with respect to the variability but not with respect to the mean Brier score, and hence we cannot conclude that the survival time of a patient can be adequately modeled based on information derived from gene expression profiling. Note that this benchmark experiment is based on a learning sample of 60 patients, 42 of which are censored. This implies that, on average, only 38 unique patients are included in one bootstrap learning sample, which explains the large variability.

17.8   Conclusion

Modeling the relationship between a clinically interesting response variable, such as tumor subtype of survival time, and gene expression levels of thousands of genes for only a small number of patients is a challenge to statistical methodology. The analyses in this chapter give an overview on ensemble methods for modeling high-dimensional gene expression data and illustrate both advantages and shortcomings. For samples involving on the order of 100 patients, ensemble methods seem appropriate for constructing models for the prediction of tumor subtypes of renal cell cancer or stages of B-cell leukemia. It should be noted though that ensemble methods as used here are not more than black-box prediction methods. Modeling censored time-to-event data in such a high-dimensional setting is even more difficult, especially when a substantial number of patients are censored. Finally, we would like to mention that much research is currently conducted in the field of ensemble methods in both statistics and machine learning, and one can expect further methodological developments and new software packages offering more functionality in the near future.

18

Browser-based Affymetrix Analysis and Annotation

C. A. Smith

Abstract

webbioc is a CGI-based interface to Bioconductor methods for preprocessing and analyzing Affymetrix data. It wraps up the functionality of a number of Bioconductor packages into a consistent environment that can be deployed for use by small groups or large departments. Without ever seeing a command prompt, it will take the user from raw data to annotated lists of the most significantly differentially expressed genes. It will optionally make use of a back-end computer cluster for batch processing. This chapter will discuss the appropriate circumstances under which webbioc should be deployed and the pros and cons of using it versus the typical command line environment of R. Installation and configuration will be fully covered. Use of the Web-based interface will be visually demonstrated. Finally, we will describe how to expand the interface by adding additional analysis modules.

18.1 Introduction

webbioc is a *Web interface* for several of the Bioconductor microarray analysis packages. It represents a significant departure from the command-line interface discussed thus far. It is designed to be installed at local sites as a shared bioinformatics resource. webbioc currently only provides Affymetrix analysis, although collaborations on a cDNA module are of some interest. The existing modules provide a workflow that takes users from CEL files to differential expression with multiple hypothesis testing control and finally to meta-data annotation of the gene lists.

A number of other projects add a Web interface to R. The Shaw Laboratory at the Baylor College of Medicine has compiled a useful list of many such projects.[1] Current Web interface projects can be subdivided based on their answers for two architectural decisions: 1) Whether to allow users to enter their own R code. 2) Whether to maintain a user's R session across multiple form submissions. webbioc answers no to both. It does not allow user entry of R code and executes a new instance of R each time the user submits a form.

As a consequence of those decisions, webbioc is much less flexible than an interactive R session. It can only perform a preset list of tasks on user data. New tasks must be explicitly programmed. However, in some cases, "less is more." A user does not need to remember any command syntax. Depending on the number of commands that must be executed to do an analysis, the Web interface may actually be faster and less error-prone. A number of other advantages are discussed in the next section.

18.1.1   Key user interface features

webbioc expands on the infrastructure and algorithmic strengths of Bioconductor in the R environment and adds a number of usability enhancements to the experience of using Bioconductor:

- *Short learning curve.* Using the Web interface, the user does not need to know how to use either a command line interface or the R language. Depending on the programming savvy of the user, R tends to take a long time to fully master. On the other hand, webbioc abstracts away much of the complexity and presents an interface that is very quickly understood by most any biologist.

- *Ease of installation.* After an initial installation by a system administrator, there is no need to install additional software on user computers. Installing and maintaining an R installation with all the Bioconductor packages can be a daunting task, often best suited to a system administrator. Using webbioc, only one such installation needs to be maintained.

- *Discoverability.* Graphical user interfaces are significantly more discoverable than command line interfaces. That is, a user browsing around a software package is much more likely to discover and use new features if they are graphically presented. Additionally, a unified user interface for the different Bioconductor packages can help show how they can be used together in a data-processing pipeline. Ideally, a user should be able to start using the Web interface without reading external documentation.

---

[1]http://franklin.imgen.bcm.tmc.edu/R.web.servers/.

- *Documentation.* Embedding context-sensitive on-line help into the interface helps first-time users make good decisions about which statistical approaches to take. Because of its power, Bioconductor includes a myriad of options for analysis. Helping the novice statistician wade through that pool of choices is an important aspect of webbioc.

18.2   Deploying webbioc

18.2.1   System requirements

webbioc requires a number of packages external to R. Here are a number of the most important requirements. A full listing can be found in the webbioc vignette.

- **Unix/Linux/Mac OS X:** webbioc was written in a Unix environment and depends on many Unix conventions including path names, directory structure, and interprocess communication. It will not run under Windows.

- **Perl 5.8:** The Web interface and much of the core logic uses Perl. webbioc uses several Perl modules where are only available in Perl 5.8 and later.

- **Netpbm:** The Netpbm series of programs handle graphics manipulation.

- **SGE or PBS (optional):** webbioc has been developed and tested with two batch queueing systems, the Sun Grid Engine and the Portable Batch System. When using a batch queueing system, webbioc requires that there be a shared filesystem mounted in the same place on the Web server as on the execution nodes. Please note that these are optional as webbioc can also run jobs by forking to the background. Adding support for other batch queueing systems is a straightforward matter.

18.2.2   Installation

To install webbioc, first verify that the webbioc package is installed. Next create a directory within the Web server's *CGI* directory that will hold all the Perl scripts. One might typically call that directory bioconductor. Copy all the files from R_LIBS/webbioc/cgi/ into that directory.

Next verify that the file permissions are correct on the copied files. Go to the directory to which you copied the files. Use the command: chmod 755 *.cgi to make the CGI scripts executable and the command: chmod 644 *.pm to make the support modules readable.

You must now decide how you want to link the Bioconductor Web interface into your existing Web site. It was designed to integrate seamlessly into an existing site design. webbioc includes a very rudimentary home page in R_LIBS/webbioc/www/. You may either use this as a starting point or create your own. Please note that you may have to change the links slightly depending on where you place the CGI scripts.

The last step of the installation is to create two directories where the Web interface can store files. Remember that if you are using a batch queueing system, these two directories must be shared between the Web server and all execution nodes via NFS (or some other file sharing mechanism). If other users have access to any of the machines running the Web interface, we recommend setting the permissions of these directories so that only the Web server user can read them.

The first directory will be for storing uploaded files. The largest type of file uploaded will probably be CEL files. They are typically around 10 MB each. Therefore, depending on expected server usage, the directory should be kept on a partition with hundreds of megabytes to gigabytes of free space. It can be stored anywhere and does not necessarily have to be Web-accessible.

The second directory will be used for storing the results of jobs that clients submit. Job results will typically be anywhere from 5 MB to 5 KB. The free space necessary for this directory is again subject to usage. This directory must be accessible via the Web server to allow results to be delivered asynchronously. Both directories should be regularly purged of old files.

To facilitate installation and updating of meta-data packages, webbioc includes a function to download and install every meta-data package from the Bioconductor Web site. The following uses reposTools to install all meta-data packages and update any out-of-date meta-data packages. (Change the path name to match your system.)

```
library("webbioc")
installReps("/library/install/path")
```

Make sure to run R with a user who has permission to write to the library directory. Depending on your site, you may wish to set up a cron job to execute this code approximately once per month to check for updates or additions to the meta-data packages. If packages are already up-to-date, it will not waste bandwidth nor CPU by re-installing them.

18.2.3 Configuration

Beyond putting the CGI scripts and HTML page in the right places and setting up directories to receive files, all configuration is done through the Site.pm file. Configuration options for specifying file locations, batch queueing system, and HTML header and footer information are located in that file. Full details can be found in the **webbioc** vignette.

18.3   Using webbioc

Although **webbioc** is designed to be modular and allow the user to use different combinations of analyses, the current selection of modules lend themselves to a linear workflow. The basic sequence involves: 1) Preprocessing raw .CEL files into a set of normalized gene expression measures. 2) Finding differentially expressed genes using basic statistical tests controlled by multiple testing procedures. 3) Annotating the gene list with meta-data linking to various on-line resources. That workflow will be demonstrated here.

18.3.1   Data Preprocessing

The first step in preprocessing Affymetrix data is uploading CEL files containing the raw intensity values from the PM/MM oligonucleotide probe pairs. To begin, the user first creates a new Upload Manager session, which is uniquely and privately identified by a token consisting of roughly 25 alphanumeric characters. The user should record the session token for future use and to avoid uploading large raw data files more than once. The session token can optionally be stored in a cookie for convenience.

For this demonstration, we will be using six CEL files from a spike-in study that used the HG-U95A chip. The study is described in Section 2.5. The six chips represent three replicates each of identical samples with 14 gene groups spiked-in at a simulated fold change of 2. The gene groups were spiked-in at varying concentrations ranging from 0 to 1024 pM. The Upload Manager with the six CEL files can be seen in Figure 18.1.

After the user selects the number of CEL files for preprocessing, **webbioc** provides a succinct but powerful interface for reordering and renaming samples, producing more coherent reports. The user has the choice of using a high-performance implementation of the RMA expression or creating a custom expression measure using different routines for background correction, normalization, PM correction, and multi-probe summarization. Log base 2 transformation of the data can be easily disabled if so desired. The main options screen can be seen in Figure 18.2.

Results of preprocessing are returned in two formats. The first is a tab-delimited text file that can be opened in Excel or any other microarray analysis software. Second, **webbioc** creates an *exprSet* that can be further utilized by the Web interface for subsequent analysis. The *exprSet* is fully compatible with R on any platform, if the user alternatively wishes to

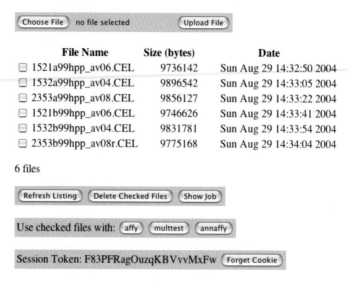

Figure 18.1. Spike-In CEL files stored with the Upload Manager.

perform further local analysis. The results output screen can be seen in Figure 18.3.

Thus webbioc may be used to provide biologists easy access to some of the advanced Affymetrix preprocessing routines, such as GCRMA, available only through Bioconductor.

18.3.2   Differential expression multiple testing

A common task is to query for significantly differentially expressed genes. This is accomplished with the multtest module of webbioc. For the most part, it leverages functionality found in that package while adding some additional statistical and graphing tools. An example of the main configuration page for the multtest module can be seen in Figure 18.4.

Before starting to use the multtest module, you must first indicate which *exprSet* to analyze and how many sample classes there are. In almost all cases, that will be two, one for a baseline and another for an experimentally varied group. You may want to use more than two groups if you were interested in analysis using multiclass tests such as the F-test. After deciding how many classes there are, you must assign each sample to a class.

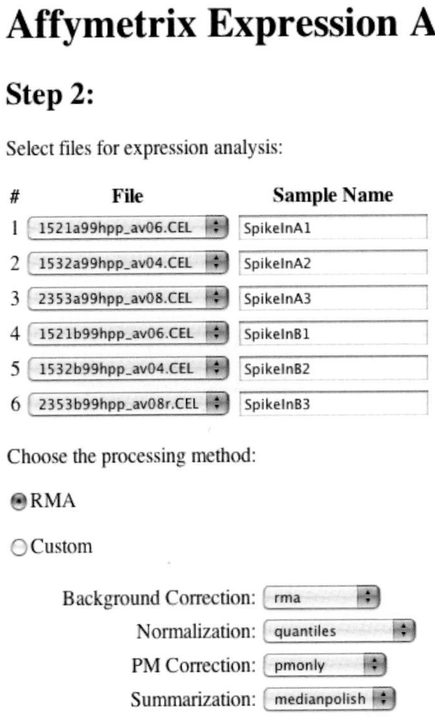

Affymetrix Expression Analysis: affy

Step 2:

Select files for expression analysis:

#	File	Sample Name
1	1521a99hpp_av06.CEL	SpikeInA1
2	1532a99hpp_av04.CEL	SpikeInA2
3	2353a99hpp_av08.CEL	SpikeInA3
4	1521b99hpp_av06.CEL	SpikeInB1
5	1532b99hpp_av04.CEL	SpikeInB2
6	2353b99hpp_av08r.CEL	SpikeInB3

Choose the processing method:

⦿ RMA

◯ Custom

Background Correction: rma
Normalization: quantiles
PM Correction: pmonly
Summarization: medianpolish

Figure 18.2. RMA preprocessing using custom sample labels.

You may also ignore individual chips, completely excluding them from the analysis.

After assigning classes to each sample, you must chose a statistical test and a suitable multiple testing procedure. There are a broad array of choices here and describing them all would be beyond the scope of this chapter. Either of the t-tests is often a good choice for a differential expression test. However, one must recognize the weaknesses of using such procedures with small sample sizes. For more in-depth information about multiple testing procedures, see Chapter 15.

A final important choice to make is how to compute the raw p-values which are then adjusted by the multiple testing procedure. Choosing parametric will use the classical methods based on continuous distributions to determine p-values. On the other hand, you my also choose to use permutation methods developed specifically for the multtest package. In most cases the permutation methods will take significantly longer to compute. The affects of both can easily be investigated using the Web interface.

In addition to producing a list of significantly differentially expressed genes, the Web interface produces a number of diagnostic plots including an M vs. A plot, Normal quantile-quantile plot, and a multiple testing

Output Files:

affy-IApLG7X2.txt
affy-IApLG7X2.exprSet

Output Archive:

affy-IApLG7X2.tar.gz

Job Summary:

Files	1521a99hpp_av06.CEL, 1532a99hpp_av04.CEL, 2353a99hpp_av08.CEL, 1521b99hpp_av06.CEL, 1532b99hpp_av04.CEL, 2353b99hpp_av08r.CEL
Sample Names	SpikeInA1, SpikeInA2, SpikeInA3, SpikeInB1, SpikeInB2, SpikeInB3
Processing	RMA
Copy back	Yes
E-Mail	

Figure 18.3. Plain text and R object output from preprocessing.

procedure selectivity plot. The Web interface distinguishes selected genes in the corresponding plots by highlighting them in red. In this case, the 100 genes with the smallest p-values are highlighted. See Figure 18.5. For more information about visualizing data see Chapter 10.

The Web interface also allows the user to select a specific subset of genes for focused analysis. In the case of the Affymetrix *spike-in* experiment, we may wish to investigate how the theoretical fold changes correlate with the observed fold changes. Figure 18.6 shows the text box where a list of probeset IDs can be entered. Also note the expression values can be included in the output HTML table.

The resulting HTML page can be seen in Figure 18.7. With the exception of probeset IDs 1708_at and 37777_at, each of the fold changes should be 2. However one observes that the fold change values are lower than their spiked-in concentrations. It is a known artifact of the RMA preprocessing method.

18.3.3   Linked annotation meta-data

Completing the analysis workflow, the Web interface includes a module that exposes the functionality of annaffy. The user can produce annotation for either their own probeset list or use a list previously identified with the multtest module. The interface allows the user to select exactly which

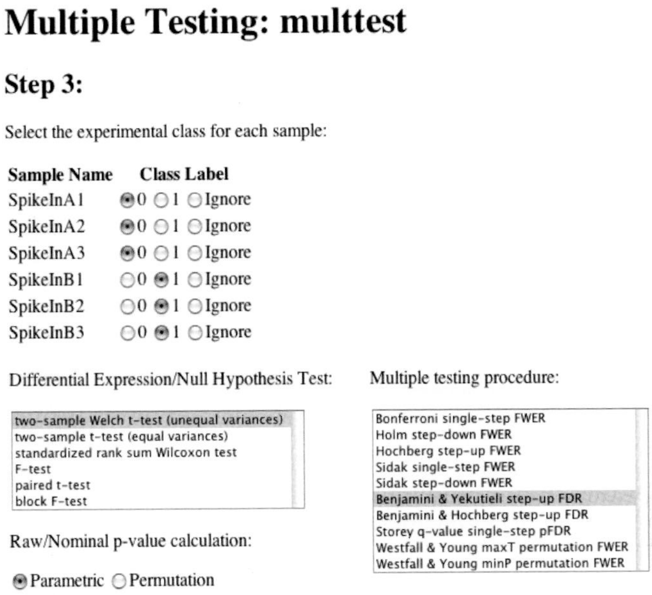

Multiple Testing: multtest

Step 3:

Select the experimental class for each sample:

Sample Name	Class Label
SpikeInA1	⦿ 0 ○ 1 ○ Ignore
SpikeInA2	⦿ 0 ○ 1 ○ Ignore
SpikeInA3	⦿ 0 ○ 1 ○ Ignore
SpikeInB1	○ 0 ⦿ 1 ○ Ignore
SpikeInB2	○ 0 ⦿ 1 ○ Ignore
SpikeInB3	○ 0 ⦿ 1 ○ Ignore

Differential Expression/Null Hypothesis Test:

```
two-sample Welch t-test (unequal variances)
two-sample t-test (equal variances)
standardized rank sum Wilcoxon test
F-test
paired t-test
block F-test
```

Multiple testing procedure:

```
Bonferroni single-step FWER
Holm step-down FWER
Hochberg step-up FWER
Sidak single-step FWER
Sidak step-down FWER
Benjamini & Yekutieli step-up FDR
Benjamini & Hochberg step-up FDR
Storey q-value single-step pFDR
Westfall & Young maxT permutation FWER
Westfall & Young minP permutation FWER
```

Raw/Nominal p-value calculation:

⦿ Parametric ○ Permutation

Figure 18.4. Finding significant differences in expression using Welch's two-sample t-test and Benjamini and Yekutieli False Discovery Rate control.

columns from the saved *aafTable* and annotation meta-data should be included in the final report. The interface for appending annotation can be seen in Figure 18.8. The resulting annotated list of the spiked in genes can be seen in Figure 18.9.

18.3.4   Retrieving results

Each webbioc module produces a single result report per analysis. Each set of results is associated with a unique job ID which that with the name of the module and is followed by pseudo-random alphanumeric characters. For convenience, users may choose to receive e-mail notification when jobs complete. In addition, the Web interface keeps a log of every job associated with a given Upload Manager session token. The log links directly to individual result pages. The log for the analyses shown here can be seen in Figure 18.10.

Output Files:

mt-dXIjW2fh.html
mt-dXIjW2fh.txt
mt-dXIjW2fh.aafTable

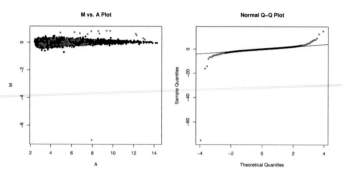

Figure 18.5. Log fold change (M) vs. mean expression level (A) and normal quantile-quantile plot. 100 most significantly differentially expressed genes highlighted in red. Multiple testing selectivity plot not shown.

☑ Limit results to [a total number of ▲▼] [100]

Test only these gene names: (optional)

37777_at, 684_at, 1597_at, 38734_at, 39058_at, 36311_at, 36889_at, 1024_at, 36202_at, 36085_at,
40322_at, 407_at, 1091_at, 1708_at, 33818_at, 546_at

☑ Include expression values in results

☑ Copy aafTable back to the upload manager for further annotation

Web Page Title:

[Multiple Testing Results]

E-mail address where you would like your job status sent: (optional)

[   ]

(Submit Job)

Figure 18.6. Testing only the spiked-in genes.

18.4   Extending webbioc

18.4.1   Architectural overview

webbioc is written in a combination of Perl, R, and shell scripts. For most processing, the Perl-driven Web interface dynamically creates an R script, which is run in batch mode. A shell script controls the execution of R and catches any errors that result.

The Web interface can be configured to execute the shell script in one of two ways. In the single-machine configuration, Perl forks an auxiliary

Multiple Testing Results

Gene Name	Fold Change	t statistic	raw p-value	BY-value	SpikeInA1	SpikeInA2	SpikeInA3	SpikeInB1	SpikeInB2	SpikeInB3
1708_at	0.00748285	-75.1783	2.83979e-07	1.53609e-05	11.6201	11.4091	11.3983	4.35112	4.35828	4.53484
36311_at	1.78041	8.01721	0.00133173	0.0351877	8.01939	8.16066	8.26515	8.82946	9.0498	9.06143
36085_at	1.56409	6.67941	0.00263475	0.0351877	11.7377	11.9323	11.942	12.4019	12.5018	12.6431
33818_at	1.44709	12.542	0.00272448	0.0351877	12.0001	12.046	12.0126	12.4734	12.5755	12.6063
36202_at	1.80831	14.9127	0.0032526	0.0351877	8.96827	8.93134	8.94066	9.69834	9.89211	9.80748
39058_at	1.522	6.29361	0.00647858	0.0518251	7.35147	7.07007	7.26847	7.76902	7.81483	7.93334
38734_at	1.68358	6.452	0.00670669	0.0518251	5.90158	5.96839	5.77727	6.43198	6.68741	6.768
1024_at	1.79783	5.08026	0.00773145	0.0522759	9.12801	9.21852	9.48133	9.94088	10.0488	10.3661
40322_at	1.24703	3.69562	0.0219794	0.116058	12.7414	12.8605	12.6707	13.0738	13.1897	12.962
37777_at	1.26935	5.12702	0.023199	0.116058	4.61404	4.61678	4.55163	4.91442	4.83865	5.05424
36889_at	1.59339	5.15807	0.0236013	0.116058	6.36458	6.45193	6.49989	6.87052	7.14858	7.28628
546_at	1.76948	4.63587	0.0303107	0.123966	7.51343	7.55785	7.38205	8.1228	8.62115	8.12614
1091_at	1.1791	4.04871	0.030583	0.123966	12.9182	13.002	12.9705	13.1154	13.2974	13.1864
684_at	1.19348	4.06014	0.0320849	0.123966	6.69316	6.56973	6.49395	6.88041	6.85299	6.79467
407_at	1.3629	2.46449	0.120473	0.434438	3.40453	3.44308	2.8699	3.66442	3.66393	3.79324
1597_at	1.14191	1.92201	0.149202	0.504412	7.03613	6.99243	6.74736	7.02577	7.1638	7.17249

16 Genes

Figure 18.7. HTML table with statistics and log expression measures. Lower than expected fold changes are observed.

thread that then runs the script. However, webbioc can also be configured to use the Portable Batch System (*PBS*) or the Sun Grid Engine (*SGE*) to send the job to another computer for processing.

If used in a cluster configuration, webbioc depends on having a shared partition between the Web server and all the compute nodes. This is typically accomplished with NFS. The shared partition allows the compute nodes to push results directly out to the user without directly interacting with the Web server.

Microarray analysis is very data intensive and tends to produce large files. Thus special care must be taken to efficiently move data back and forth between the Web-client and the server. webbioc uses two different systems for exchanging data with the client, one for input and the other for output.

The Upload Manager handles all files to be processed by Bioconductor. When users start a session with the upload manager, they are granted a unique token that identifies the session. Using that short random string of letters and numbers, they may access their uploaded files from any of the Bioconductor tools. In that way, users must only upload files a single time. Upload manager sessions are meant to be temporary, with any files being automatically deleted after a given amount of time.

For output of results, webbioc creates static HTML pages on the fly that contain relevant images or files. Like the upload manager, job results are uniquely identified and are only temporarily stored. Because saving old results can potentially fill up the disk, those should be automatically purged on a regular basis.

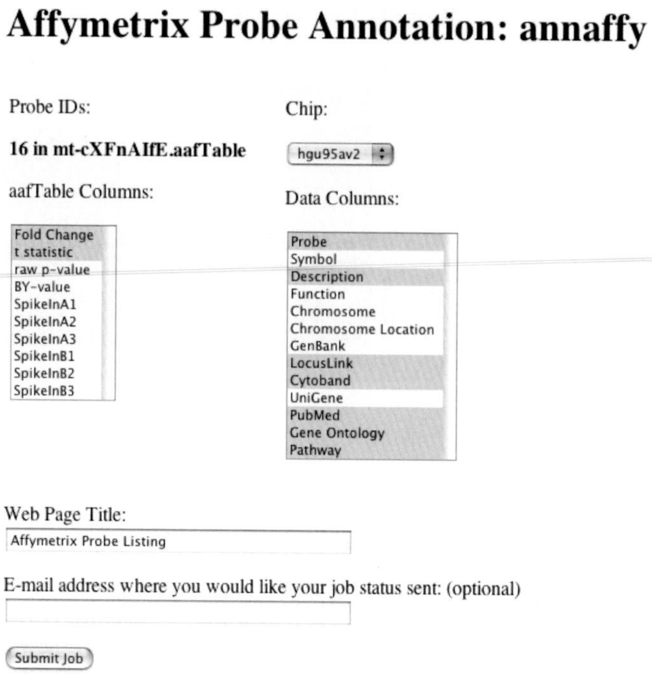

Affymetrix Probe Annotation: annaffy

Probe IDs: Chip:

16 in mt-cXFnAlfE.aafTable   [hgu95av2 ⬍]

aafTable Columns: Data Columns:

```
Fold Change                      Probe
t statistic                      Symbol
raw p-value                      Description
BY-value                         Function
SpikeInA1                        Chromosome
SpikeInA2                        Chromosome Location
SpikeInA3                        GenBank
SpikeInB1                        LocusLink
SpikeInB2                        Cytoband
SpikeInB3                        UniGene
                                 PubMed
                                 Gene Ontology
                                 Pathway
```

Web Page Title:
[Affymetrix Probe Listing   ]

E-mail address where you would like your job status sent: (optional)
[   ]

(Submit Job)

Figure 18.8. Annotating statistical data with columns of linked meta-data.

The results of one job may need to be the input of another. For instance, expression summary data created by preprocessing using affy needs to be supplied to multtest for detection of differentially expressed genes. To facilitate that data exchange, the Web interface will optionally copy results back to the upload manager for processing by other packages.

The data interchange format used by webbioc is the R binary file format. By convention, each file contains only one object. Additionally, instead of using the common .Rda or .Rdata extensions, webbioc uses the class of the stored object as the file extension. This provides a useful abstraction that many computer users have come to understand and expect. The extensions are merely for the benefit of the user as the Web interface ignores them.

18.4.2 Creating a new module

In general, each of the modules in webbioc attempt to encapsulate the functionality of a single Bioconductor package. That design decision can be seen in the affy, multtest, and annaffy modules, that expose much of the functionality of their respective packages. The affy module leverages the modular nature of the affy package and includes preprocessing components from gcrma and vsn.

Affymetrix Probe Listing

Probe	Description	LocusLink	Cytoband	PubMed	Gene Ontology	Pathway	Fold Change	t statistic
1708_at	mitogen-activated protein kinase 10	5602	4q22.1-q23	8	JUN kinase activity MAP kinase kinase activity MAP kinase activity ATP binding protein serine/threonine kinase activity JNK cascade protein amino acid phosphorylation signal transduction nucleus transferase activity	MAPK signaling pathway	0.00748285	-75.1783
36311_at	phosphodiesterase 1A, calmodulin-dependent	5136	2q32.1	7	calmodulin-dependent cyclic-nucleotide phosphodiesterase activity calmodulin binding signal transduction hydrolase activity	Purine metabolism	1.78041	8.01721

Figure 18.9. Final HTML table with linked annotation, fold change, and *t*-statistics.

Job Log

Job Name	Title	Date
aaf-afazatgn	Affymetrix Probe Listing	Mon Aug 30 17:25:48 2004
mt-cXFnAIfE	Multiple Testing Results	Mon Aug 30 17:15:59 2004
mt-dXIjW2fh	Multiple Testing Results	Mon Aug 30 17:14:48 2004
affy-IApLG7X2	Affymetrix Expression Analysis	Mon Aug 30 17:09:12 2004

Figure 18.10. Log linking to results of every job associated with a given session token.

An important part of the design of webbioc is that modules take standardized input formats in the form of R objects. For instance, multtest takes an *exprSet* object as input, the Bioconductor standard for storing microarray data sets. annaffy takes an *aafTable* object as input, that includes all the information it will need to add annotation. Thus when you create your own packages, it is important to determine what the inputs and outputs will be. If you are creating a cDNA preprocessing module, you would likely want to create an *exprSet* as your output. If you were implementing some other method for finding differentially expressed genes, you might want to produce an *aafTable*. Remember that all data is exchanged through files stored in the Upload Manager, with a single R object per file.

Once you have determined the functionality and input/output of your module, you need to plan the steps of the user interface. If your module takes a varying number of input parameters, you may first need a screen where the user selects that number and then produce the appropriate HTML form based on the user input. (An alternative could be to use JavaScript to dynamically create an appropriately sized form without having the initial page.) Examples of that can be seen by looking at the source of the `affy.cgi` and `multtest.cgi` scripts.

The main structure of the currently implemented modules is as follows: 1) Initialize instances of the `CGI` and `FileManager` objects to be used during script instances, linking the `FileManager` object to the current session. 2) Next decide which processing step the user is working on and call the corresponding step function. 3) Each step is handled by a function that handles all the processing and HTML display. The initial form is also handled by a function, making a separate HTML page unnecessary. 4) A `generate_r` function creates the text R code that is executed by the job control system. That text is passed to the `create_files` function that generates the necessary scripts for batch execution. 5) Finally, after any module-specific functions, there is an `error` function that displays problems with processing.

18.5   Conclusion

The Web interface in the webbioc package offers a complete solution for doing statistical analysis of Affymetrixbased microarray experiments. Through the affy package, it gives the user the ability to explore many combinations of background correction, normalization, and summarization in probe-level preprocessing. Second, it leverages the multtest package for multiple comparisons testing and global control of the error rate. Several visualizations are also provided. Finally, the results of both analysis stages can be fully annotated using annaffy. Though limited to a largely linear workflow, enough flexibility is programmed in to provide many possibilities for interactive analysis.

Beyond its current abilities, webbioc provides a template and infrastructure for building interfaces for other types of microarray analysis including clustering, classification, and other novel methods for assessing differential expression. In order to handle the large file sizes inherent in microarray data, it provides an easy mechanism to transfer and store raw data on the server. Additionally, it allows users to easily backtrack and determine exactly which parameters led to a given result. Finally, it is built from the ground up to make use of a back-end computing cluster for batch data processing and will thus scale to a large number of users.

Part IV

Graphs and networks

19

Introduction and Motivating Examples

R. Gentleman, W. Huber, and V. J. Carey

19.1   Introduction

Graphs are fundamental structures of discrete mathematics and have found applications in many scientific disciplines that consider networks of interacting elements (Strogatz, 2001). A graph consists of a set of nodes and a set of edges that connect nodes. The nodes are entities of interest and the edges represent relationships between the entities. For example, the entities may be a set of proteins in a cell, and the relationship modeled may be the existence of a physical interaction between two proteins. We will use the notation $G = (V, E)$ to specify a graph G, with V denoting the node set, and E denoting the edge set. Elements of E relate pairs of elements of V. Edges can be assigned weights, directions, and types. Some applications make use of specialized forms of graphs such as *multigraphs*, *bipartite graphs*, and *hypergraphs* (Gallo et al., 1993; Berge, 1973). These are defined in Section 20.2.1 below.

It is now common to encounter publications in the biosciences that make explicit use of graphs and graph theory. In this part of the book, we discuss a body of software that is available for working with graphs to build mathematical and statistical models for experimental data. This chapter provides a general discussion of practicalities of using graphs for data analysis, presenting three motivating examples with some illustrations of software functionality. We address the formal concepts of graph theory and algorithms on graphs in Chapter 20 and provide a detailed description of the relevant software resources available in Bioconductor in Chapter 21. Finally, in Chapter 22 we present some case studies.

19.2   Practicalities

Using graphs as models for data analysis and data representation poses a number of challenges. In many cases, the data that are available to construct the graphs of interest are imperfect. The process of constructing a graphical model on the basis of experimental data must address three potential complications: *false positives*, relationships that appear in the experimental data, but are not real; *false negatives*, relationships that are real and were probed experimentally, but were not detected; and *untested relationships*, where no information is available. In order to make appropriate use of the data, we will need to keep these issues in mind as we explore the resultant graphs. Uncertainty is usually not part of a purely mathematical approach to graph theory, but it cannot be ignored in the context of experimental data. Uncertainty affects how we use and think about graphs or networks. Uncertainty of relationships being modeled also impacts the design of software, the choice of algorithms, and the interpretation of the output.

We caution the reader against over-interpretation of graphical models. Differences between graphs may be entirely due to artifacts, such as the bias of biomedical research toward matters related to human disease. For example, if one type of gene or gene product is well studied and there are many experimental tools that can be used to study it, then graphs based on it are likely to contain many edges. On the other hand less is known about genes that are hard to study, or that are not directly implicated in diseases, and the resultant graphs tend to be sparse. Hence an observed difference may be merely a reflection of how extensively the genes were studied.

19.2.1   Representation

An abstract graph can be represented for computational purposes in many different ways. Bioconductor supports representations based on node and edge lists, adjacency matrices, and from-to matrices. Software has been developed to translate between representations, a process sometimes referred to as "coercion." The representation used for a graph can have a profound effect on the running time of algorithms that are applied to it. It is advisable to make timing comparisons on different representations before committing to a particular one. The most appropriate or efficient strategy for representing the graph will depend on many factors such as the size of the graph and the types of operations that are going to be applied to it. More details are provided in Section 21.2.

19.2.2   Algorithms

We emphasize the reuse of existing, tested implementations of graph algorithms. Bioconductor provides interfaces to many of the algorithms coded in the open source Boost graph library (Siek et al., 2002). A prototype interface to LEDA (Melhorn and Näher, 1999) is also available.

Good implementations for many algorithms required in bioinformatics are still needed. Algorithms adapted to deal with incompleteness and uncertainty are of particular interest. For example, Scholtens and Gentleman (2004) developed a special form of *clique* that is appropriate for protein complex data where different forms of uncertainty are prevalent. For hypergraphs, Krishnamurthy et al. (2003) describe an extension of DFS, and Klamt and Gilles (2004) developed an analog of the mincut algorithm for biochemical reaction networks.

19.2.3 Data Analysis

Graphs play roles in three complementary areas related to data analysis. First, graphs provide a data structure for *knowledge representation.* Examples include metabolic and signal transduction networks that are stored in graph form. This provides the user with a computational object that can easily and naturally be used and that reflects, in software, the physical objects and relationships of interest. Graphs are used for knowledge representation in the Gene Ontology (GO), and bipartite graphs between genes and scientific papers that cite them are another form of knowledge representation.

A second application of graphs to statistical methodology is their use in *exploratory data analysis (EDA)*. A knowledge-representation graph can be juxtaposed with observed data to guide the discovery of interesting phenomena in the observations. Examples include the mapping of gene expression data onto static knowledge-based graphs (Storch et al., 2002; Zhou et al., 2002; Doniger et al., 2003).

A further role for graphs is in *statistical inference*. For example, one might want to make inferential statements such as that two genes are related due to significantly frequent co-citation, or that gene expression is related to protein complex co-membership (Ge et al., 2001). In this context, we have found that random graphs play an important role and we consider a variety of random graph models, such as the Erdös-Rényi model, as well as simulation models that randomly permute the labels on the nodes.

19.3   Motivating examples

19.3.1 Biomolecular Pathways

The behavior of a biological system depends on the properties of its individual components as well as on their interaction networks. Graphs are a natural tool for the analysis and modeling of such networks. Figure 19.1 shows an illustration of the integrin-mediated cell-adhesion pathway, as

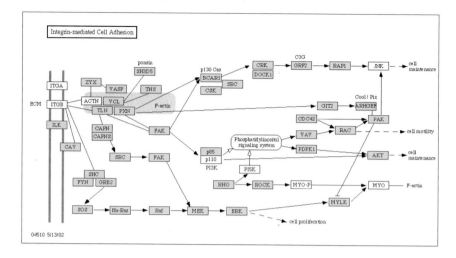

Figure 19.1. The integrin-mediated cell adhesion pathway as rendered by KEGG.

provided by KEGG (Kanehisa and Goto, 2000). Taken as a graph, V is the set of proteins, complexes, and processes comprising this description of the integrin-mediated cell adhesion pathway, and E is their set of interactions. In order to be able to perform queries and computations on this pathway, we need to have it in a machine-readable format. The package graph contains a manually curated version of this graph.

```
> library("graph")
> data("integrinMediatedCellAdhesion")
> class(IMCAGraph)
```

The function acc(g,n) is a method from the graph package that returns all nodes that are *accessible* in the graph g through a path from node n. This allows us to ask, for example, which nodes are downstream of the "son of sevenless" (SOS) protein.

```
> s <- acc(IMCAGraph, "SOS")
```

$SOS

Ha-Ras	Raf	MEK
1	2	3
ERK	MYLK	MYO
4	5	6
F-actin	cell proliferation	
7	5	

We can also ask which of the nodes has the highest number of out-going edges.

```
> deg <- degree(IMCAGraph)$outDegree
> deg[which.max(deg)]
```

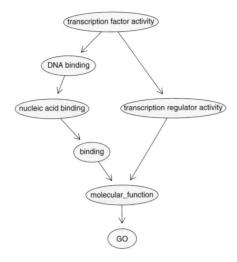

Figure 19.2. Graph of GO relationships for the term "transcription factor activity."

ITGB
 6

Although the graph object `IMCAGraph` was curated manually from the visual representation of the pathway in Figure 19.1, there are now public databases that represent biological pathways in a formal, ontology-based manner. Examples include KEGG (Kanehisa and Goto, 2000), *reactome* (Joshi-Tope et al., 2005), BioCarta (biocarta.com), and the National Cancer Institute cMAP (see Section 7.6.2).

19.3.2   Gene ontology: A graph of concept-terms

In the parlance of computer science, an *ontology* is an explicit formal specification of how to represent objects, concepts, and other entities that exist in some subject area of interest and the relationships that hold among them. It provides a structured vocabulary for that subject area. GO (Chapter 7) is an ontology whose subject area is the description of gene products. The relationships between terms are structured as a directed acyclic graph (DAG) in which each term may be a *child* of one or more *parents*. Child terms are stipulated to be more specific than their parents. Child terms may have more than one parent term and they may have different relationships with the different parents.

As an example, the term "transcription factor activity" is in the molecular function (MF) ontology and has the GO label `GO:0003700`. We can learn about GO's treatment of this term by extracting the corresponding record from the `GOTERM` environment.

```
> library("GO")
> library("GOstats")

> GOTERM$"GO:0003700"

GOID = GO:0003700
Term = transcription factor activity
Definition = Any protein required to initiate or
    regulate transcription; includes both gene
    regulatory proteins as well as the general
    transcription factors.
Ontology = MF
```

The induced graph for this term, based on the MF hierarchy, can be produced using the GOGraph function of the package GOstats.

```
> tfG <- GOGraph("GO:0003700", GOMFPARENTS)
```

The graph is shown in Figure 19.2. We can also look at the children of GO:0003700, these are terms about transcription factor activity that are more specific.

```
> tfch <- GOMFCHILDREN$"GO:0003700"

[1] "GO:0003705"

> tfchild <- mget(tfch, GOTERM)

$"GO:0003705"
GOID = GO:0003705
Term = RNA polymerase II transcription factor
    activity, enhancer binding
Definition = Functions to initiate or regulate RNA
    polymerase II transcription by binding a
    promoter or enhancer region of DNA.
Ontology = MF
```

Mappings between manufacturer identifiers for microarray probes and GO terms are available in Bioconductor meta-data packages. These packages, organized by chip model numbers (e.g., hgu95av2 for the Affymetrix HG-U95Av2 GeneChip). The mapping data resources include the evidence codes that characterize the reasons for linking a gene product to a GO term (see Section 7.5.4 for more details).

19.3.3   Graphs induced by literature references and citations

The National Library of Medicine's PubMed resource provides access to more than 11 million citations to medical and biological literature in the form of Web-accessible detailed provenance information, abstracts, and hyperlinks. We have described the basic functioning of this resource in

Figure 19.3. A bipartite graph between genes (LocusLink identifiers, nodes starting with "LL") and articles (PubMed identifiers, nodes starting with "PM").

Chapter 7. Here we have a look at its graph structures. Each paper abstracted by PubMed contains a list of the specific genes that are mentioned in the corresponding paper. The genes are identified by LocusLink IDs.

Our analysis employs an *affiliation network* in the parlance of social network analysis. We form a bipartite graph, where one set of nodes correspond to papers indexed in PubMed and the other set of nodes correspond to LocusLink identifiers. Edges in this graph relate LocusLink identifiers (genes) to PubMed identifiers (papers). An example for such a graph, representing a subgraph of the neighborhood of the DKC1 gene (LocusLink identifier 1736), is shown in Figure 19.3. In a general approach, these edges may have weights and other attributes associated with them, but we can consider them as merely representing a citation relationship between a gene and a paper.

Different transformations of this bipartite graph are interesting. Let $\mathbf{A}$ denote its adjacency matrix, where the rows represent genes, the columns represent papers, and a one (zero) in the (i, j) entry indicates that gene i was cited (not cited) in paper j. Two other matrices are of immediate interest, the so-called *one-mode graphs* that represent relationships between genes or between papers. In the following, we use Boolean matrix multiplication, in which non-zero outputs of ordinary matrix multiplication are converted to ones. The matrix $\mathbf{G} = \mathbf{A}\mathbf{A}'$ is the adjacency matrix of the graph representing gene-gene co-citation. If $\mathbf{G}_{i,j}$ is non-zero then genes i and j are jointly cited in at least one paper. The matrix $\mathbf{H} = \mathbf{A}'\mathbf{A}$ is the adjacency matrix of the graph representing relationships among papers. If

$\mathbf{H}_{i,j}$ is non-zero then there is some gene that papers i and j both mention. We will consider the properties of these graphs and describe software and different operations that can be performed in Section 22.4.

19.4   Discussion

There are many problems in computational biology that can be represented and tackled in terms of graphs and algorithms on graphs. There is a great need for software, and in particular for software that integrates data analytic capabilities with methods for querying and manipulating graphs. We have produced an approach to such an environment in Bioconductor, however much remains to be done.

We have discussed three examples. Pathways, where the interest is in better understanding functional relationships between the components. The ontology example indicates the importance of embedding terminology graphs and their mappings to bioinformatic objects in an environment that provides ready access to statistical and graph-theoretic algorithms. The literature example indicates the importance of integrating graph structures and algorithms with text mining software, to support the interpretation of data analytic results through links to the biomedical literature.

20

Graphs

W. Huber, R. Gentleman, and V. J. Carey

Abstract

In this chapter, we describe and discuss various definitions and algorithms for graphs, their representation, and uses. The presentation is formal and we leave references to software and usage for the later chapters. Our goal is to use graphs to explore, navigate, represent, and model biological data. Hence, we must often specialize general concepts and ideas to the tasks at hand. Some of our motivation is taken from the area of social network analysis where many similar problems have been considered and there is a rich history of both concepts and methods.

20.1 Overview

Graphs provide natural models for systems of related entities. The entities and their relationships define the interpretation of the graph. The bulk of this chapter concerns the mathematical notation, definitions, and abstractions needed to discuss graphs. We begin by reviewing basic definitions of graph structures, connectivity properties and measures, and operations on graphs that are useful in various bioinformatic contexts. The treatment is not comprehensive, and we refer readers to more complete references such as Gross and Yellen (1999), Sedgewick (2002), and Berge (1973). We have based our presentation and notation on that used in Gross and Yellen (1999).

Graphs will be employed as models for systems of objects and relationships between these objects. The relationships are modeled as two-place relations: for objects a and b and relationship R in the model, the notation $R(a, b)$ is interpreted as "a has relation R to b." When $R(a, b)$ is true, the representing graph will possess an edge between nodes a and b; otherwise, a and b are not directly connected by any edge.

Relationships modeled by edges may be dichotomous [$R(a, b)$ holds or does not hold] or we may consider a more general interpretation of R as a two-place function with discrete or continuous range. Such valued relations are often represented by using weights on the edges in the graph. Additional complexity can be accommodated by introducing formal types as features of modeled relations. For example, a graph with genes as nodes can simultaneously model homologies among genes using edges of one type and co-citation in medical literature using edges of another type. Regulatory networks can also be modeled using graphs. In this case, the nodes represent the genes or gene products involved in the process, and the directed edges represent different actions such as activation, enhancement, or inhibition of the target gene by the regulatory element.

In some cases, such as transcription factor networks, the relationships between nodes in the graph are directed. There is no conceptual difficulty in mixing directed relationships and undirected relationships within the same graph.

20.2   Definitions

The *graph* $G = (V, E)$ is a structure that consists of two sets. The elements of V are called nodes (the term *vertex* is also commonly used) and the elements of E are referred to as edges. Each edge has either one or two nodes associated with it; they are called its endpoints. Both nodes and edges can have types and various other attributes associated with them. We use the notation $|S|$ to denote the cardinality of the set S. Thus $|V|$ denotes the the number of nodes in G. We will also use the notation $V(G)$ to denote the node-set of graph G, and $E(G)$ to denote the edge-set of graph G.

We will use a backslash to denote set difference, so that $S\backslash T$ denotes those elements in the set S that are not in T. When dealing with graphs G and U, we use $G\backslash U$ to denote the graph H, satisfying $V(H) = V(G)\backslash V(U)$ and $E(H)$ given by $E(G)\backslash F(G, U)$, where $F(G, U)$ is the set of edges in G possessing endpoints in $V(U)$.

Edges are binary relations that join two nodes. An edge is said to be *incident at* a node if the node is an endpoint for the edge. A *self-loop* is an edge that joins one node to itself. A *proper edge* is an edge that is not a self-loop, and a *multi-edge* is a set of two or more edges that have the same endpoints. A *directed edge* is an edge where one endpoint is designated the *head* and the other the *tail*. Directed edges join the tail node to the head node but not vice versa. A *directed graph*, or *digraph*, is a graph where all edges are directed. The *underlying* graph of a digraph is the graph that results from making all directed edges undirected edges.

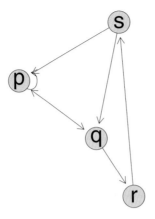

Figure 20.1. A simple graph.

Two nodes are said to be *adjacent* if they are joined by an edge. Two edges are adjacent edges if they are joined by a node. The *degree* of node v is denoted $\deg(v)$ and is equal the number of proper edges incident at v plus twice the number of v's self-loops. For directed graphs we define *in degree* to be the number of edges directed at the node and *out degree* to be the number of edges that go out from the node (edges whose tail is at the node of interest). A *complete graph* is a graph such that every pair of nodes is joined by an edge.

In Figure 20.1 node p has a self-loop, there is no edge between nodes p and r. The other edges are all directed, as there are arrowheads only on one end.

In a graph a *walk* from node v_0 to node v_n is an alternating sequence of nodes and edges, $W = \langle v_0, e_1, v_1, \cdots, v_{n-1}, e_n, v_n \rangle$ such that the endpoints of e_i are v_{i-1} and v_i for $i = 1, \ldots, n$. In a digraph we refer to the analogous structure as a directed walk. The length of a walk when no edge weights are defined is the number of edges traversed. If edge weights are defined, the length will be computed by summing the edge weights. A walk is said to be *closed* if $v_0 = v_n$, so that it starts and ends at the same node. Software for working with walks is discussed in Section 21.3.2 and Figure 21.7 demonstrates some of the concepts.

A node v is said to be reachable from node u if there is a walk from u to v. A graph is said to be *connected* if there is a walk between every pair of nodes in the graph, otherwise it is said to be disconnected. A digraph is said to be *weakly connected* if its underlying graph is connected. Two nodes w and z in a digraph are said to be mutually reachable if there is a directed walk from w to z **and** a directed walk from z to w. A digraph is said to be *strongly connected* if every pair of nodes in the digraph are mutually reachable. Figure 21.6 demonstrates some of the differences between strong and weak connectivity.

The distance between two nodes u and v in a graph is the length of the shortest walk joining them. For a digraph the directed distance is the length of the shortest directed walk. Note that the distance function so defined for digraphs may not be symmetric in its arguments. We define a *trail* to be a walk with no repeated edges and a *path* to be a walk with no repeated nodes, except possibly the first and last. A non-trivial closed path is called a *cycle*.

For a graph $G = (V, E)$ the *connectivity* is defined to be the minimum number of edges whose removal results in a disconnected graph. This number is denoted $k(G)$. If $k(G) = l$, then G is said to be l–connected. A *cut* in G is a set of edges whose removal disconnects the graph. A minimum cut is a cut with the minimum number of edges. If C is a minimum cut set of a non-trivial graph G, then $|C| = k(G)$. The connectivity of the graph in Figure 20.1 is 2.

Connectivity properties can also be described in terms of nodes. Interest often accrues to those nodes whose deletion from a connected graph G results in a disconnected graph. A *cut-set* is a node set U such that $G \backslash U$ has more components (formally defined below) than G does. A *cut-node* is a cut-set consisting of a single element.

A *subgraph* of $G = (V, E)$ is a graph $H = (W, F)$ where W is subset of V, and F is a subset of E, and all edges in F have their endpoints in W. A subgraph H is said to *span* G if $V = W$, that is, if their node sets are the same. An *induced subgraph* is a subgraph that is defined in terms of a node set S and contains all edges from E that have both endpoints in S. If G is a directed graph, then so are all subgraphs. Subgraphs can also be induced by edge sets in an analogous manner. The *boundary* of a subgraph H is the set of nodes S in $V \backslash W$ such that every element s of S has an edge to at least one node in W. We denote by T the set of edges in E that have one endpoint in S and the other in W.

A *clique* is a subset of the nodes in V such that every pair of nodes in the subset is joined by an edge. If the clique is a proper subset of no other clique, then we call it a *maximal clique*. Any node adjacent to a node v is said to be a neighbor of v. A *component* of a graph G is a maximal connected subgraph. In a graph G we refer to the component of a node v as the set of nodes that are reachable from v and denote this $C(v)$. Cliques are one type of cohesive subgroup. That is, they are sets of nodes for which there is a high degree of relatedness as demonstrated by the existence of many edges. For our purposes, such a notion of cohesive subgroup will be too restrictive, and we will consider these ideas in more detail in Section 20.3. A *spanning tree* of a graph is a subgraph that both spans the original graph and is itself a tree. A graph is connected if and only if it contains a spanning tree.

We now diverge somewhat from the standard graph theoretic terminology. Our reason for doing so is that the concepts we are about to address,

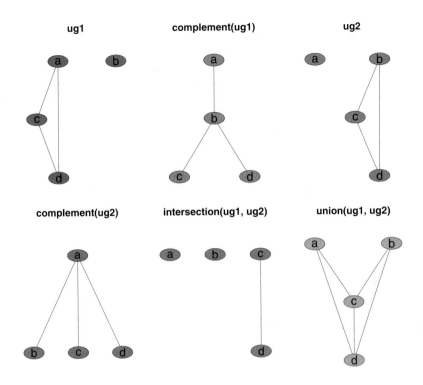

Figure 20.2. Set operations on graphs.

including graph unions and intersections, will need to be tailored to our particular situation. In many of our analyses we will be considering a graph defined in terms of a fixed and scientifically relevant set of nodes (for example all genes in a genome). Our experiment or analysis will be working with different subsets of that large well defined set of nodes and edges.

In this general context our graphs, G and H have node sets that are subsets of the larger universal node set. We define the *union* of two graphs G and H to be the graph F satisfying $V(F) = V(G) \cup V(H)$ and $E(F) = E(G) \cup E(H)$. The *intersection* graph is defined analogously, substituting $\cap$ for $\cup$ in the previous definition. The *complement* of a graph $G = (V, E)$ is the graph $G' = (V, E')$ where E' are those edges in the complete graph on V that are not in E. These concepts are presented in Figure 20.2.

20.2.1 Special types of graphs

There are a few special types of graphs that deserve a bit more attention because they play important roles in some of the applications that we will consider. The main ones are bipartite graphs, hypergraphs, and directed acyclic graphs (DAGs).

Bipartite Graphs. If the nodes of a graph $G = (V, E)$ can be partitioned into two sets U and W such that every edge in E is an undirected relationship between one node in U and one node in W, then G is said to be a *bipartite graph*. Note that there can be no edges between the elements of U or between the elements of W. Thus relationships between nodes in U are mediated through the nodes in W and vice versa.

Two graphs called *one mode graphs* can be derived from a bipartite graph. If U and W form the node partition of a bipartite graph G, then the edges in the one mode graph on U (resp. W) are determined by whether or not the two nodes both have edges in G to a common element of W (resp. U).

A common representation for bipartite graphs is in terms of the $n = |U|$ by $m = |W|$ adjacency matrix, $\mathbf{A}$, where $\mathbf{A}_{i,j} = 1$ if there is an edge from u_i to w_j and otherwise $\mathbf{A}_{i,j} = 0$. We note that usual notions of valued edges or edges with types can also be easily accommodated in this representation.

The *mode* of a network is the number is the number of partitions of the node set determined by some general node property. For example, a two-mode network can be used to describe the relationships between genes and scientific papers, or between proteins and protein complexes. In each of these cases, the node set can be partitioned by node type. Two-mode graphs are often referred to as *affiliation networks*.

In social network analysis, the two types of nodes are often referred to as *actors* and *events*. Among the basic ideas that are represented by such graphs is the concept that relationships between actors is mediated by the events that they attend (or the clubs or social groups that they belong to). In Chapter 22, we consider the relationships between genes due to co-citation and the relationship between proteins due to co-membership in protein complexes can be represented by a bipartite graph.

It is worth noting that *adjacency* in the one-mode graphs means that both nodes have an edge to (at least one) common node in the other node set. However, accessibility is less easy to interpret. Two nodes u_i and u_j that are *accessible*, but not adjacent have a connection or relationship between them is less direct – they are connected by a sequence of related actors and events but do not themselves share memberships directly.

Bipartite graphs can be directed . In this case the edges could, for example, represent temporal relationships. If we consider a biochemical reactions, then the different molecules involved form one node set while the reactions form the other. An edge from a molecule to a reaction denotes an input and an edge from a reaction to a molecule would denote a product. Because chemical reactions typically involve multiple inputs and produce multiple outputs, these graphs can be viewed as multivariate generalizations of the usual graphs (where edges represent binary relationships). This concept is formalized in the next section.

Hypergraphs. Hypergraphs are closely related to bipartite graphs (Berge, 1973; Gallo et al., 1993). Hypergraphs generalize the graph concept, allowing for the specification of relationships that are one to many and many to many.

A *hypergraph* G can be defined as a pair (V, E), where V is a set of nodes, and E is a set of hyperedges between the vertices. Each hyperedge is a set of vertices, $E_i = \{u, v, \ldots\}$, where $u, v, \ldots \in V$.

A directed hypergraph G can be represented by a pair $H = (V, E)$, where V is the set of nodes and $E = \{E_1, E_2, \ldots, E_m\}$ is the set of hyperedges. A directed hyperedge is an ordered pair, $E_i = (X, Y)$, of disjoint subsets of nodes; X is the tail of E_i while Y is the head. A *path* P from a node s to a node t is a sequence $s = (V_0, E_1, V_1, \ldots, E_n, V_n = t)$ of alternating nodes and hyperedges where (1) each hyperedge E_i is distinct, (2) for any $i \in \{0, 1, \ldots, n - 1\}$, $V_i = tail(E_{i+1})$, and (3) for any $i \in \{1, \ldots, n - 1\}$, $V_i = head(E_i)$.

Directed acyclic graphs. An important class of directed graphs are the *directed acyclic graphs*, which are simply directed graphs with no cycles. We note that a *tree* is a connected graph that has no cycles, however, we will not consider trees in any great detail.

DAGs have found very many uses in statistics. They form the basis for graphical models (Lauritzen, 1996; Edwards, 2000). They also play important roles in structuring concepts, both GO and MeSH are represented as DAGs. In Chapter 22, we demonstrate some of their uses in different specific problems.

20.2.2   *Random graphs*

Support of random graph generation in Bioconductor currently operates under the constraint that the node set is fixed and edges are randomly generated in a user-selectable fashion.

The most commonly explored formulation is based on randomly selecting edges from a complete graph (Erdös and Rényi, 1959) and is often referred to as the Erdös-Rényi model. After the user specifies the number of nodes and desired number of edges, a graph is generated by sampling edges, without replacement, from the set of $\binom{n}{2} = n(n - 1)/2$ possible edges. An equivalent version allows the user to specify a fixed probability that any two nodes are joined by an edge.

Random graphs can also be modeled in terms of node degree distributions (Newman et al., 2001). An algorithm for simulating random graphs with an arbitrary degree distribution was given by Newman et al. (2002).

A third type of random graph is generated by assuming the existence of some node attributes or latent variables that determine the presence, and possibly the weight, of edges between them. In this model, the nodes are associated with a set of identically distributed, independent random variables. The occurrence and weight of edges is determined by a function f that maps the pairs of random variables into some output space. For example, suppose that there are M binary attributes that can be associated with each node. In one scenario edges are deemed to exist between nodes u and v if they share the presence of at least one attribute. In other cases, the edge may be valued to represent the number of shared attributes or, in some cases, the attributes may themselves have real values and the edge weight could be the sum of the values of the shared attributes. The scheme is quite general. Random realizations of the node attributes can be generated using standard simulation procedures.

Although such random graph algorithms can be valuable for testing algorithms and for demonstration, their use for generating a reference distribution or a null model in a statistical analysis of a real-world graph should be approached with caution. In applications, we have noted that graphs realized from these models possess features that seem incompatible with the observed graph, and hence basing inference on simulations from such models is problematic. In many cases a permutation-type test, for example permuting the node labels, may be more sensible.

20.2.3   Node and edge labeling

A standard (1-based) node labeling of a graph $G = (V, E)$ is a one-to-one mapping between the integers from 1 to $|V|$ and the nodes in G. A standard edge labeling of G is the one-to-one mapping between the integers from 1 to $|V|$ choose 2 such that the edge labeled 1 is between nodes 1 and 2, the edge labeled 2 is between nodes 1 and 3 and so on. This definition does not handle multi-edges and self-loops, but can be extended. Standard node and edge labelings are useful for the implementation of algorithms on graphs.

20.2.4   Searching and related algorithms

Breadth first search (BFS) and depth first search (DFS) are two strategies for traversing a graph to discover properties of nodes or paths of substantive interest. Conceptually each algorithm starts with a specified node and then traverses all edges and nodes in a specific way. For DFS, at each node the next node to visit is one that goes "deeper" into the graph. For BFS, all nodes adjacent to the current node are visited before any of their descendant nodes are. Examples are shown in Section 21.3.1 and Figure 21.5.

Algorithms for finding connected components and cut sets are often based on DFS. Flow maximization problems are typically addressed using BFS.

20.3   Cohesive subgroups

Finding collections or subsets of nodes that have a close relationship to each other is often of fundamental importance. Identification of maximal cliques has limited interest as the maximality criterion is so restrictive. When dealing with imperfect systems or experimental data, we will need to deal with the problems that arise due to various types of missingness, false positives, false negatives, and unexplored relationships. In this setting, we will need a notion of cohesive subgroup that is appropriate for these sorts of problems.

Our development here follows that of Wasserman and Faust (1994). They consider a number of different notions of cohesive subgroups that include n-cliques, k-plexes and λ-sets.

In this section, we let G_s denote a subgraph of $G = (V, E)$ with nodes V_s and edges E_s defined as members of E with both end-points in V_s.

n-**cliques.** An n-clique is a sub-graph with nodes V_s such that the distance $d(v, u)$ between nodes v and u satisfies $d(v, u) \leq n$ for all nodes $v, u \in V_s$.

k-**plexes.** A k-plex is a maximal subgraph, V_s, containing v_s nodes, in which each node is adjacent to no fewer than $v_s - k$ nodes. Let $deg_s(u)$ denotes the degree of node u within the subgraph V_s. Then a k-plex is a subgraph V_s such that $deg_s(u) \geq v_s - k$, for all $u \in V_s$, and such that there is no node w in $V \backslash V_s$ such that $deg_s(w) \geq v_s - k$. For valued relationships, the requirement may be changed to require the existence of edges with value greater than δ.

One way to view this definition is that we are allowing up to k false negative edges per node. False positive edges, if infrequent, are unlikely to cause problems, because the probability that all nodes within a subgraph have a false positive edge to the same node tends to be negligible. There are exceptions, however, and in some cases the experimental technology being used may induce correlated false positive, or false negative, edges.

k-**core.** A k-core is defined similarly to a k-plex, with the main difference being that for a k-core, the minimum number of edges that must exist is specified, rather than the maximum number that can be missing. Again a slight, but obvious, modification is needed to address graphs with valued relationships.

Within-to-without comparisons. Another way to think of a cohesive subgroup is as a set of nodes that are more similar (or related) to each other than they are to the other nodes. When viewed in this manner, one might look for regions of the graph in which the concentration of edges between nodes in that region is larger than the concentration of edges from that region to the rest of the graph.

Some of these ideas have been embodied in the notions of λ-sets (Borgatti et al., 1990). Let $\lambda(w, u)$ denote the minimum number of edges that must be cut (or removed) so that there is no path between nodes w and u. For any graph $G = (V, E)$, a set of nodes $W \subset V$ is a λ-set if for all $u, v, w \in W$ and $l \in V \backslash W$ $\lambda(u, v) \geq \lambda(w, l)$. Borgatti et al. (1990) note that the members of a λ-set do not need to be adjacent; they can be quite distant from each other.

20.4   Distances

The path-length between any two nodes in a graph induces a distance between the nodes. In many cases, the shortest path will be used, but other alternatives may be appropriate for specific applications. If the graph has weighted edges, then these can easily be accommodated. Multi-graphs (graphs with multiple types of edges) can have different distances determined by the different types of edges. Other notions of distance, such as the number of paths that exist between two points, or the number of edge-cuts required to separate two nodes, can also be used. As noted in Chapter 12, the distance measure used can have a large effect, and the appropriate distance measure will usually be problem-specific.

In some cases, the natural structure of the graph will suggest different distances. For example, GO (Chapter 7) is represented as three different directed acyclic graphs (DAGs), each with a root. In some cases, the three are linked by a common root. Various methods for assessing similarity based on GO have been used (Balasubramanian et al., 2004):

- the similarity between subgraph g_i and subgraph g_j, $s(g_i, g_j)$ is computed as the length of the shortest shared path to the root node.

- the similarity between subgraph g_i and subgraph g_j, $s(g_i, g_j)$ is computed as $|g_i \cap g_j|$ divided by $|g_i \cup g_j|$.

These measures can easily be combined across ontologies. For example, one might want to require similarity both in terms of function and in terms of cellular location. For microarray experiments, distances between genes can be computed based on these measures by first finding, for each gene, the induced GO graph and then applying the above measures. Significance can often be assessed using a resampling scheme.

Once a decision has been made about a distance measure for objects organized in a graph, standard tools for cluster analysis or multidimensional scaling can be applied to the inter-object distances. See Chapter 12 for more details on distances and Chapter 10 for more details visualizing distances.

We end on a cautionary note. Just because one can easily compute distances and paths in a graph does not mean that one should do this. If we consider the co-citation graphs as exemplars of *affiliation networks*, one must interpret relationships in the one-mode graphs quite carefully. Whereas the affiliation network describes relationships between sets of *actors* and sets of *events*, the one-mode networks describe relationships between *pairs* of actors and *pairs* of events. Inference should not be made about groups that are larger than two.

21

Bioconductor Software for Graphs

V. J. Carey, R. Gentleman, W. Huber, and J. Gentry

Abstract

We describe software tools for creating, manipulating, and visualizing graphs in the Bioconductor project. We give the rationale for our design decisions and provide brief outlines of how to make use of these tools. The discussion mirrors that of Chapter 20 where the different mathematical constructs were described. It is worth differentiating between packages that are mainly infrastructure (sets of tools that can be used to create other pieces of software) and packages that are designed to provide an end-user application. The packages graph, RBGL, and Rgraphviz are infrastructure packages. Software developers may use these packages to construct tools aimed at specific applications areas, such as the GOstats package.

21.1   Introduction

Computing with graphs for bioinformatics requires attention to three basic problems. First, data structures that represent graphs and that can be readily employed in bioinformatic modeling and statistical computing are required. Second, algorithms for graph traversal and analysis (including shortest-path and cut-set determination, connectivity measurement, and decompositions) must be adapted to the structures and modeling activities that we want to address. Third, methods for layout, annotation, and visualization of graph structures are needed. Because graphs in bioinformatics will tend to be large and complex, dynamic visualization tools that facilitate interactive focus and change of view will be at a premium.

The data structure problem is addressed by the Bioconductor package graph. Package RBGL is currently the primary source of software for graph algorithms. Package Rgraphviz is the primary graph visualization resource. The graph package is entirely a creation of the Bioconductor core. Packages RBGL and Rgraphviz are interfaces to third party projects. The Boost Graph Library (Siek et al., 2002) is a C++ library devoted to portable implementation of Standard Template Library (STL) concepts for graph computations, and is at the heart of RBGL. Graphviz (Gansner and North, 1999) is a C/C++ library devoted to layout and visualization of graphs encountered in telecommunications research. We greatly appreciate the fact that the Boost and Graphviz groups have produced high-quality software with sufficiently open licenses to meet our requirements.

21.2   The graph package

The graph package contains the fundamental classes and methods that are needed to manipulate graphs and that the other two packages rely on. Interested readers are referred to the vignettes included in that package for more a more detailed discussion of some of the issues raised here.

Graphs can have distinct but equivalent representations. Different representations will be efficient for different sorts of problems, and some experimentation or exploration may be needed to determine what is an efficient representation for a given problem. Many different texts (Cormen et al., 1990; Sedgewick, 2002) provide more than one implementation. The most appropriate or efficient strategy for representing the graph will depend on many factors such as the size of the graph and the sorts of operations that are going to be applied to it. Sedgewick (2002, Table 17.1) provides details of the efficiency of different operations on graphs depending on their representation. We envisage an extremely wide range of sizes to be of interest to the users of our software, from graphs that contain a few tens of nodes up to those containing millions or more. We also envisage that a very wide range of algorithms and other manipulations will be of interest. Some users will want to have very efficient representations for very large graphs, while others, perhaps exploring different cellular pathways, will want an extremely rich set of classes for nodes and edges. They will want to be able to represent in a graph objects of different types (proteins, molecules, RNA) and different relationships between objects and processes (enhances, inhibits, modifies).

Representation of a graph by its node and edge lists is a natural approach in R. Another representation is the adjacency matrix, which encodes the existence of an edge by a non-zero entry in an $n \times n$ matrix. This representation can be inefficient for undirected graphs (and is usually replaced by the upper triangular portion), but is suitable for directed graphs. The SparseM and Matrix packages, from CRAN, allow compact representation of large, sparse graphs using a sparse matrix encoding of the adjacencies.

For bipartite graphs, an adjacency matrix is commonly used. The graph package supports these and other representations and provides facilities for converting between representations. There are also tools for serializing graphs to the GXL (Winter, 2001) and Tulip (David and Mathieu, 2004) formats. The Graphviz (Gansner and North, 1999) project supplies a gxl2dot translator, so R graphs can be converted to the Graphviz dot format.

The package graph defines the virtual class *graph*. It provides a uniform user interface to the various concrete classes that implement particular graph representations. Currently, graph provides the classes *graphNEL* for a node and edge-lists representation and *graphH* for a hash table representation. There are also tools to coerce to and from a sparse matrix representation.

In working with genomic data, we have found that two special types of graphs arise often enough that special implementations are advantageous.

The *clusterGraph* class represents the output of clustering algorithms. For this type of graph, all nodes that are assigned to a cluster have edges to all other nodes within that cluster, but there are no between-cluster edges. Edges can have weights, and these could for example depend on the distance between the nodes. The graphs are undirected and the edges are not typed.

The *distGraph* class represents a graph explicitly by inter-node distances. The representation for this class is essentially as a distance matrix, which holds the n times $n - 1$ distinct inter-node distances. Simple operations are defined on these graphs such as thresholding so that all edges whose distance is larger than some specified limit are removed. We note that this is not the same as using a graph to derive distances (see Chapter 20) but rather that some relevant distances (perhaps based on physical attributes of the nodes, such as mRNA expression levels) are used to define the edge weights in a graph where the objects are represented as nodes.

21.2.1   Getting started

We begin by loading the graph library. Next we can create a graph from scratch by specifying its nodes and edges. We have also indicated that we want the graph to be a directed graph, using the edgemode argument.

```
> library("graph")
> myNodes <- c("s", "p", "q", "r")
> myEdges <- list(s = list(edges = c("p", "q")),
+       p = list(edges = c("p", "q")), q = list(edges = c("p",
+           "r")), r = list(edges = c("s")))
> g <- new("graphNEL", nodes = myNodes, edgeL = myEdges,
+       edgemode = "directed")
> g
```

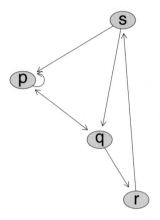

Figure 21.1. Visualization of the directed graph **g**.

```
A graph with  directed  edges
Number of Nodes = 4
Number of Edges = 7
```

The graph that we have created is of class *graphNEL* and uses the node-edge lists representation. A plot is shown in Figure 21.1. We can find out about the nodes, edges, and node degrees of **g**:

```
> nodes(g)
```

```
[1] "s" "p" "q" "r"
```

```
> edges(g)
```

```
$s
[1] "p" "q"
```

```
$p
[1] "p" "q"
```

```
$q
[1] "p" "r"
```

```
$r
[1] "s"
```

```
> degree(g)
```

```
$inDegree
s p q r
1 3 2 1
```

```
$outDegree
```

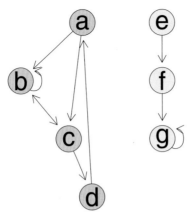

Figure 21.2. Visualization of the graph g2, which is used in the examples for the functions adj and acc.

```
s p q r
2 2 2 1
```

The functions adj and acc provide the names of the *adjacent* and *accessible nodes*. For the graph, g2, that is displayed in Figure 21.2, the result of applying these functions is

```
> adj(g2, "c")

$c
[1] "b" "d"

> acc(g2, c("b", "f"))

$b
a c d
3 1 2

$f
g
1
```

Note that the acc method returns information both on the identities of accessible nodes and the lengths of the walks to these nodes from the specified source nodes.

Functions that perform basic manipulations on graphs carry out the specified changes on a copy of the input graph and return the copy. They do not alter the input graph. These include clearNode, which removes all edges (of all types) from the specified node or nodes. combineNodes lets you group a set of nodes into a single new node. The function join combines its two arguments into a new graph. Nodes with identical labels are amalgamated, as are edges with identical labels. For directed graphs, the function `ugraph` returns the underlying undirected graph.

We can also explore operations such as the union, intersection, and complement of graphs using the correspondingly named functions. These operations are defined more fully in Chapter 20 and are illustrated in Figure 20.2.

21.2.2   Random graphs

Generating *random graphs* can be useful for testing algorithms. Random graphs are also used as a null model in statistical inference. The `graph` package provides three methods of random graph generation. `randomEGraph` for random edge graphs, `randomNodeGraph` for graphs with specified node degree distributions, and `randomGraph` for random graphs based on latent variables. These are discussed in more detail in Section 20.2.2. In the next example we create an undirected random edge graph,

```
> set.seed(123)
> nodeNames <- sapply(0:99, function(i) sprintf("N%02d",
+     i))
> rg <- randomEGraph(nodeNames, edges = 50)
```

It is shown in Figure 21.3. The degree of each node can be obtained through the `degree` function. Figure 21.4 shows the overall degree distribution, obtained through the commands

```
> deg <- degree(rg)
> hist(deg)
```

Notice that as a consequence of the generation method, the resulting node degree distribution is Binomial, with parameters n, the number of possible edges incident at each node, and p, the ratio of the number of edges in the graph to the total number of possible edges. We have superimposed the exact Binomial calculations over the histogram in Figure 21.4.

```
> size <- numNodes(rg) - 1
> prob <- numEdges(rg)/choose(numNodes(rg), 2)
```

21.3   The RBGL package

The RBGL package provides a direct interface to many of the graph algorithms available in the Boost Graph Library (Siek et al., 2002). Table 21.1 lists the main functionalities currently provided. Most of these interfaces were contributed by Li Long of the Swiss Institute for Bioinformatics.

The next two subsections review tasks involved with using RBGL to do computations on graph connectivity and shortest paths.

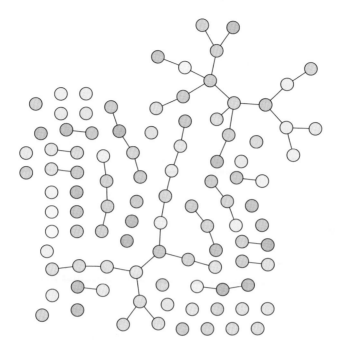

Figure 21.3. A random edge graph with 100 nodes and 50 edges.

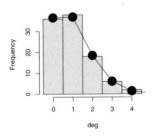

Figure 21.4. Histogram of the degree distribution of a random edge graph with 100 nodes and 50 edges. The black dots represent the density of a binomial distribution with `size`=99 and `prob`=0.01.

RBGL functions	Comments
Traversals	
bfs	BFS
dfs	DFS
Shortest paths	
dijkstra.sp	Single-source, nonnegative weights
bellman.ford.sp	Single-source, general weights
dag.sp	Single-source, DAG
johnson.all.pairs.sp	Returns distance matrix
Minimal spanning trees	
mstree.kruskal	Returns edge list and weights
prim.minST	As above
Connectivity	
connectedComp	Returns list of node-sets
strongComp	As above
edgeConnectivity	Returns index and minimum disconnecting set
init.incremental.- components	Special processing for evolving graphs
incremental.components	
same.component	Boolean in the incremental setting
Maximum flow algorithms	
edmunds.karp.max.flow	List of max flow, and edge-
push.relabel.max.flow	specific flows
Vertex ordering	
tsort	Topological sort
cuthill.mckee.ordering	Reduces bandwidth
sloan.ordering	Reduces wavefront
min.degree.ordering	Heuristic
Other functions	
transitive.closure	Returns from-to matrix
isomorphism	Boolean
brandes.betweenness.- centrality	Indices and dominance measure
circle.layout	Returns vertex coordinates
kamada.kawai.spring.layout	Returns vertex coordinates

Table 21.1. Names of key functions in RBGL. Working examples for all functions are provided in the package manual pages.

21.3.1   Connected graphs

We can obtain the connected components of the graph `rg` generated in the previous section using the `connComp` function.

```
> cc <- connComp(rg)
> table(listLen(cc))

 1  2  3  4 15 18
36  7  3  2  1  1
```

cc is a list whose elements are the individual connected components (represented by their node names). We see that the graph has one large component of size 18 and another one of size 15. It also has 36 singletons. Let us have a closer look at the largest connected component, see Figure 21.5.

```
> wh <- which.max(listLen(cc))
> sg <- subGraph(cc[[wh]], rg)
```

We can perform a depth first search (DFS) on this graph.

```
> dfs.res <- dfs(sg, node = "N14", checkConn = TRUE)
> nodes(sg)[dfs.res$discovered]

 [1] "N14" "N94" "N40" "N69" "N02" "N67" "N45" "N53" "N28"
[10] "N46" "N51" "N64" "N07" "N19" "N37" "N35" "N48" "N09"
```

The result of a call to `dfs` is a list with two elements, `discovered` and `finish`. The numeric vector `discovered` contains the discovery order of the nodes, and `finish` is the order of completion. We see that N94 is the first node visited after N14, N40 is the second, and so on. In Figure 21.5, the graph is rendered with both the node labels and the order in which they are visited. A complementary search algorithm is breadth first search (BFS):

```
> bfs.res <- bfs(sg, "N14")
> nodes(sg)[bfs.res]

 [1] "N14" "N94" "N64" "N37" "N48" "N40" "N69" "N67" "N07"
[10] "N19" "N35" "N09" "N02" "N45" "N28" "N53" "N46" "N51"
```

Here, the output is simply the vector of node names in BFS order.

For *directed graphs*, the function `connComp` calculates the *weakly connected components*. The function `strongComp` implements Tarjan's algorithm to obtain the *strongly connected components* (Tarjan, 1975). To illustrate this, we use graph g2 from Figure 21.2, and the results are shown in Figure 21.6.

```
> sc <- strongComp(g2)
> wc <- connComp(g2)
```

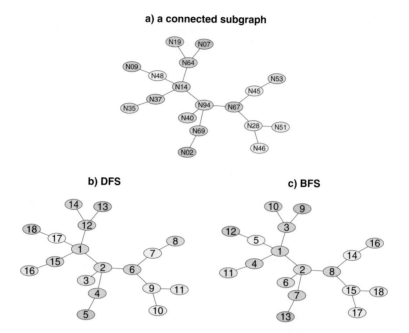

Figure 21.5. a) The graph sg, the largest connected component of the graph in Figure 21.3. b) Nodes of the graph are labeled with respect to visiting order in a depth first search (DFS). c) Visiting order in a breadth first search (BFS).

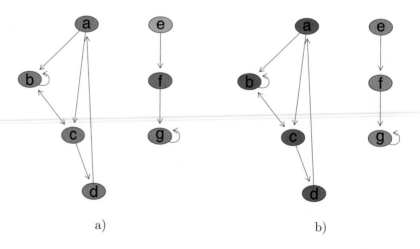

Figure 21.6. a) Colors represent the four strongly connected components of graph g2. b) The weakly connected components are colored the same.

21.3.2 Paths and related concepts

An important class of algorithms that we will often use are shortest path algorithms. The efficiency of shortest path algorithms depends on the properties of the edge weights. It is relevant to distinguish graphs that have a common edge weight, edge weights that are nonnegative integers, edge weights that are nonnegative reals, and edge weights that are arbitrary reals.

There are several variants of the shortest path problem: in the *single-pair* problem, we look for the shortest path between two given nodes. In the *single-source* or equivalently in the *single-destination* problem, we look for all shortest paths from one node to every other node in the graph. Finally, in the *all-pairs* problem, we look for all shortest paths between all pairs of nodes. There is no algorithm for solving the single-pair problem that is asymptotically faster than algorithms that solve the single-source problem.

Let us first look at the single-source shortest path algorithms that are provided in the Boost Graph Library. If all edge weights are the same, the shortest path is most efficiently found by a breadth first search. If the edge weights are different, but all positive, Dijkstra's shortest path algorithm (Dijkstra, 1959; Cormen et al., 1990) is preferred, `dijkstra.sp`. The function `sp.between` is a convenience wrapper that allows users to specify multiple start and end nodes. In that case, the function returns a list whose elements are themselves lists with elements: `path`, `length`, and `pweights`. If two nodes are not connected, i.e. there is no *path* between them, then the distance is reported as infinity, `Inf`.

In the next example we create a random edge graph that has 100 nodes N00, ..., N99 and `nEdges=100` edges. The graph is shown in Figure 21.7.

```
> set.seed(123)
> rg2 <- randomEGraph(nodeNames, edges = nEdges)
> fromNode <- "N43"
> toNode <- "N81"
> sp <- sp.between(rg2, fromNode, toNode)
> sp[[1]]$path

 [1] "N43" "N08" "N88" "N73" "N50" "N89" "N64" "N93" "N32"
[10] "N12" "N81"

> sp[[1]]$length

[1] 10
```

For the *single-source* problem, we can call the function `dijkstra.sp` directly.

```
> allsp <- dijkstra.sp(rg2, start = fromNode)
> sum(!is.finite(allsp$distances))

[1] 26
```

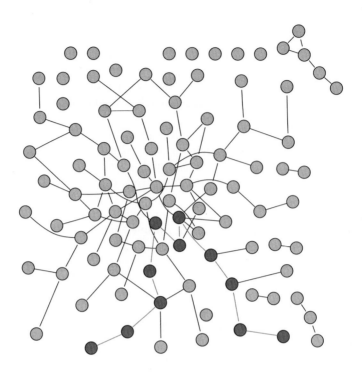

Figure 21.7. The shortest path between nodes N43 and N81 in the graph `rg2` is colored in red.

The return value `allsp` is a list with three elements: `start`, the name of the start node, `distances`, the vector of distances to all other nodes in the graph, and `penult`, the vector of predecessor nodes on the path from `start`. We see that 26 nodes are not connected to N43, and for these the value of `distance` is infinity (`Inf`). The histogram of the finite values is shown in Figure 21.8. The vector of penultimate nodes `penult` can be used to reconstruct all shortest paths,

```
> i1 <- match(fromNode, nodes(rg2))
> i2 <- match("N15", nodes(rg2))
> pft <- RBGL::extractPath(i1, i2, allsp$penult)
> nodes(rg2)[pft]

[1] "N43" "N08" "N88" "N83" "N61" "N21" "N15"
```

If we need to calculate the shortest path lengths between all pairs of nodes, an efficient algorithm is provided by the function `johnson.all.pairs.sp`.

```
> ap <- johnson.all.pairs.sp(rg2)
> table(signif(ap, 3))
```

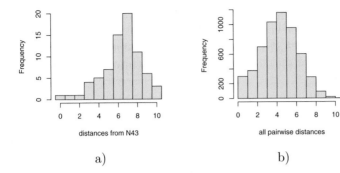

Figure 21.8. a) Histogram of the distances from N43 to all other accessible nodes in the graph **rg2** from Figure 21.7. b) Histogram of all pairwise distances.

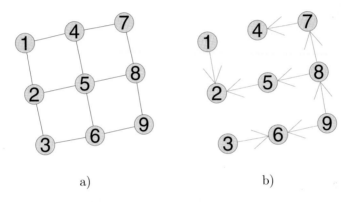

Figure 21.9. a) The regular lattice **gr**, a graph with nine nodes and twelve undirected edges. b) A minimal spanning tree for this graph.

```
   0     1     2     3     4     5     6     7     8     9    10    11
 100   200   378   698  1032  1162   956   604   290    92    20     2
 Inf
4466
```

A *minimal spanning tree* can be calculated with `mstree.kruskal`.

```
> mst <- mstree.kruskal(gr)
> mst$edgeList

     [,1] [,2] [,3] [,4] [,5] [,6] [,7] [,8]
[1,]    1    3    9    9    8    8    7    5
[2,]    2    6    6    8    5    7    4    2
```

The function returns a list with three elements: `nodes` contains the node names, `edgeList` is a from-to matrix representation of the tree's edges, and weights contains their weights. Figure 21.9 shows the example graph gr and one of its minimal spanning trees. There is a direct relationship between minimal spanning trees and single linkage hierarchical clustering. We note that algorithms for computing minimal spanning trees are also provided in the R packages ade4 (Chessel et al., 2004) and ape (Paradis et al., 2004). The latter focuses on phylogenetic trees, which are an important special case of graphs.

21.3.3   RBGL summary

Table 21.1 indicates the scope of current Boost algorithms covered by RBGL. Maximum flow algorithms will be interfaced in the near future. Because bipartite graphs and hypergraphs are likely to play a substantial role, we will also need specialized algorithms for them. Both Klamt and Gilles (2004) and Krishnamurthy et al. (2003) discuss algorithms specialized for hypergraphs.

Further experience is needed to establish the structures of outputs that most effectively support use of graph algorithms in bioinformatic workflows. Currently RBGL functions return a variety of list and matrix structures, and reduction in diversity of output formats will be undertaken as experience accumulates.

21.4   Drawing graphs

Drawing a graph can be an important aid to understanding the structure of the relationships encoded in the graph. Graph drawing consists of *graph layout*, embedding the nodes and edges in a two-dimensional plane, and *graph rendering*, which embellishes the layout with symbols that identify and describe nodes and edges, and *graph display*, which puts the rendering on a viewable surface. A nice overview of the problems and some of the solutions is given in Battista et al. (1999).

We note that graph layout is not, in general, automatic. Users that want publication-quality graphs will need to expend some time and energy trying different parameter settings and manually adjusting the node and edge attributes until they are satisfied. The figures in this section were laid out using the Graphviz (Gansner and North, 1999) library for rendering graphs. Graph layout is a substantial and difficult problem, and Graphviz offers a number of different layout algorithms. Our interface to Graphviz, the R package Rgraphviz, was designed to allow access to as much of the Graphviz functionality as possible, while retaining a user interface that is familiar to users of R and that allows users to freely combine graph visualization with the computational and visualization capabilities of R.

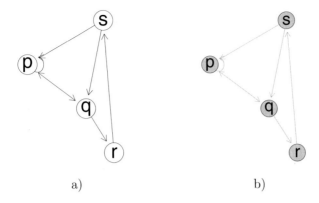

Figure 21.10. a) Result of a call to `plot(g)`, where `g` is the graph that we constructed on page 349. b) Result of a call to `plot` with a non-default value for the global attributes parameter `attr`, see page 363.

The strategy is as follows. An instance of the *graph* class is passed to Graphviz. It returns an object that contains the *layout* information for that graph. The layout indicates which nodes go where, and how the edges are drawn between them. Essentially, the layout consists of the two-dimensional (*x*- and *y*-) coordinates of the graph's nodes and a parameterization of the trajectories of the edges. In simple layouts, the edges are straight lines, specified by their start and end points. In more complex layouts, the edges are *Bezier curves*, which are parameterized by planar locations of knots. The layout can then be *displayed* through R's graphics devices. In the simplest case, these steps can be performed through one single call to the function `plot` with an object of class *graph*:

```
> library("Rgraphviz")
> plot(g)
```

The result is shown in Figure 21.10.

For more fine-grained control of the layout, the rendering, or the display, it is possible to access the functions and data structures that are associated with each intermediate step. The available options may be grouped in different categories:

- The choice of the layout algorithm.
- Global properties, such as size, aspect ratio, and background color of the plot.
- Properties of the nodes, such as labels, shape, fill and outline color. These may be set for all nodes or on a per node basis.
- Properties of the edges, such as labels, line style, color. Again, these may be set for all edges or per edge.

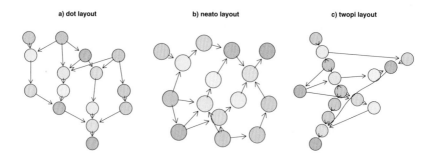

Figure 21.11. Three different graph layout algorithms applied to the same directed graph. a) *dot* aims at visualizing the hierarchies in a directed graph. b) *neato* tries to arrange the nodes in a way that, as much as possible, the edges do not overlap and have the same length. c) *twopi* aims at visualizing the radial structure of the graph.

Rgraphviz allows users to access any of the three different layout algorithms available in *Graphviz*, see Figure 21.11. *dot* is a hierarchical layout algorithm for directed graphs with four main phases: cycles are broken, nodes are assigned to layers, nodes are rearranged in layers to minimize edge crossings, and finally edges are computed as splines. *neato* is a layout algorithm for undirected graphs and is closely related to statistical multidimensional scaling. It creates a virtual physical model and optimizes for low energy configurations. It was recently augmented with a scalable stress majorization algorithm. *twopi* is a circular layout (Wills, 1997). Basically, one node is chosen as the center and put at the origin. The remaining nodes are placed on a sequence of concentric circles centered about the origin, each a fixed radial distance from the previous circle. All nodes distance 1 from the center are placed on the first circle; all nodes distance 1 from a node on the first circle are placed on the second circle; and so forth. Each of these layouts can be viewed in Figure 21.12.

Although these layout algorithms provide a flexible set of alternatives many problems and issues remain. In particular, there are rather substantial differences between drawing graphs and visualization. Many graph layout algorithms can be viewed as multi-objective optimization problems, and their solution is generally handled using optimization software. Visualization is different and typically has as one of its objectives the notion of accurately conveying information to the intended audience. In a number of papers H. Purchase (Purchase, 2000; Ware et al., 2002) has begun investigations into effective information visualization from graphs. We propose extending the available layout algorithms to the types of graphs that are prevalent in bioinformatics and computational biology. In carrying out those extensions we will pay particular attention to the issues of visualization in a biological context.

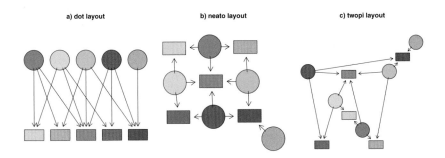

Figure 21.12. Three different graph layout algorithms applied to the same bipartite graph.

21.4.1 Global attributes

Global attributes are set via a list and passed in as the attrs argument to the plot function, or, as we will see later, to the agopen function. Default values for all attributes can be generated through the function getDefaultAttrs.

```
> defAttrs <- getDefaultAttrs()
> names(defAttrs)

[1] "graph"   "cluster" "node"    "edge"

> names(defAttrs$node)

[1] "shape"     "fixedsize" "fillcolor" "label"
[5] "color"     "fontcolor" "fontsize"  "height"
[9] "width"

> defAttrs$node$fillcolor

[1] "transparent"
```

The function getDefaultAttrs takes two arguments, a partial global attribute list, whose entries override the defaults, and the layout algorithm to be used (dot, neato, or twopi).

```
> nodeA <- list(fillcolor = "lightblue")
> edgeA <- list(color = "goldenrod")
> attrs <- getDefaultAttrs(list(node = nodeA, edge = edgeA))
> plot(g, attrs = attrs)
```

The result is shown in Figure 21.10b.

21.4.2 Node and edge attributes

Graphviz allows users to set a wide range of attributes on nodes and edges. It is possible to define subsets of nodes that are to be treated as a group.

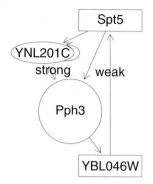

Figure 21.13. The graph g from Figure 21.1, rendered with customized node and edge attributes.

Much, but not all, of this flexibility has been carried over into Rgraphviz. In addition, because Rgraphviz uses R's graphics for the rendering, users can use any of that functionality to control the actual plotting. As we will see in some of the examples, this provides an extremely flexible plotting system. The full details of the different layout parameters can be obtained from the Graphviz documentation and the Rgraphviz documentation.

The attributes of individual nodes and edges are set via the `nodeAttrs` and `edgeAttrs` arguments to the `plot` (or `agopen`) function. For example, the visualization of the graph g2 in Figure 21.2 was plotted using the commands

```
> cc <- connectedComp(g2)
> colors <- rep(c("#D9EF8B", "#E0F3F8"), listLen(cc))
> names(colors) <- unlist(cc)
> plot(g2, nodeAttrs = list(fillcolor = colors))
```

In the first line, we calculate cc, the list of the connected components of g2. In the second and third line, different colors are assigned to different components, and then these colors are used for the `plot` command.

By default, nodes use the node name as their label and edges do not have a label. However, both can have custom labels supplied via attributes. In the next code chunk, we set several attributes and then render the graph g, which we have already seen in previous examples, with these attributes.

```
> globA <- list(node = list(width = "3", height = "1",
+     shape = "box"))
> nodeA <- list(label = c(p = "YNL201C", q = "Pph3",
+     r = "YBL046W", s = "Spt5"), shape = c(p = "ellipse",
+     q = "circle"))
> edgeA <- list(label = c("p~q" = "strong", "r~s" = "weak"),
+     color = c("p~q" = "red", "r~s" = "blue"))
> plot(g, attrs = globA, edgeAttrs = edgeA, nodeAttrs = nodeA)
```

The attributes here are solely for demonstration and have no further meaning. The result is shown in Figure 21.13.

21.4.3   The function agopen and the Ragraph class

Often, it is useful to separate the tasks of graph layout, rendering, and display. For example, one may want to use the node coordinates in the layout for further calculations in R, or produce several versions of a graph visualization with the same layout. Performance can also be an issue: the layout of large graphs can take considerable time, and often it is convenient to store the layout in an R object for reuse.

Layout is performed through the function agopen. It takes arguments layoutType, attrs, nodeAttrs, and edgeAttrs, which are described above, as well as a number of additional arguments that control the layout. For details, we refer to the manual page of agopen. It returns an object of class *Ragraph*, which contains the *laid out* graph, that is, the x- and y-coordinates of each nodes and parameterized curves for the edges. This needs to be distinguished from the class *graph*, which contains the abstract, mathematical graph.

```
> lg <- agopen(g, attrs = globA, edgeAttrs = edgeA,
+      nodeAttrs = nodeA, name = "ex1")
```

The layout is contained in two lists of *AgNode* and *AgEdge* objects. We can access these through functions of the same name:

```
> ng <- AgNode(lg)
> length(ng)

[1] 4

> class(ng[[1]])

[1] "AgNode"
attr(,"package")
[1] "Rgraphviz"

> slotNames(ng[[1]])

 [1] "center"    "name"      "txtLabel"  "height"
 [5] "rWidth"    "lWidth"    "color"     "fillcolor"
 [9] "shape"     "style"
```

The slot named center of ng[[1]] is an object of class *xyPoint*, and we can access the 2-vector of x- and y- coordinates through the function getPoints.

```
> sapply(ng, function(x) getPoints(x@center))

     [,1] [,2] [,3] [,4]
[1,]  568  235  374  572
[2,]  524  414  218   36
```

Similarly, for the edges:

```
> eg <- AgEdge(lg)
> slotNames(eg[[1]])

 [1] "splines"  "sp"        "ep"        "head"
 [5] "tail"     "arrowhead" "arrowtail" "arrowsize"
 [9] "color"    "lty"       "lwd"       "txtLabel"

> sapply(eg, function(x) (x@color))

[1] "black" "black" "black" "red"   "black" "blue"
```

The slot `splines` contains a list of *BezierCurve* objects that parameterize the trajectory of the edge. This class is used to describe Bezier curves and there is a `lines` method to facilitate drawing.

```
> eg[[1]]@splines[[1]]

460,488 415,473 364,457 322,442
```

21.4.4   *User-defined drawing functions*

As we have seen, **Rgraphviz**'s `plot` method for *graph* and *Ragraph* objects provide built-in node drawing facilities which are simple to use and will be adequate in many cases. But one of the real strengths of the *Ragraph* `plot` method lies in the possibility of specifying user-defined node drawing functions. This permits generation of arbitrarily complex node displays. These displays can make use of the full computational and graphical versatility of R. An example is shown in the code chunk below, and the result is displayed in Figure 21.14. The argument `drawNode` of the `plot` method can be either a single function (which is then called for each node in turn) or a list of functions (one for each node). In the next code chunk, we define a function `drawThermometerNode` that draws a thermometer symbol at each node location. Thermometer symbols can be used to represent numbers between 0 and 1. The result is shown in Figure 21.14.

```
> prop <- seq(0.2, 0.8, length = numNodes(g))
> colors <- brewer.pal(numNodes(g), "Set2")
> names(prop) <- names(colors) <- nodes(g)
> drawThermometerNode <- function(node) {
+       x <- getX(getNodeCenter(node))
+       y <- getY(getNodeCenter(node))
+       w <- getNodeLW(node) + getNodeRW(node)
+       h <- getNodeHeight(node)
+       nm <- name(node)
+       symbols(x, y, thermometers = cbind(w, h, prop[nm]),
+           fg = colors[nm], inches = FALSE, add = TRUE,
+           lwd = 3)
+ }
```

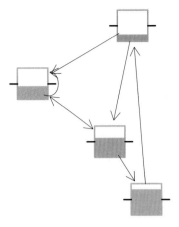

Figure 21.14. The graph **g** plotted with a custom node drawing function
`drawNode`.

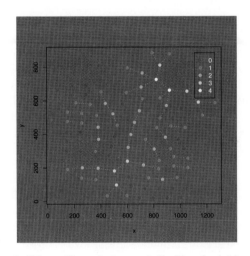

Figure 21.15. "Manual" rendering and display of a laid out graph.

```
> lg <- agopen(g, name = "")
> plot(lg, drawNode = drawThermometerNode)
```

Complete flexibility can be obtained by extracting the x- and y- coordinates of the nodes, the parameterizations of the edges, and relevant other
attributes from the *Ragraph* object, and then displaying the graph through
R's generic plotting facilities or through another specialized visualization
package. In this way, it is possible to combine the graph layout capabilities of Graphviz with the versatility of R and its add-on packages in
computation and visualization.

The layout `layoutRandomEGraph` that is used in the next example was calculated on page 352. The result is shown in Figure 21.15.

```
> par(bg = "#606060")
> nr <- AgNode(layoutRandomEGraph)
> x <- sapply(nr, function(j) getX(getNodeCenter(j)))
> y <- sapply(nr, function(j) getY(getNodeCenter(j)))
> dg <- degree(rg) + 1
> colors <- rev(colorRampPalette(brewer.pal(9, "GnBu"))(max(dg)))
> plot(x, y, pch = 16, col = colors[dg])
> for (ed in AgEdge(layoutRandomEGraph)) for (s in splines(ed)) {
+     lines(bezierPoints(s), lty = 3, col = "#f0f0f0")
+ }
> legend(max(x), max(y), paste(0:max(degree(rg))),
+     col = colors, xjust = 1, text.col = "white",
+     pch = 16)
```

21.4.5 Image maps on graphs

A graph visualization provides a global overview over the relationships between nodes. Often, it is useful to be able to *drill down* and to get more specific information about particular nodes or edges of interest.

A simple but often effective method to provide drill-down facilities is based on the ability of Web browsers to display images with *tooltips* and *hyperlinks*. Tooltips are short bits of explanatory text that are displayed when the mouse pointer moves into a certain region, and disappear when the pointer leaves the region. By clicking on a hyper-link, one can display another graphic or HTML page in a different browser window or browser frame.

Although the production of the "drill down information" is up to the user, the method `imageMap` in the Rgraphviz package offers some help in the generation of the HTML image map that can be read by the browser together with the bitmap image containing the graph visualization.

The function takes as arguments a graph layout object of class *Ragraph*, a *connection* con, a named list `tags` of tool tips and/or hyper-link tags, the name of the bitmap image file `imgname`, and its dimensions `width height`.

```
> imageMap(object, con, tags, imgname, width, height)
```

Using this approach, all possible drill down queries need to be precomputed and stored, and the responses are then simply navigated in the graphical user interface of a Web browser. This can be compared to an approach where an R process dynamically accepts and responds to queries. Both approaches have their advantages and disadvantages. It is possible to build interactive mouse-driven user interfaces to R programs, either using R's own graphics or the tcltk or RGtk packages.

22

Case Studies Using Graphs on Biological Data

R. Gentleman, D. Scholtens, B. Ding, V. J. Carey, and W. Huber

Abstract

In this chapter we consider four specific data-analytic and inferential problems that can be addressed using graphs. We demonstrate the use of the software and methods described in Chapters 20 and 21 on real problems in computational biology. We will show how one can investigate relationships between gene expression and protein-protein interaction data, how GO annotations can be used to analyze gene sets, how literature citations can be related to experimental data, and how gene expression data can be mapped on pathways.

22.1 Introduction

In our first example, we demonstrate how graphs can be used to perform an analysis that relates gene expression data to protein complex co-membership data. The question of interest was whether genes in a protein complex are more likely to have a similar pattern of gene expression than genes in different complexes. More details are reported by Balasubramanian et al. (2004), which in turn was based on the work of Ge et al. (2001). Balasubramanian et al. (2004) used two graphs defined on a common set of nodes: the genes present in yeast. The relationship represented by the edges in the first graph is co-membership in a cluster of correlated expression, while the edges in the second graph represent co-membership in a protein complex.

In our second example, we consider sets of genes and use the Hypergeometric distribution to identify GO terms that have an over-representation of the selected genes. Other categorizations, such as pathways, or chromosomal location (e.g., cytochrome band), can be analyzed similarly.

In the third example, data from the National Library of Medicine (NLM) are used to provide links between genes and scientific articles. We note that these relationships can be phrased in terms of a bipartite graph and use that observation together with standard techniques from social networks analysis to identify interesting relationships between genes and papers.

In the fourth example, we explore pathway data and demonstrate one way of relating gene expression data to pathway information. The analysis is mainly exploratory and demonstrates some of the benefits that accrue from linking R and Graphviz.

22.2   Comparing the transcriptome and the interactome

Our title for this section is largely the same as that of Ge et al. (2001); and we will demonstrate how to carry out the bulk of the analysis that they report, using tools in the packages graph, Rgraphviz, and RBGL. We will make use experimental data from the yeastExpData package.

The methods that we will consider can be implemented in many other ways but the advantage to using a graph-based approach is the abstraction that it provides. The models are similar to those discussed by Balasubramanian et al. (2004) and we refer the interested reader to the GraphAT package which can be used to reproduce their results.

Ge et al. (2001) assembled gene expression data from a yeast cell-cycle experiment (Cho et al., 1998), literature protein-protein interaction (PPI) data, and yeast two-hybrid data. We have curated the data slightly to make it simpler to carry out the analyses. In particular, we reduced the data to the 2885 genes that were common to all experiments.

The relevant data sets are ccyclered, which is a dataframe with 11 columns and 2885 rows describing the set of common genes, and litG, which is a graph representing the curated set of literature predicted protein-protein interactions. We note that this data set is not up to date, but retain it because it provides answers that coincide with those of Ge et al. (2001).

The information about which cluster a gene is in can be obtained from ccyclered. We use that to create a *cluster graph* (see Section 21.2). In the cluster graph, edges are between all genes that are in the same cluster, and no edges connect genes from different clusters. The graph ccClust has 30 complete subgraphs.

```
> library("yeastExpData")
> data(ccyclered)
> clusts <- split(ccyclered[["Y.name"]], ccyclered[["Cluster"]])
> cg1 <- new("clusterGraph", clusters = clusts)
> ccClust <- connectedComp(cg1)
```

We next turn our attention to a brief exploration of the literature based collection of protein-protein interactions. We make use of the data in litG and examine the *connected components* found therein.

```
> data(litG)
> ccLit <- connectedComp(litG)
> cclens <- listLen(ccLit)
> table(cclens)
```

```
cclens
    1     2     3     4     5     6     7     8    12    13    36    88
 2587    29    10     7     1     1     2     1     1     1     1     1
```

We see that most of the proteins, 2587, do not have edges to others, and that there are a few, rather large sets of connected proteins. The largest one contains 88, the next largest 36. We plot these in Figures 22.1 and 22.2.

```
> ord <- order(cclens, decreasing = TRUE)
> sG1 <- subGraph(ccLit[[ord[1]]], litG)
> sG2 <- subGraph(ccLit[[ord[2]]], litG)
```

22.2.1 Testing associations

It is now easy to determine how many pairs of genes have both a protein-protein interaction and are found in the same expression cluster. To compute this, we simply find the intersection of the cluster-graph and the literature graph.

```
> commonG <- intersection(cg1, litG)
```

```
A graph with  undirected  edges
Number of Nodes = 2885
Number of Edges = 42
```

We see there are 42 edges in common. This might seem like a small number, but in fact it is significantly larger than what would be expected by chance. There are several ways to test this. One way is to generate an appropriate null distribution and to compare the observed value, 42, to the values from this distribution. To generate the null distribution, there are some reasons to consider random edge graphs (Erdös and Rényi, 1959), and this is what Ge et al. (2001) did. However, if one examines the random graphs generated using the random edge model, they seldom resemble the structure in the graph based on the observed data. We propose generating the null distribution by permuting the node labels on the observed data graph.

In the next code chunk, we show a small function that performs the node label permutation test. Notice, from Figure 22.3, that the maximum number of edges in the intersection of the permuted graphs is much smaller than that observed in our data, 42. This justifies our assertion that there

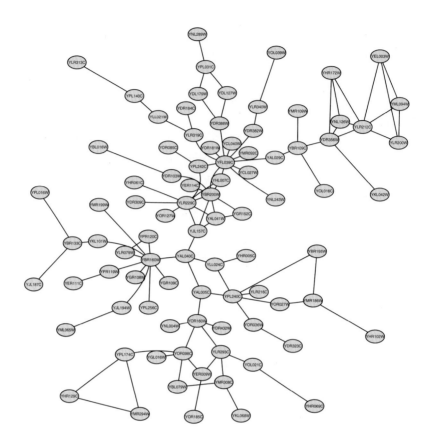

Figure 22.1. The largest PPI connected component.

is a significant relationship between gene expression pattern and protein complex co-membership, consistent with the findings of Ge et al. (2001).

```
> nodePerm <- function(g1, g2, B = 1000) {
+     n1 <- nodes(g1)
+     sapply(1:B, function(i) {
+         nodes(g1) <- sample(n1)
+         numEdges(intersection(g1, g2))
+     })
+ }
> set.seed(123)

> nPdist <- nodePerm(litG, cg1)
```

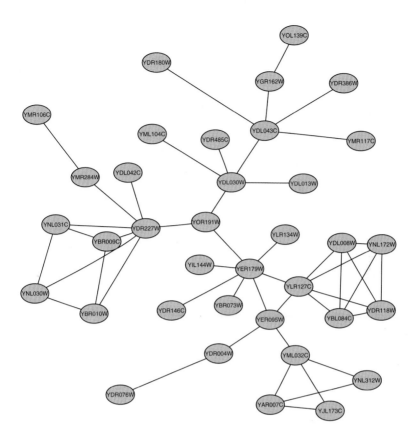

Figure 22.2. Another large PPI connected component.

22.2.2   Data analysis

Now that we have satisfied our testing curiosity, we might want to carry out a little exploratory data analysis. There are clearly some questions that are of interest including:

- Which of the expression clusters have intersections and with which of the literature clusters?

- Are there expression clusters that have a number of literature cluster edges going between them (and hence suggesting that the expression clustering was too fine or that the genes involved in the literature cluster are not cell-cycle regulated).

- Are there known cell-cycle regulated protein complexes, and do the genes involved tend to cluster together in both graphs?

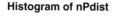

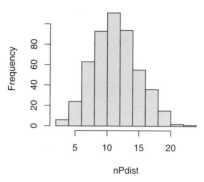

Figure 22.3. A histogram of the number of common edges as computed by a node label permutation model.

- Is the expression behavior of genes that are involved in multiple protein complexes different from that of genes that are involved in only one complex?

Many of these questions require access to more information. For example, we need to know more about the pattern of expression related to each of the gene expression clusters so that we can try to interpret them better. We need to have more information about the likely protein complexes from the literature data so that we can better identify reasonably complete protein complexes and given them, then identify those genes that are involved in more than one complex. But, the most important fact to notice is that all of the substantial calculations and computations (given the meta-data) can be phrased in terms of operations on graphs. This makes it both simple to think about what to do as well as to carry out the operations.

22.3   Using GO

In this section, we consider some of the ways in which data from GO can be used. A fairly extensive description of GO is given in Chapter 7 and we will presume that the reader is familiar with that material. Other more detailed examples involving GO and the analysis of genomic data are available through the vignettes in the **GOstats** package and in reference (Gentleman, 2004).

We make use of the ALL data (Chiaretti et al., 2004) to provide examples of how to make use of GO data in different data analytic situations. We select the B-cell leukemia cases, and from these, we will compare those with BCR/ABL to those with no observed cytogenetic abnormalities (labeled NEG). To reduce the set of genes for consideration, we applied two different sets of filters. Gene filtering is considered in more detail in Chapter 14 and by von Heydebreck et al. (2004). A non-specific filter was used to remove genes that showed little or no change in expression level across experiments. The resulting data set had 2391 probes remaining. To select genes whose expression values were associated with the phenotypes of interest (BCR/ABL and NEG), we used the mt.maxT function from the multtest package, which computes a permutation based t-test for comparing two groups.

After adjustment for multiple testing, there were only 19 probes (which correspond to 16 genes) with an adjusted p-value below 0.05. Using those genes, we obtain the set of most-specific GO terms in the MF ontology that they are annotated at. We then use these terms, together with the parent-child relationships, to find the GO graph that contains all less specific terms and we refer to that graph as the *induced* GO graph. This graph is rendered in Figure 22.4. Nodes are labeled by the most specific four digits in their GO label, that is GO:0005125 is labeled as 5125. The most specific terms are at the top of the graph and arrows go from more specific nodes to less specific ones. The node in the bottom center is the MF node. Clearly some sort of interactivity would be beneficial and you might consider using the imageMap function from the Rgraphviz package.

22.3.1 Finding interesting GO terms

In our example, we have selected a set of genes that are thought to be expressed differently in two subgroups of interest but these same methods apply equally to sets of genes that have been obtained in other ways, say by some form of clustering. Then questions that arise are: whether genes that comprise a cluster have a common function; are involved in common processes; or perhaps, are co-located in some compartment of the cell.

The test is quite straightforward. Given a set of genes and a categorization of those genes, say using one of the three ontologies, we find the set of all unique GO terms within the ontology that are associated with one or more of the genes of interest (i.e., the induced GO graph). Next, for each term, we count the interesting genes annotated at that node and obtain the number of genes assayed that are annotated at the node. Basically, we form the two-way table that identifies a gene as interesting, or not, and as being annotated at the node, or not. The unique LocusLink identifiers, and not the manufacturers identifiers, should be used because there are often multiple probes for a single LocusLink identifier on each chip.

We can ask if there are more interesting genes at the node than one might expect by chance. If that is true, then that term can be thought of as being overrepresented in the data. This question can be answered using a Hypergeometric distribution. The function GOHyperG, available in

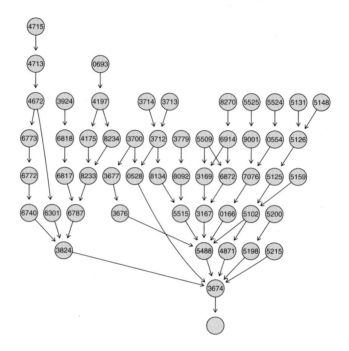

Figure 22.4. The induced GO graph for the selected genes; truncated GO identifiers are used as labels.

the GOstats package, takes as input a set of LocusLink identifiers, finds the induced GO graph and performs the Hypergeometric test at each node.

There are some issues that arise in the interpretation of the resultant p-values. First, we note that often very many hypotheses will have been tested and that some form of p-value correction will be needed. However, there is no simple or straightforward way to do that. The different hypotheses are not independent by virtue of the way that GO is structured and even with this difficulty addressed, we are most likely interested in patterns of p-values that correspond to structure in GO rather than single p-values that exceed some threshold. For these reasons, we prefer to report unadjusted p-values and leave corrections to the discretion of the user. These and other issues were considered in more detail by Gentleman (2004), however, much more research in this area is needed.

A second issue that arises is the fact that nodes of the induced GO graph with few genes annotated at them will typically have small p-values. This phenomenon occurs due to the way that we selected nodes for evaluation and the structure of GO. Recall that a gene annotated any node is also annotated at all less specific nodes in the GO hierarchy. Many genes are annotated out quite far into the leaves of the GO graph and hence at

	GO ID	Term	p	n
1	GO:0005131	growth hormone receptor b...	0.002	1
2	GO:0005148	prolactin receptor bindin...	0.004	2
3	GO:0005159	insulin−like growth facto...	0.011	6
4	GO:0003924	GTPase activity	0.014	101
5	GO:0008270	zinc ion binding	0.014	557
6	GO:0030693	caspase activity	0.021	12
7	GO:0004715	non−membrane spanning pro...	0.021	12
8	GO:0046914	transition metal ion bind...	0.026	663
9	GO:0043169	cation binding	0.029	1034
10	GO:0005488	binding	0.04	4825
11	GO:0005525	GTP binding	0.041	181
12	GO:0019001	guanyl nucleotide binding	0.043	187
13	GO:0004713	protein−tyrosine kinase a...	0.043	187
14	GO:0043167	ion binding	0.048	1185
15	GO:0046872	metal ion binding	0.048	1185
16	GO:0005126	hematopoietin/interferon−...	0.065	37
17	GO:0017076	purine nucleotide binding	0.087	976
18	GO:0000166	nucleotide binding	0.091	990

Table 22.1. GO terms, p-values, and numbers of genes for a selection of GO categories.

nodes that have relatively few other genes annotated there. Calculation of the Hypergeometric p-values for these nodes results in very small p-values. Others have dealt with this issue by defining the concept of depth in the GO graph (the number of edges to the root node) and then only using nodes that are neither too deep nor too shallow.

In the next code chunk, we show how to take an induced GO graph, gGO, and a set of interesting genes, gNsLL, and find the Hypergeometric p-values. This is done using GOHyperG. Because the data come from a HG-U95Av2 chip, we use the set of genes on that chip as the set of all genes in the Hypergeometric test. We then make use of the resultant p-values to provide colors for the nodes.

```
> gNsLL <- unique(unlist(mget(names(gde), env = hgu95av2LOCUSID,
+       ifnotfound = NA)))
> gGhyp <- GOHyperG(gNsLL)
```

In Figure 22.5, we reproduce the plot from Figure 22.4 except that we have now colored the nodes according to the p-value obtained from the Hypergeometric test described above. The nodes in Figure 22.5 are colored either dark red or light blue depending on whether the unadjusted Hypergeometric p-value was less than 0.1 or not. The GO terms for the nodes colored red are printed below. The relevant biology suggests that these are quite reasonable.

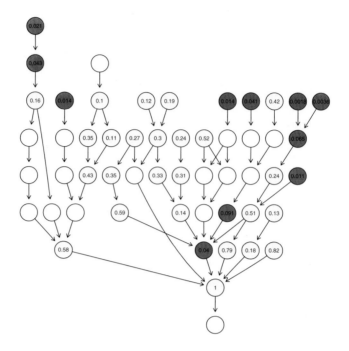

Figure 22.5. The induced GO graph colored according to unadjusted Hypergeometric p-values, whose values are given in the nodes.

We see in Table 22.1 that the nodes with the smallest p-values do tend to be the nodes with few genes annotated at them. However, there are also some nodes with quite small p-values and large counts, such as GO:0008270 and GO:0003924, and these would surely be of some interest in subsequent explorations.

It is interesting to note that we can also ask, and answer, the question about underrepresented GO terms. That is, we can find nodes in the GO graph that, given their size, should have contained one or more interesting genes, under the null hypothesis.

22.4 Literature co-citation

In this section, we consider the graph structure of literature co-citation data and explore some of the ways it can be used to help add meaning to a data analysis. The basic statistical models and paradigm will be presented first, and subsequently we apply them to co-citation via PubMed; see Chapter 7 for more details on PubMed. There are many different problems that can be addressed using these data, but we will consider only a few of them.

One of the problems in providing concrete recommendations is the lack of a gold standard against which to measure the performance of the various tools. We have used a number of examples where we believe one can make a reasonable statement about whether two genes are related and then contrast the different measures and adjustments with respect to their agreement with this point of view. Of course, your opinion might be different, and in that case you would naturally select a test statistic to use accordingly. More details on the approach and more extensive examples were given by Ding and Gentleman (2004).

One can consider citation in terms of a bipartite graph. The genes represent one type of node and the scientific papers represent the other type of node. An edge exists between a gene and a paper if the gene is cited in the paper. In this graph, there are no edges between papers and no edges between genes. The relationships between genes are mediated by the papers and the relationships between papers are mediated by the genes. From this bipartite graph, we can generate two *one mode* graphs. One is the graph whose nodes are genes and an edge exists between two genes if they are co-cited in one or more papers. Edge weights can be used in this graph to count the number of co-citations. The second type of one mode graph that is of some interest is the graph whose nodes represent papers, and an edge exists between two papers if they co-cite at least one gene. Edge weights can be used to represent the number of genes that have been co-cited.

In the context of a co-citation graph (see Section 20.2.1 for more details), the *actor size* is the number of papers that cite the gene of interest, while the *event size* is the number of genes that are cited by a specific paper. We note that some adjustment for either actor or event size can improve the inference and should be considered; we discuss this in Section 22.4.2. For example co-citation in a paper such as that by Strausberg et al. (2002), which cites more than 15,000 genes, has very little information and one would not generally treat co-citation in this paper as indicating any relationship between genes. On the other hand, co-citation in a paper that discusses only two or three genes is a much stronger indication of an intrinsic biological relationship. Interested readers are referred to Section 22.4.1 below, Chapter 8 of the book by Wasserman and Faust (1994), and the article of Ding and Gentleman (2004) for further considerations.

The concepts of adjacency, reachability, and connectedness can all be applied to bipartite graphs, and hence to affiliation networks. Of these, the strongest and most interpretable property will be adjacency. If we consider the co-citation network, the notion of a relationship based on reachability seems very vague and would be difficult to interpret. Similarly, it will be difficult to place much meaning on the path length between two genes, or two papers. We also note that that notions reachability, diameter, and connectedness in the one mode networks are likely to be of little biological interest. In a co-citation graph, only the direct co-citations are likely to be important. Two genes that are co-cited will share an edge, and it is not clear

	$gene_2$		
$gene_1$	n_{11}	n_{12}	$n_{1.}$
	n_{21}	n_{22}	$n_{2.}$
	$n_{.1}$	$n_{.2}$	n

Table 22.2. Notational conventions for a two-way table.

that the existence of a path, say through some third gene, is any evidence of a relationship. One might be willing to argue that the existence of many, very short paths between two genes of interest constitutes evidence of a relationship, but that requires a different approach.

22.4.1   Statistical development

For many of the two-way tables that arise in bioinformatics, one of the entries in the table is much much larger than the other ones. For example, with co-citation comparisons, or when comparing annotation at a particular GO term with some other property, we find that most genes have neither property and so one entry in the two-way table is very large compared to the other three. To facilitate discussion, we will use the convention that the n_{22} entry in the two-way table is the one with the very large number. To further ease the exposition, we will base some of the discussion on the notion that we want to compare two genes, $gene_1$ and $gene_2$, on the basis of their co-citations in the medical literature.

When one entry in the table (Table 22.2) is much larger than the others, the actual distribution of the test statistics may be quite far from the asymptotic distributions that are commonly used to assess significance. It may be prudent to rely on test statistics that either do not use n_{22} or which do not depend heavily on it. Some of these were studied by Ding and Gentleman (2004) and we discuss their findings here. Ding and Gentleman (2004) considered a wide range of statistics and recommended the three following statistics as performing well, under different situations.

- **Concordance measure**

$$n_{11}$$

- **Jaccard Index** (Jaccard, 1912)

$$\frac{n_{11}}{n_{11} + n_{12} + n_{21}}$$

- **Hubert's Γ** (Hubert, 1987; Good, 1994)

$$\frac{n_{11}n_{22} - n_{12}n_{21}}{\sqrt{(n_{11} + n_{12})(n_{21} + n_{22})(n_{11} + n_{21})(n_{12} + n_{22})}}$$

The range of the Jaccard Index is [0,1] and for Hubert's Γ the range is [-1,1]. Hubert's Γ is equivalent to the fourfold correlation coefficient.

Ding and Gentleman (2004) carried out an empirical study of the performance of the different test statistics together with adjustments for both event size and actor size. Although there is no gold standard by which to compare these different test statistics, it is nonetheless important to attempt to understand the properties of the different test statistics. Notions of power and size are therefore approximate, and based on comparisons where a biological association between genes could be determined, and other cases using genes where it was very unlikely that any true biological association exists. That is, Ding and Gentleman (2004) considered genes that are likely to have a biologically meaningful relationship, as well as those that, despite frequent co-citation, are not likely to have a biologically meaningful relationship. They found that the χ^2 and odds ratio based statistics do not, in general, perform as well as the Concordance, Jaccard Index and Hubert's Γ based statistics. They also found that actor size adjustment tends to make tests too conservative, whereas event-size adjusted Concordance and Jaccard Index tend to be too anti-conservative. The necessary software for carrying out these tests is provided in the CoCiteStats package.

Let $\mathcal{N}$ denote the set of actors, with cardinality n, and let $\mathcal{M}$ denote the set of events, with cardinality m. We denote the affiliation matrix as $\mathbf{A}$, where $\mathbf{A}_{i,j}$ is 1 if actor i was present at event j. The corresponding one mode networks can be then be found as as $\mathbf{X}^{\mathcal{N}} = \mathbf{AA}'$ and $\mathbf{X}^{\mathcal{M}} = \mathbf{A}'\mathbf{A}$. Note that in $\mathbf{X}^{\mathcal{M}}$ the i, j entry is the number of events that both actor i and actor j attended. In some cases we will be interested in Boolean versions of these matrices, that is versions of $\mathbf{X}^{\mathcal{M}}$ and $\mathbf{X}^{\mathcal{N}}$ that have entries that are zero or one, which indicate whether actors i and j attended one or more events together.

We return to the subject of actor and event size adjustments. We note that a very large event (a paper that cites very many genes) is likely to co-cite two genes, but the information about their relationship is weaker than if they were co-cited in a paper that cited only a small number of genes. Considering, instead, actors we see that an actor (or gene) that attends many events is much more likely to be affiliated with other actors than an actor that attends few events. In the context of co-citation, this says that a well studied gene is more likely to be associated with other genes than a recently discovered, or recently studied, gene.

One argument that is often made in social network theory is that the measure of association between actors should be logically independent of the event size. When the data are presented in the form of a two-way table, the odds ratio is one measure of association that is logically independent of group size. An alternative discussed in Wasserman and Faust (1994) is to normalize either of $\mathbf{X}^{\mathcal{M}}$ or $\mathbf{X}^{\mathcal{N}}$ so that all row and column totals are equal; this idea will not be explored here.

The use of $\mathbf{X}^{\mathcal{N}}$ intrinsically assumes equal weighting of papers. The size of the papers, however, may also play a key role in deciding significance of association between genes and some adjustment may be needed. There are various ways of doing this, but in principle one should down-weight large papers as their information content is less. We consider a weight equal to the inverse of the number of genes cited for each paper, i.e., paper size.

22.4.2   Comparisons of interest

Now we revise the gene-gene contingency table, Table 22.2, for the case where the comparison of interest is between two genes, $gene_1$ and $gene_2$, on the basis of co-citation. We let $n'_{ij} = \sum_{l \in Pub_{ij}} 1/N_l$, $i, j = 1, 2$ where N_l is the size of paper l, i.e., the number of genes cited by PubMed l, Pub_{11} is the set of papers citing both genes; Pub_{12} citing $gene_1$ but not $gene_2$, Pub_{21} citing $gene_2$ but not $gene_1$, and Pub_{22} are those citing neither genes. Hence n'_{ij} is a weighted version of n_{ij} where the weight depends on the number of genes cited by each paper. We can then use the three statistics proposed in Section 22.4.1, Concordance, Jaccard's index, and Hubert's Γ, with n_{ij} replaced by n'_{ij}.

22.4.3   Examples

We begin with a small example to clarify some of the relevant issues. TRO and BYSL form a complex mediating cell adhesion. Suzuki et al. (1999) studied expression of these two genes in human placenta. These two genes are the only two human gene products referred to in this paper (PMID: 10026108). Conversely, they were also co-cited in Strausberg et al. (2002) (PMID: 12477932) where ESTs were generated from libraries enriched for full-length cDNAs; there is no direct association between the genes they have cited other than the fact that their cDNA sequences can be obtained. So we can see that the paper by Suzuki et al. (1999) is very informative about these two genes, and their potential relationship, while that by Strausberg et al. (2002) is not.

We consider the Concordance measure, Hubert's Γ, and the Jaccard Index. For all three we also consider gene size adjustments, paper size adjustments and both gene and paper size adjustments, thus yielding four statistics for each of these.

Example 1

We first look at the association between two genes, BYSL with LocusLink ID 705 and TRO with LocusLink ID 7216. As noted above, they have been co-cited twice (PMID: 12477932,10026108) where the second paper cited only these two genes and the first one cited 14596 genes. Even though one of the papers citing both is general (PMID: 12477932), the other (PMID: 10026108) is a very specific paper discussing the two genes. Moreover, the two genes were cited in only 4 and 8 papers respectively, hence we believe that there is an association between them and we would like to use a test statistic that is capable of detecting that relationship.

$$
\begin{array}{c}
\quad 7216 \\
705\;\begin{array}{|cc|c}
2 & 2 & 4 \\
6 & 74666 & 74672 \\
\hline
8 & 74668 & 74676
\end{array}
\end{array}
$$

	Concordance	Jaccard	Hubert
None	2.0000	0.2000	0.3535
	(0.0000)	(0.0600)	(0.0800)
GS	0.9911	0.9824	0.9822
	(0.1000)	(0.1000)	(0.1000)
PS	0.5001	0.0832	0.1579
	(0.0000)	(0.0000)	(0.0000)
BOTH	0.9855	0.9715	0.9710
	(0.0800)	(0.0800)	(0.0800)

Table 22.3. PubMed co-citation: Locuslink ID 705 and 7216.

Using a Hypergeometric distribution the exact p-value for testing the null hypothesis that gene 705 and 7216 are not related is 0.377 when no edge weights are considered, indicating no significant association between them. Failure to account for the edge weights may offer an explanation.

Table 22.3 reports the results for the three statistics from Section 22.4.1. For each statistic, we also considered four versions: no adjustment (None), gene size adjustment (GS), paper size adjustment (PS) and both gene and paper size adjustment (Both). The numbers listed in each entry are the score and p-value (in parentheses).

Results from Concordance, Jaccard Index, and Hubert's Γ are quite consistent, the original Concordance statistic and paper size adjusted Concordance, Jaccard Index, and Hubert's Γ are significant at 0.05 level. This suggests that paper size adjustment is useful especially as one of the papers under investigation is extremely large in size. The adjustments for gene size all lead to non-significant results.

An analysis using GO by Ding and Gentleman (2004) indicated that the two genes are highly significantly related in their biological processes.

Example 2

The previous example suggests that both the number of co-citations and the paper size are important in determining the level of significance. To see this more clearly, we consider genes 10038 (ADPRTL2) and 10039 (ADPRTL3) which are co-cited four times. The sizes of the papers citing 10038 and 10039 are 3,2,2,2, all relatively small compared with previous examples. Moreover, the genes were cited 7 and 8 times respectively.

		10039	
10038	4	3	7
	4	74665	74669
	8	74668	74676

	Concordance	Jaccard	Hubert
None	4.0000	0.3636	0.5345
	(0.0000)	(0.0000)	(0.0000)
GS	0.9937	0.9875	0.9874
	(0.0000)	(0.0000)	(0.0000)
PS	1.8333	0.3771	0.5476
	(0.0000)	(0.0000)	(0.0000)
BOTH	0.9956	0.9913	0.9913
	(0.0000)	(0.0000)	(0.0000)

Table 22.4. PubMed co-citation: Locuslink ID 10038 and 10039.

All results reported in Table 22.4 are significant. This suggests that if paper size is small then there is no obvious need for paper size adjustment; almost all the statistics, with or without adjustment, yield similar results.

Application to gene lists. Here we use the test statistics, suggested above, but aggregate them over the set of genes in the gene list or over the boundary of the gene list.

Given a list of genes, D, one can find the boundary of that list, with respect to the one mode co-citation graph $\mathbf{X}^{\mathcal{N}}$. This boundary is simply the set of genes that were co-cited one or more times with the genes in D. Because there are many papers that cite thousands of genes, the boundary itself will not be very interesting, and we will typically restrict our attention to those genes where the sum of the edge weights exceeds some threshold. This cut-off can be determined empirically.

Once the boundary has been determined, we might want to find those genes that have a particularly strong association with the genes in D. While parametric tests are not generally available, a resampling test can be used to assess significance. Alternatively, we can compute pairwise relationships between the members of D itself. These distances, could then be analyzed, using multidimensional scaling or they could form the basis for yet another graph.

We return to the ALL example begun in Section 22.3. In that example, we selected genes whose expression values were associated with the phenotypes of interest (BCR/ABL and NEG) using a permutation-based t-test to compare the two groups. We found 19 probes, corresponding to 16 genes, that had adjusted p-values below 0.05. Suppose that we wanted to find out whether there are subsets of these genes that are closely related, according to co-citation. We can also ask if there are other genes that are closely related to the selected genes that we did not find. We first obtain the unique LocusLink identifiers and then map these to the set of papers that cite the genes. We begin with the data object `intLLc` that contains the LocusLink identifiers for the selected genes. For each of these we first obtain the number of citations for each gene.

```
> papersByLL <- mget(intLLc, humanLLMappingsLL2PMID,
+       ifnotfound = NA)
> ncit <- sapply(papersByLL, length)
> ncit
```

```
  25   687   195  2534 23145  7277   841  4599  2273    87
  68     5     7    28     3    10    94    24    10     6
6935  9697  9900  3937  1396  8835
  10     5     4    11     4    12
```

We see that the number of citations ranges from 94 to 3. Next, we can construct a simple co-citation graph, on these genes and here we need only concern ourselves with this rather small set of papers. The paper sizes were also computed and they range from 14596 to 1.

```
> num <- length(papersByLL)
> grels <- vector("list", length = num)
> names(grels) <- names(papersByLL)
> for (i in 1:num) {
+     curr <- papersByLL[[i]]
+     grels[[i]] <- lapply(papersByLL, function(x) {
+         mt <- match(x, curr, 0)
+         if (any(mt > 0))
+             curr[mt]
+         else NULL
+     })
+ }
> for (i in 1:num) grels[[i]] <- grels[[i]][-i]
```

We have now computed the edges that are present in our graph. Next we want to see which papers co-cite genes from among our list.

```
> gr2 <- lapply(grels, function(x) {
+     slen <- sapply(x, length)
+     x[slen > 0]
+ })
> table(unlist(gr2))
```

```
12477932 14702039
     132       30
```

We notice that all of the co-citations between the genes we have selected are due to two papers, one by Strausberg et al. (2002) and a similar one by Ota et al., and hence there is no information about relationships between these genes to be gleaned from the currently available medical literature.

We can take a more exploratory approach. For instance, starting with the
same set of genes, the boundary of their co-citation graph can be examined.
That is, we are looking for all genes that have a co-citation with one or
more of the genes in our list. We will need to discount the very large papers,
and hence we will make use of edge weights in constructing our graph and
subsequently will trim those elements of the boundary with edge weights
that are small.

Finding the boundary is relatively straightforward. Given our list of
genes, we first find their citations, and using those citations we find the
information on genes cited in those papers. In the next code chunk, a simple function, LL2wts, that carries out this computation is provided. Given a
set of LocusLink IDs it finds all papers that cite these genes. Then, taking
those papers, it finds all genes they cite and creates a weight vector, where
the weights are 1 over the papers sizes. Finally, a list of the named weight
vectors is output.

```
> LL2wts <- function(inList) {
+     pBLL <- mget(inList, humanLLMappingsLL2PMID,
+         ifnotfound = NA)
+     numL <- length(inList)
+     ans <- NULL
+     for (i in 1:numL) {
+         lls <- mget(as.character(pBLL[[i]]),
+             humanLLMappingsPMID2LL,
+             ifnotfound = NA)
+         lens <- sapply(lls, length)
+         names(lens) <- NULL
+         wts <- rep(1/lens, lens)
+         wtsbyg <- split(wts, unlist(lls, use.names = FALSE))
+         ans[[i]] <- sapply(wtsbyg, sum)
+     }
+     ans
+ }
> vv <- LL2wts(intLLc)
```

Given vv, we can answer a number of questions. For example, we can
find which of the elements of vv have the largest weights, we can see which
genes are connected to more than one gene in our list of interesting genes,
and of those, which have relatively high weights.

```
> allLL <- unique(unlist(sapply(vv, names)))
> bdrywts <- rep(0, length(allLL))
> names(bdrywts) <- allLL
> for (wvec in vv) bdrywts[names(wvec)] <- bdrywts[names(wvec)] +
+     wvec
> wts <- bdrywts[!(allLL %in% intLLc)]
> sum(wts > 1)
```

[1] 20

```
> range(wts[wts > 1])
```

```
[1] 1.08 9.00
```

We can see that there are 20 genes that have weights that are larger than 1 and hence might warrant further study. We can find those that are on the HG-U95Av2 chip by using the chip-specific annotation pacakge, hgu95av2.

```
> LL95 <- unlist(as.list(hgu95av2LOCUSID))
> bdryLL <- names(wts[wts > 1])
> onC <- match(bdryLL, LL95, 0)
> unlist(mget(names(LL95[onC]), hgu95av2SYMBOL))
```

517_at	1084_at	2043_s_at	1441_s_at	2024_s_at
"SHFM3P1"	"ABL2"	"BCR"	"FAS"	"LYN"
879_at	32725_at	38350_f_at	40567_at	34448_s_at
"MX2"	"BID"	"TUBA2"	"TUBA3"	"CASP2"
36143_at	38281_at	486_at	1765_at	38755_at
"CASP3"	"CASP7"	"CASP9"	"CASP10"	"FADD"
1867_at	40969_at	35681_r_at		
"CFLAR"	"SOCS3"	"ZFHX1B"		

22.5   Pathways

In this section, we consider some uses of pathway information in the analysis of gene expression data. Although the concept of a pathway does not have a rigorous definition, the general concept is widely used. For example, the biological process ontology from GO describes itself as being less than a pathway.

Associating gene expression data with pathways has been considered by many others, including Doniger et al. (2003). In some applications, one might render a pathway and color the nodes (genes) according to changes in expression across experimental conditions. Although this approach has some appeal, there are other uses for pathway data. Pathways can be used to perform subgroup analysis where interest is restricted to a set of genes that are associated with a particular pathway. However, there are many situations where one would not expect the expression levels to change. For example, many signal transduction pathways are known to end in the activation of a transcription factor. Thus, to know if the pathway is active, it seems more reasonable to study the targets of the transcription factor than the constituent elements of the pathway.

In our first example, we consider the network structure of the pathways themselves. We make use of the bipartite graph that relates genes and pathways and study the one mode network on pathways that results from it. In our second example, we take a single pathway, the integrin-mediated cell-adhesion pathway, and render it in different ways, using gene expression data to modify the outputs.

22.5.1   The graph structure of pathways

Consider the bipartite graph where one set of nodes are genes and the other set of nodes are pathways. We are interested in understanding the relationships between pathways due to shared genes, or shared sets of genes. We represent the bipartite graph in terms of an incidence matrix; see Section 20.2.1 for more details.

We construct the graph based on the data available from the HG-U95Av2 GeneChip array from Affymetrix. It might be of more interest to consider the construction of this graph based on all mappings for a given organism rather than restricting our attention to a particular chip, but this restriction makes the computations manageable. The construction is considered in some detail as readers are likely to find it useful for creating their own bipartite graphs. There are two relevant mappings, those from probes to pathways, and the converse, from pathways to probes. For the HG-U95Av2 chips these are available as `hgu95av2PATH`, which holds the mappings from probesets to the pathways, and `hgu95av2PATH2PROBE`, which contains the mappings from pathways to probesets. We first load the necessary libraries and then look to see how many pathways different genes are annotated at.

```
> library("hgu95av2")
> library("annotate")
> gene1 <- unlist(eapply(hgu95av2PATH, length))
> table(gene1)

gene1
    1     2     3     4     5     6     7     8    10    11
11264   635   363   208    71    33     6    10     7    10
   12    13    15
    7     3     8
```

We see that some genes are annotated at many pathways, while most are annotated at only one. Since genes are annotated at pathways using LocusLink identifiers we next reduce the data by removing any duplicate probes.

```
> pathLL <- eapply(hgu95av2PATH2PROBE, function(x) {
+     LLs <- getLL(x, "hgu95av2")
+     unique(LLs)
+ })
> pLens <- sapply(pathLL, length)
> range(pLens)

[1]   1 219

> uniqLL <- unique(unlist(pathLL, use.names = FALSE))
```

We see that pathway sizes are between 1 and 219 for LocusLink identifiers from this chip. We note that these sizes are with respect to the set of genes that we have information on. The actual size (number of genes) in a pathway could be quite different, and for some calculations we will want the actual set of genes, but for others we will need to focus on those genes for which we have data.

Now that we have computed `pathLL`, that is really all that is needed. We can find out how many pathways there are (136), and how many unique LocusLink identifiers there are (2297). In the incidence matrix representation of our bipartite graph, we let LocusLink identifiers denote the rows and pathways denote the columns. The data in `pathLL` are easily transformed to an adjacency matrix where the pathways are the columns, and the genes are the rows.

```
> Amat <- sapply(pathLL, function(x) {
+       mtch <- match(x, uniqLL)
+       zeros <- rep(0, length(uniqLL))
+       zeros[mtch] <- 1
+       zeros
+ })
```

Now that we have an incidence matrix for the pathways, we can construct the one mode graphs for genes and for pathways. We leave the gene graph for the reader to explore and instead consider the pathway graph. The diagonal entries of `pwGmat` will be the counts of the number of genes in each pathway. We set these to zero so that they do not get interpreted as self-loops.

```
> pwGmat <- t(Amat) %*% Amat
> diag(pwGmat) <- 0
> pwG <- as(pwGmat, "graphNEL")
```

Although we could use Rgraphviz to lay out the graph, it has too many nodes and edges to provide a meaningful visualization using standard layout methodologies. Further research is needed to develop good layout strategies for this graph. However, we can examine some of the basic characteristics of the graph.

We can find the connected components.

```
> ccpwG <- connectedComp(pwG)
> sapply(ccpwG, length)

  1   2   3   4   5
132   1   1   1   1
```

We see that there are four singletons, and otherwise all the pathways are connected by the genes that are assayed on the HG-U95Av2 chip. In the next code chunk we find and print the names of the singletons.

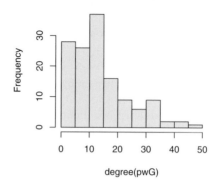

Figure 22.6. The degree distribution of the pathway graph.

```
> library("KEGG")
> for (i in ccpwG) {
+     if (length(i) == 1)
+         cat(get(i, KEGGPATHID2NAME), "\n")
+ }

Basal transcription factors
Retinol metabolism
Proteasome
Chondroitin / Heparan sulfate biosynthesis
```

These pathways might be connected to each other, or to other pathways, through genes that were not assayed.

We computed the degree distribution of the pathway graph and plotted a histogram in Figure 22.6. Pathways are the nodes in this graph, and so we see that some pathways have many edges to other pathways, and hence are quite central. It might be useful to use edge weights to indicate the number of shared genes, and this could then be used in coloring the edges or perhaps in thresholding them.

Other analyses might focus on finding shared components, for example finding out whether one pathway is wholly contained within another. We will need good layout algorithms for single pathways. We will also need layout mechanisms for joining together different pathways.

22.5.2   Relating expression data to pathways

We now consider a method for relating gene expression data to pathways. Other approaches have been considered, in particular by the GenMAPP project (Doniger et al., 2003), and some of our own work has been reported in R News (Gentry et al., 2004). We consider the integrin-mediated cell-adhesion pathway, as represented at *KEGG*. The KEGG pathway label is `hsa04510` and the graphical representation from KEGG was shown in Figure 19.1. Users can either access the KEGG Web site directly, or they can use the KEGGSOAP package to obtain more information about this pathway. For any microarray experiment, Bioconductor meta-data packages can be used to find associations between probes and the genes involved in different KEGG pathways.

To obtain the pathway graph, you have several different options. You can construct one yourself, based on the available data and potentially expert biological advice, or you can make use of the information from the cMAP project, which is available in the cMAP package. For this particular pathway, we have already taken the information available in KEGG and used that to construct a graph representation of the pathway. The relevant data structures are constructed from two objects in the graph package. The object `IMCAGraph` is an instance of the graphNEL class, representing the pathway as a mathematical graph with named nodes and directed edges. The object `IMCAAttrs` is a list of plotting attributes for each node in the graph, such as the color.

We return to the ALL data and ask whether or not there are differences between the two groups (BCR/ABL and NEG) with respect to expression levels of genes in this pathway. We use the subset of the ALL data computed in Section 22.3. However, we do not carry out any gene selection, instead we consider the expression levels of the different genes in this pathway, and how those levels depend on phenotype (whether the samples are BCR/ABL or NEG).

Next, we obtain the mapping between the probes on the Affymetrix array and the genes in the pathway.

```
> hsa04510 <- hgu95av2PATH2PROBE$"04510"
> hsaLLs <- getLL(hsa04510, "hgu95av2")
```

There are 52 nodes in this pathway, and of these 45 represent genes. We find that there are 114 probesets for these genes on the HG-U95Av2 chip. There are many different ways to deal with the duplicate probesets, and here we take the simplistic approach of just selecting the first match. We note that an appropriate investigation of these data would involve a more detailed consideration of how to deal with multiple probes per gene.

In the next code chunk, we extract the LocusLink identifiers associated with each node in the graph and then for each of these take the first probeset that maps to it. We also check to see which of the genes in the pathway have no probes associated with them; these will have a value of `NA` in `whProbe`.

```
> LLs <- unlist(sapply(IMCAAttrs$LocusLink, function(x) x[1]))
> whProbe <- match(LLs, hsaLLs)
> probeNames <- names(hsaLLs)[whProbe]
```

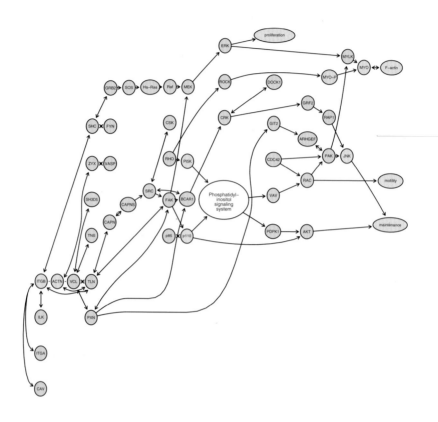

Figure 22.7. The integrin-mediated cell-adhesion network.

```
> names(probeNames) <- names(LLs)
> pN <- probeNames[!is.na(probeNames)]
```

We lay out the graph using agopen, as we want to render the same graph several times.

```
> IMCg <- agopen(IMCAGraph, "", attrs = IMCAAttrs$defAttrs,
+     nodeAttrs = IMCAAttrs$nodeAttrs, subGList = IMCAAttrs$subGList)
> plot(IMCg)
```

In Figure 22.7 we see the pathway laid out, with nodes that represent genes colored green. Now that we have found a set of probes that map to each gene in the pathway, we split the data into those with BCR/ABL and those that have no abnormalities and render the pathway, once for each group. For each group, we will plot a pie chart for each node. The pie chart will reflect a split, across the gene, of the samples for that gene. We will use splits of $(0, 6]$, for low, $(6, 8.5]$ for moderate and $(8.5, \infty]$ for high, levels of expression. This visualization is different from one that colors nodes according to whether the genes are more highly expressed in one group than the other. It allows the reader to compare the distribution of expression, for each gene, between the two phenotypes.

Now that we have found the expression levels and computed the counts for each of the probes, we are ready to layout the graph and then render it, once for each phenotype we are interested in. The resulting plots are shown in Figure 22.8. Using pie charts for the nodes in the graph is easily done, and the procedure is documented in the Rgraphviz package. We note that due to the modular nature of the graph drawing procedures in Rgraphviz, virtually any R plot can be used for the nodes in a graph; see also Section 21.4.4. It is also easy to simply color the nodes according to which group has higher levels of expression, as is done by many others.

The graphs themselves are quite interesting. The similarity in distribution of expression levels, especially for those genes on the right half of the graph is remarkable. On the left side, we draw your attention to FYN, which has about 3/4 of the samples in the high range for BCR/ABL while for the NEG samples about 3/4 of the samples are moderate.

22.6   Concluding remarks

In this chapter, we have presented four case studies that made use of the tools that were introduced in the earlier chapters of this section. Our purpose was not to promulgate the examples themselves, but rather to demonstrate the flexibility of the software tools that are available and to emphasize that virtually any analysis can be undertaken, with a small amount of additional programming. You should only be limited by your ideas and the available data.

There are still many questions to answer, and much software needs to be written. We will need specialized graph algorithms to deal with the fact that many biological relationships are measured with error, and hence usual constructs and algorithms may fail or be unusable when false negative and false positive relationships exist. Visualizing graphs, as opposed to layout, is a difficult problem and one that is starting to get some attention. We hope that the tool kit of graph algorithms and methods described here, linked to the R statistical computing framework, will foster many new developments.

a) pie chart graph for BCR/ABL

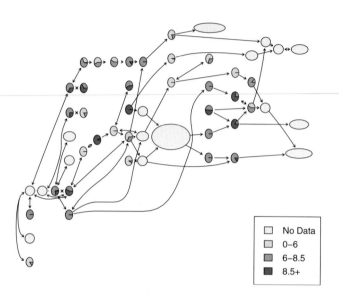

b) pie chart graph for NEG

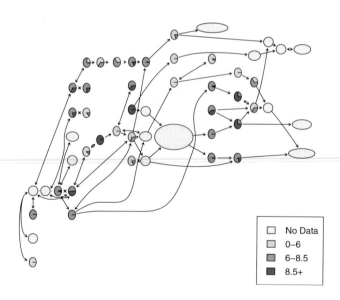

Figure 22.8. Pie chart graphs representing gene expression data for a) BCR/ABL samples, b) NEG samples.

Part V

Case studies

23

limma: Linear Models for Microarray Data

G. K. Smyth

Abstract

A survey is given of differential expression analyses using the linear modeling features of the limma package. The chapter starts with the simplest replicated designs and progresses through experiments with two or more groups, direct designs, factorial designs and time course experiments. Experiments with technical as well as biological replication are considered. Empirical Bayes test statistics are explained. The use of quality weights, adaptive background correction and control spots in conjunction with linear modelling is illustrated on the $\beta 7$ data.

23.1   Introduction

limma is a package for *differential expression* analysis of data arising from microarray experiments. The package is designed to analyze complex experiments involving comparisons between many RNA targets simultaneously while remaining reasonably easy to use for simple experiments. The central idea is to fit a linear model to the expression data for each gene. The expression data can be log-ratios, or sometimes log-intensities, from two-color microarrays or log-intensity values from one-channel technologies such as Affymetrix. Empirical Bayes and other shrinkage methods are used to borrow information across genes making the analyses stable even for experiments with small number of arrays (Smyth, 2004; Smyth et al., 2005).

limma is designed to be used in conjunction with the affy or affyPLM packages for Affymetrix data as described in Chapters 2 and 25. With two-color microarray data, the marray package may be used for preprocessing as described in Chapter4. limma itself also provides input and normalization functions that support features especially useful for the linear modeling approach.

This chapter gives a survey of differential expression analyses starting with the simplest replicated designs and progressing through experiments with two or more groups, factorial designs, and time course experiments. For the most part, this chapter does not analyze specific data sets but gives instead generic code which can be applied to any data set arising from the designs described. Analyses of specific data sets are given in Chapters 4, 14, 16, and 25. One purpose of this chapter is to place these analyses in context and to indicate how the methods would be extended to more complex designs. This chapter was prepared using limma version 1.8.18.

23.2   Data representations

The starting point for this chapter and many other chapters in this book is that an experiment has been performed using a set of microarrays hybridized with two or more different RNA sources. The arrays have been scanned and image-analyzed to produce output files containing raw intensities, usually one file for each array. The arrays may be *one-channel* with one RNA sample hybridized to each array or they may be *two-channel* or *two-color* with two RNA samples hybridized competitively to each array.

Expression data from experiments using one-channel arrays can be represented as a data matrix with rows corresponding to probes and columns to arrays. The `rma` function in the affy package produces such a matrix for Affymetrix arrays. The output from `rma` is an *exprSet* object with the matrix of log-intensities in the `exprs` slot. See Chapters 2 and 25 for details.

Experiments using two-color arrays produce two data matrices, one each for the green and red channels. The green and red channel intensities are usually kept separate until normalization, after which they are summarized by a matrix of log-ratios (*M*-values) and a matrix of log-averages (*A*-values). See Chapter 4 for details.

Two-color experiments can be divided into those for which one-channel of every array is a common reference sample and those that make direct comparisons between the RNA samples of interest without the intermediary of a common reference. Common reference experiments can be treated similarly to one-channel experiments with the matrix of log-ratios taking the place of the matrix of log-intensities. Direct two-color designs require some special techniques. Many features of limma are motivated by the desire to obtain full information from direct designs and to treat all types of experiment in a unified way.

Sections 23.3 to 23.11 will assume that a normalized data object called MA or `eset` is available. The object `eset` is assumed to be of class *exprSet* containing normalized probeset log-intensities from an Affymetrix experiment, while `MA` is assumed to contain normalized M and A-values from an experiment using two-color arrays. The data object `MA` might be an *marrayNorm* object produced by `maNorm` in the `marray` package or an *MAList* object produced by `normalizeWithinArrays` or `normalizeBetweenArrays` in the `limma` package, although *marrayNorm* objects usually need some further processing after normalization before being used for linear modeling as explained in Section 23.4. The examples of Sections 23.4 to 23.11 remain valid if `eset` or `MA` is just a matrix containing the normalized log-intensities or log-ratios.

Apart from the expression data itself, microarray data sets need to include information about the *probes* printed on the arrays and information about the *targets* hybridized to the arrays. The targets are of particular interest when setting up a linear model. In this chapter, the target labels and any associated covariates are assumed to be available in a *targets frame* called `targets`, which is just a `data.frame` with rows corresponding to arrays in the experiment. In an *exprSet* object, this data frame is often stored as part of the *phenoData* slot, in which case it can be extracted by `targets <- pData(eset)`. Despite the name, there is no implication that the covariates are phenotypic in nature, in fact they often indicate genotypes such as wild-type or knockout. In an *marrayNorm* object, the targets frame is often stored as part of the `maTargets` slot, in which case it can be extracted by `targets <- maInfo(maTargets(MA))`. `limma` provides the function `readTargets` for reading the targets frame directly from a text file, and doing so is often the first step in a microarray data analysis.

23.3   Linear models

`limma` uses *linear models* to analyze designed microarray experiments (Yang and Speed, 2003; Smyth, 2004). This approach allows very general experiments to be analyzed nearly as easily as a simple replicated experiment. The approach requires two matrices to be specified. The first is the *design matrix*, which provides a representation of the different RNA targets that have been hybridized to the arrays. The second is the *contrast matrix*, which allows the coefficients defined by the design matrix to be combined into contrasts of interest. Each contrast corresponds to a comparison of interest between the RNA targets. For very simple experiments, the contrast matrix may not need to be specified explicitly.

The first step is to fit a linear model using `lmFit`, which fully models the systematic part of the data. Each row of the design matrix corresponds to an array in the experiment and each column corresponds to a coefficient. With one-channel data or common reference data, the number of coefficients will be equal to the number of distinct RNA sources. With direct-design two-color data, there will be one fewer coefficients than distinct RNA targets, or the same number if a dye-effect is included. One purpose of this step is to estimate the variability in the data.

In practice, one might be interested in more or fewer comparisons between the RNA targets than there are coefficients. The contrast step, which uses the function `contrasts.fit`, allows the fitted coefficients to be compared in as many ways as there are questions to be answered, regardless of how many or how few these might be.

Mathematically, we assume a linear model $E[\mathbf{y}_j] = \mathbf{X}\boldsymbol{\alpha}_j$ where $\mathbf{y}_j$ contains the expression data for the gene j, $\mathbf{X}$ is the design matrix, and $\boldsymbol{\alpha}_j$ is a vector of coefficients. Here $\mathbf{y}_j^T$ is the jth row of the expression matrix and contains either log-ratios or log-intensities. The contrasts of interest are given by $\boldsymbol{\beta}_j = \mathbf{C}^T \boldsymbol{\alpha}_j$ where $\mathbf{C}$ is the contrasts matrix. The `coefficients` component of the fitted model produced by `lmFit` contains estimated values for the $\boldsymbol{\alpha}_j$. After applying `contrasts.fit`, the `coefficients` component now contains estimated values for the $\boldsymbol{\beta}_j$.

With one-channel or common reference microarray data, linear modeling is much the same as ordinary ANOVA or multiple regression except that a model is fitted for every gene. With data of this type, design matrices can be created in the same way that one would do when modeling univariate data. With data from two-color direct designs, linear modeling is very flexible and powerful but the formation of the design matrix may be less familiar. The function `modelMatrix` is provided to simplify the construction of appropriate design matrices for two-color data.

23.4   Simple comparisons

The simplest possible microarray experiment is one with a series of replicate two-color arrays all comparing the same two RNA sources. For a three-array experiment, comparing wild-type (wt) and mutant (mu) RNA, the targets frame might contain the following entries:

FileName	Cy3	Cy5
File1	wt	mu
File2	wt	mu
File3	wt	mu

A list of the top genes that show evidence of differential expression between the mutant and wild-type might be found for this experiment by

```
> fit <- lmFit(MA)
> fit <- eBayes(fit)
> topTable(fit, adjust = "fdr")
```

where `MA` holds the normalized data. The default design matrix used here is just a single column of ones. This experiment estimates the fold change of mutant over wild-type. Genes that have positive M-values are more highly expressed in the mutant, whereas genes with negative M-values are more highly expressed in the wild-type. The analysis is analogous to the classical single-sample t-test except that empirical Bayes methods have been used to borrow information between genes.

A simple modification of the above experiment would be to swap the dyes for one of the arrays. The targets frame might now be

FileName	Cy3	Cy5
File1	wt	mu
File2	mu	wt
File3	wt	mu

and the analysis would be

```
> design <- c(1, -1, 1)
> fit <- lmFit(MA, design)
> fit <- eBayes(fit)
> topTable(fit, adjust = "fdr")
```

Alternatively, the design matrix could be constructed, replacing the first of the above code lines, by

```
> design <- modelMatrix(targets, ref = "wt")
```

where `targets` is the targets frame.

If there are at least two arrays with each dye orientation, for example

FileName	Cy3	Cy5
File1	wt	mu
File2	mu	wt
File3	wt	mu
File4	mu	wt

then it may be useful to estimate probe-specific *dye-effects*. The dye-effect is estimated by an intercept term in the linear model. Including the dye-effect uses up one degree of freedom, which might otherwise be used to estimate the residual variability, but may be valuable if many genes show non-negligible dye-effects.

Integrin $\alpha 4\beta 7$ experiment. Chapter 4 introduces a data example with six replicate arrays including three dye-swaps in which integrin $\beta 7+$ memory T help cells play the role of "mutant" and $\beta 7-$ cells play the role of "wild-type." Here, we continue from the "quick start" section of that chapter where an *marrayNorm* object `normdata` is created and a top table gene listing is presented. For this data it proves important to include a dye effect.

```
> design <- cbind(Dye = 1, Beta7 = c(1, -1, -1,
+       1, 1, -1))
> fit <- lmFit(normdata, design, weights = NULL)
> fit <- eBayes(fit)
```

Now

```
> topTable(fit, coef = "Dye", adjust = "fdr")
```

reveals significant dye effects for many genes.

Note the use of `weights=NULL` in the above fit. This is needed because the function `read.GenePix` used to input the data populates the maW slot of the data object with GenePix® spot quality flags rather than with weights. The flags are just indicators that take on various negative values to indicate suspect spots with zero representing a normal spot, whereas functions in R that accept "weights" expect them to be numeric non-negative values with zero indicating complete unreliability. For this reason, it was necessary to use `weights=NULL` to tell `lmFit` to ignore the weights slot in `normdata`. Much better however is to convert the flags into quantitative weights. The code below gives weight zero to all spots with negative flags and weight one to all unflagged spots. This improves the power of the analysis and increases the number of apparently differentially expressed genes.

```
> w <- 0 + (maW(normdata) >= 0)
> fit <- lmFit(normdata, design, weights = w)
> fit <- eBayes(fit)
> tab <- topTable(fit, coef = "Beta7", adjust = "fdr")
> tab$Name <- substring(tab$Name, 1, 20)
> tab
```

```
               ID                Name       M      A      t
6647   H200017286 GPR2 - G protein-cou -2.451   7.79 -14.72
11025  H200018884 Homo sapiens cDNA FL -1.598   6.62 -12.12
6211   H200019655 KIAA0833 - KIAA0833  -1.610   7.98 -10.91
11431  H200015303 CCR9 - Chemokine (C-  1.501  10.11   9.94
4910   H200003784 SEMA5A - Sema domain -1.349   6.81  -8.69
3152   H200012024 ITGA1 - Integrin, al  1.316   6.98   9.67
22582  H200017325 IFI27 - Interferon,   1.319   7.15   7.74
7832   H200004937 Homo sapiens cDNA FL -1.226   6.30  -8.96
20941  H200008015 PTPRJ - Protein tyro -0.885  11.01  -7.55
9314   H200006462     LMNA - Lamin A/C -1.213   7.93  -8.65
          P.Value     B
6647      0.0112  5.56
11025     0.0236  4.65
6211      0.0337  4.11
11431     0.0494  3.60
4910      0.1006  2.84
3152      0.1006  2.50
22582     0.1045  2.15
7832      0.1045  2.15
20941     0.1045  2.00
9314      0.1045  1.99
```

In this table, M is the $\log_2$ fold change, with positive values indicating higher expression in the $\beta7+$ cells. For the meaning of the other columns, see Section 23.12.

23.5   Technical Replication

In the previous sections, we have assumed that all arrays are biological replicates. Now consider an experiment in which two wild-type and two mice from the same mutant strain are compared using two arrays for each pair of mice. The targets might be

FileName	Cy3	Cy5
File1	wt1	mu1
File2	wt1	mu1
File3	wt2	mu2
File4	wt2	mu2

The first and second and third and fourth arrays are *technical replicates*. It would not be correct to treat this experiment as comprising four replicate arrays because the technical replicate pairs are not independent, in fact they are likely to be positively correlated.

One way to analyze these data is the following:

```
> biolrep <- c(1, 1, 2, 2)
> corfit <- duplicateCorrelation(MA, ndups = 1,
+      block = biolrep)
> fit <- lmFit(MA, block = biolrep, cor = corfit$consensus)
> fit <- eBayes(fit)
> topTable(fit, adjust = "fdr")
```

The vector `biolrep` indicates the two blocks corresponding to biological replicates. The value `cofit$consensus` estimates the average correlation within the blocks and should be positive. This analysis is analogous to *mixed model* analysis of variance (Milliken and Johnson, 1992, Chapter 18) except that information has been borrowed between genes. Information is borrowed by constraining the within-block correlations to be equal between genes and by using empirical Bayes methods to moderate the standard deviations between genes (Smyth et al., 2005).

If the technical replicates were in dye-swap pairs as

FileName	Cy3	Cy5
File1	wt1	mu1
File2	mu1	wt1
File3	wt2	mu2
File4	mu2	wt2

then one might use

```
> design <- c(1, -1, 1, -1)
> corfit <- duplicateCorrelation(MA, design, ndups = 1,
+      block = biolrep)
> fit <- lmFit(MA, design, block = biolrep, cor = corfit$consensus)
```

```
> fit <- eBayes(fit)
> topTable(fit, adjust = "fdr")
```

In this case, the correlation `corfit$consensus` should be negative because the technical replicates are dye-swaps and should vary in opposite directions.

This method of handling technical replication using `duplicateCorrelation` is somewhat limited. If for example one technical replicate was dye-swapped and the other not,

FileName	Cy3	Cy5
File1	wt1	mu1
File2	mu1	wt1
File3	wt2	mu2
File4	wt2	mu2

then there is no way to use `duplicateCorrelation` because the technical replicate correlation will be negative for the first pair but positive for the second. An alternative strategy is to include a coefficient in the design matrix for each of the two biological blocks. This could be accomplished by defining

```
> design <- cbind(MU1vsWT1 = c(1, -1, 0, 0), MU2vsWT2 = c(0,
+      0, 1, 1))
> fit <- lmFit(MA, design)
```

This will fit a linear model with two coefficients, one estimating the mutant vs. wild-type comparison for the first pair of mice and the other for the second pair of mice. What we want is the average of the two mutant vs. wild-type comparisons, and this is extracted by the contrast (MU1vsWT1+MU2vsWT2)/2:

```
> cont.matrix <- makeContrasts(MUvsWT = (MU1vsWT1 +
+      MU2vsWT2)/2, levels = design)
> fit2 <- contrasts.fit(fit, cont.matrix)
> fit2 <- eBayes(fit2)
> topTable(fit2, adjust = "fdr")
```

The technique of including an effect for each biological replicate is well suited to situations with a lot of technical replication. Here is a larger example from a real experiment. Three mutant mice are to be compared with three wild-type mice. Eighteen two-color arrays were used with each mouse appearing on six different arrays:

```
> targets

      FileName Cy3 Cy5
1391 1391.spot wt1 mu1
1392 1392.spot mu1 wt1
1340 1340.spot wt2 mu1
1341 1341.spot mu1 wt2
```

```
1395 1395.spot wt3 mu1
1396 1396.spot mu1 wt3
1393 1393.spot wt1 mu2
1394 1394.spot mu2 wt1
1371 1371.spot wt2 mu2
1372 1372.spot mu2 wt2
1338 1338.spot wt3 mu2
1339 1339.spot mu2 wt3
1387 1387.spot wt1 mu3
1388 1388.spot mu3 wt1
1399 1399.spot wt2 mu3
1390 1390.spot mu3 wt2
1397 1397.spot wt3 mu3
1398 1398.spot mu3 wt3
```

The comparison of interest is the average difference between the mutant
and wild-type mice. duplicateCorrelation could not be used here because
the arrays do not group neatly into biological replicate groups. In any case,
with six arrays on each mouse it is much safer and more conservative to fit
an effect for each mouse. We could proceed as

```
> design <- modelMatrix(targets, ref = "wt1")
> design <- cbind(Dye = 1, design)
> colnames(design)

[1] "Dye" "mu1" "mu2" "mu3" "wt2" "wt3"
```

The above code treats the first wild-type mouse as a baseline reference so
that columns of the design matrix represent the difference between each of
the other mice and wt1. The design matrix also includes an intercept term
which represents the dye effect of Cy5 over Cy3 for each gene. If no dye
effect is expected then the second line of code can be omitted.

```
> fit <- lmFit(MA, design)
> cont.matrix <- makeContrasts(muvswt = (mu1 + mu2 +
+     mu3 - wt2 - wt3)/3, levels = design)
> fit2 <- contrasts.fit(fit, cont.matrix)
> fit2 <- eBayes(fit2)
> topTable(fit2, adjust = "fdr")
```

The contrast defined by the function makeContrasts represents the average
difference between the mutant and wild-type mice, which is the comparison
of interest.

This general approach is applicable to many studies involving biological
replicates. Here is another example based on a real example conducted
by the Scott Lab at the Walter and Eliza Hall Institute (WEHI). RNA is
collected from four human subjects from the same family, two affected by
a leukemia-inducing mutation and two unaffected. Each of the two affected
subjects (A1 and A2) is compared with each of the two unaffected subjects
(U1 and U2):

FileName	Cy3	Cy5
File1	U1	A1
File2	A1	U2
File3	U2	A2
File4	A2	U1

Our interest is to find genes that are differentially expressed between the affected and unaffected subjects. Although all four arrays compare an affected with an unaffected subject, the arrays are not independent. We need to take account of the fact that RNA from each subject appears on two different arrays. We do this by fitting a model with a coefficient for each subject and then extracting the contrast between the affected and unaffected subjects:

```
> design <- modelMatrix(targets, ref = "U1")
> fit <- lmFit(MA, design)
> cont.matrix <- makeContrasts(AvsU = (A1 + A2 -
+     U2)/2, levels = design)
> fit2 <- contrasts.fit(fit, cont.matrix)
> fit2 <- eBayes(fit2)
> topTable(fit2, adjust = "fdr")
```

23.6   Within-array replicate spots

Robotic printing of spotted arrays can be set up to print more than one spot from each well of the DNA plates. Such printing means that each probe is printed two or more times a fixed distance apart. The most common printing is in duplicate, with the duplicates being either side-by-side or in the top and bottom halves of the array. The second option means that the arrays are printed with two sub-arrays each containing a complete set of probes.

Duplicate spots can be effectively handled by estimating a common value for the intra-duplicate correlation (Smyth et al., 2005). Suppose that each probe is printed twice in adjacent positions, side-by-side by columns. Then the correlation may be estimated by

```
> corfit <- duplicateCorrelation(MA, design, ndups = 2,
+     spacing = "columns")
```

Here `spacing="rows"` would indicate replicates side-by-side by rows and `spacing="topbottom"` would indicate replicates in the top and bottom halves of the arrays. The spacing may alternatively be given as a numerical value counting the number of spots separating the replicate spots.

The estimated common correlation is `corfit$consensus`. This value should be a large positive value, say greater than 0.4. The correlation is then specified at the linear modeling step:

```
> fit <- lmFit(MA, design, ndups = 2, spacing = 1,
+     cor = corfit$consensus)
```

The object `fit` contains half as many rows as does `MA`, i.e., results for the multiple printings of each probe have been consolidated. See Chapter 14 for an example of this analysis applied to the Kidney cancer study. The analysis given there demonstrates the greater power of the duplicate correlation approach compared to simply averaging the log-ratios from the replicate spots.

23.7   Two groups

Suppose now that we wish to compare two wild-type (Wt) mice with three mutant (Mu) mice using two-color arrays hybridized with a common reference RNA (Ref):

FileName	Cy3	Cy5
File1	Ref	WT
File2	Ref	WT
File3	Ref	Mu
File4	Ref	Mu
File5	Ref	Mu

The interest is to compare the mutant and wild-type mice. There are two major ways in which this comparison can be made. We can (1) create a design matrix that includes a coefficient for the mutant vs. wild-type difference, or (2) create a design matrix which includes separate coefficients for wild-type and mutant mice and then extract the difference as a contrast.

For the first approach, the design matrix might be

```
> design
```

```
       WTvsREF MUvsWT
Array1     1      0
Array2     1      0
Array3     1      1
Array4     1      1
Array5     1      1
```

This is sometimes called the *treatment-contrasts* parameterization. The first coefficient estimates the difference between wild-type and the reference for each probe, while the second coefficient estimates the difference between mutant and wild-type. For those not familiar with model matrices in linear regression, it can be understood in the following way. The matrix indicates which coefficients apply to each array. For the first two arrays, the fitted values will be just the WTvsREF coefficient. For the remaining arrays, the fitted values will be WTvsREF + MUvsWT, which is equivalent to mutant vs. reference. Differentially expressed genes can be found by

```
> fit <- lmFit(MA, design)
> fit <- eBayes(fit)
> topTable(fit, coef = "MUvsWT", adjust = "fdr")
```

There is no need here to use contrasts.fit because the comparison of interest is already built into the fitted model. This analysis is analogous to the classical *pooled two-sample t-test* except that information has been borrowed between genes.

For the second approach, the design matrix should be

```
       WT MU
Array1  1  0
Array2  1  0
Array3  0  1
Array4  0  1
Array5  0  1
```

We will call this the *group-means* parameterization. The first coefficient represents wild-type vs. the reference and the second represents mutant vs. the reference. Our interest is in the difference between these two coefficients. Differentially expressed genes can be found by

```
> fit <- lmFit(MA, design)
> fit2 <- contrasts.fit(fit, c(-1, 1))
> fit2 <- eBayes(fit2)
> topTable(fit2, adjust = "fdr")
```

The genelist will be the same as for the first approach.

The design matrices can be constructed manually or using the built-in R function model.matrix. Let Group be the factor defined by

```
> Group <- factor(c("WT", "WT", "Mu", "Mu", "Mu"),
+      levels = c("WT", "Mu"))
```

For the first approach, the treatment-contrasts parameterization, the design matrix can be computed by

```
> design <- cbind(WTvsRef = 1, MUvsWT = c(0, 0,
+      1, 1, 1))
```

or by

```
> design <- model.matrix(~Group)
> colnames(design) <- c("WTvsRef", "MUvsWT")
```

For the second approach, the group-means parameterization, the design matrix can be computed by

```
> design <- cbind(WT = c(1, 1, 0, 0, 0), MU = c(0,
+      0, 1, 1, 1))
```

or by

```
> design <- model.matrix(~0 + Group)
> colnames(design) <- c("WT", "Mu")
```

Suppose now that the experiment had been conducted with one-channel arrays such as Affymetrix rather than with a common reference, so the targets frame might be

FileName	Target
File1	WT
File2	WT
File3	Mu
File4	Mu
File5	Mu

The one-channel data can be analyzed exactly as for the common reference experiment. For the treatment-contrasts parameterization, the design matrix is as before

```
> design
```

```
       WT MUvsWT
Array1  1      0
Array2  1      0
Array3  1      1
Array4  1      1
Array5  1      1
```

except that the first coefficient estimates now the mean log-intensity for wild-type mice rather than the wild-type versus reference log-ratio. For the group-means parameterization, the design matrix is as before but the coefficients now represent mean log-intensities for wild-type and mutant rather than log-ratios versus the wild-type. Design and contrasts matrices are computed exactly as for the common reference experiment.

See Chapter 25 for a complete data analysis of a two group experiment with six Affymetrix arrays, three in each group, from the Affymetrix spike-in experiment.

23.8   Several groups

The above approaches for two groups extend easily to any number of groups. Suppose that three RNA targets are to be compared using Affymetrix arrays. Suppose that the three targets are called "RNA1," "RNA2," and "RNA3" and that the column `targets$Target` indicates which one was hybridized to each array. An appropriate design matrix can be created using

```
> f <- factor(targets$Target, levels = c("RNA1",
+      "RNA2", "RNA3"))
```

```
> design <- model.matrix(~0 + f)
> colnames(design) <- c("RNA1", "RNA2", "RNA3")
```

To make all pair-wise comparisons between the three groups, one could proceed

```
> fit <- lmFit(eset, design)
> contrast.matrix <- makeContrasts(RNA2 - RNA1,
+     RNA3 - RNA2, RNA3 - RNA1, levels = design)
> fit2 <- contrasts.fit(fit, contrast.matrix)
> fit2 <- eBayes(fit2)
```

A list of top genes for RNA2 versus RNA1 can be obtained from

```
> topTable(fit2, coef = 1, adjust = "fdr")
```

Acceptance or rejection of each hypothesis test can be decided by

```
> results <- decideTests(fit2)
```

A Venn diagram showing numbers of genes significant in each comparison is obtained from

```
> vennDiagram(results)
```

The statistic `fit2$F` and the corresponding `fit2$F.p.value` combine the three pair-wise comparisons into one F-test. This is equivalent to a one-way ANOVA for each gene except that the residual mean squares have been moderated across genes. Small p-values identify genes which vary in any way between the three RNA targets. The following code displays information on the top 30 genes:

```
> o <- order(fit2$F.p.value)
> fit2$genes[o[1:30], ]
```

Now suppose that the experiment had been conducted using two-color arrays with a common reference instead of Affymetrix arrays. For example, the targets frame might be

FileName	Cy3	Cy5
File1	Ref	RNA1
File2	RNA1	Ref
File3	Ref	RNA2
File4	RNA2	Ref
File5	Ref	RNA3

For this experiment, the design matrix could be formed by

```
> design <- modelMatrix(targets, ref = "Ref")
```

and everything else would be as for the Affymetrix experiment.

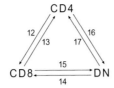

Figure 23.1. A direct design to compare three DC populations using six two-color microarrays. Each arrow represents an array, the head pointing toward the target labeled Cy5. Figure by Suzanne Thomas and James Wettenhall.

23.9   Direct two-color designs

A direct design is one in which there is no single RNA source that is hybridized to every array. As an example, we consider an experiment conducted by Dr. Mireille Lahoud at the WEHI to compare gene expression in three different populations of dendritic cells (DC). This experiment involved six cDNA microarrays in three dye-swap pairs, with each pair used to compare two DC types (Figure 23.1). The targets frame was:

SlideNumber	FileName	Cy3	Cy5
12	ml12med.spot	CD4	CD8
13	ml13med.spot	CD8	CD4
14	ml14med.spot	DN	CD8
15	ml15med.spot	CD8	DN
16	ml16med.spot	CD4	DN
17	ml17med.spot	DN	CD4

There are many valid choices for a design matrix for such an experiment. We chose the design matrix as

```
> design <- modelMatrix(targets, ref = "CD4")
> design
```

```
        CD8 DN
ml12med   1  0
ml13med  -1  0
ml14med   1 -1
ml15med  -1  1
ml16med   0  1
ml17med   0 -1
```

In this design matrix, the CD8 and DN populations have been compared back to the CD4 population. The coefficients estimated by the linear model will correspond to the log-ratios of CD8 vs. CD4 (first column) and DN vs. CD4 (second column). A linear model was fit using

```
> fit <- lmFit(MA, design)
```

All pairwise comparisons between the three DC populations were made by

```
> cont.matrix <- cbind("CD8-CD4" = c(1, 0), "DN-CD4" = c(0,
+     1), "CD8-DN" = c(1, -1))
> fit2 <- contrasts.fit(fit, cont.matrix)
> fit2 <- eBayes(fit2)
```

23.10   Factorial designs

Factorial designs are those where more than one experimental dimension is being varied and each combination of treatment conditions is observed. Suppose that cells are extracted from wild-type and mutant mice and these cells are either stimulated (S) or unstimulated (U). RNA from the treated cells is then extracted and hybridized to a microarray. We will assume for simplicity that the arrays are one-channel arrays such as Affymetrix. This section explains the form of the analysis for a hypothetical experiment. A detailed analysis of an actual factorial experiment, the Estrogen data, is given in Chapter 14. Consider the following targets frame:

FileName	Strain	Treatment
File1	WT	U
File2	WT	S
File3	Mu	U
File4	Mu	S
File5	Mu	S

The two experimental dimensions or *factors* here are Strain and Treatment. Strain specifies the genotype of the mouse from which the cells are extracted, and Treatment specifies whether the cells are stimulated or not. All four combinations of Strain and Treatment are observed, so this is a factorial design. It will be convenient for us to collect the Strain/Treatment combinations into one vector as follows:

```
> TS <- paste(targets$Strain, targets$Treatment,
+     sep = ".")
> TS
```

```
[1] "WT.U" "WT.S" "Mu.U" "Mu.S" "Mu.S"
```

It is especially important with a factorial design to decide what are the comparisons of interest. We will assume here that the experimenter is interested in 1) which genes respond to stimulation in wild-type cells, 2) which genes respond to stimulation in mutant cells, and 3) which genes respond differently in mutant compared to wild-type cells. These are the questions that are most usually relevant in a molecular biology context. The first of these questions relates to the WT.S vs. WT.U comparison and the second to Mu.S vs. Mu.U. The third relates to the difference of differences, i.e., (Mu.S-Mu.U)-(WT.S-WT.U), which is called the *interaction* term.

We describe first a simple way to analyze this experiment using limma commands in a similar way to that in which two-sample designs were analyzed. Then we will go on to describe a traditional statistical approach using factorial model formulas. The two approaches are equivalent and yield identical bottom-line results. The most basic approach is to fit a model with a coefficient for each of the four factor combinations and then to extract the comparisons of interest as contrasts:

```
> TS <- factor(TS, levels = c("WT.U", "WT.S", "Mu.U",
+     "Mu.S"))
> design <- model.matrix(~0 + TS)
> colnames(design) <- levels(TS)
> fit <- lmFit(eset, design)
```

This fits a model with four coefficients corresponding to WT.U, WT.S, Mu.U and Mu.S, respectively. Our three contrasts of interest can be extracted by

```
> cont.matrix <- makeContrasts(WT.SvsU = WT.S -
+     WT.U, Mu.SvsU = Mu.S - Mu.U, Diff = (Mu.S -
+     Mu.U) - (WT.S - WT.U), levels = design)
> fit2 <- contrasts.fit(fit, cont.matrix)
> fit2 <- eBayes(fit2)
```

We can use topTable to look at lists of differentially expressed genes for each of three contrasts, or else

```
> results <- decideTests(fit2)
> vennDiagram(results)
```

to look at all three contrasts simultaneously.

The analysis of factorial designs has a long history in statistics, and a system of factorial *model formulas* has been developed to facilitate the analysis of complex designs. It is important to understand though that the above three molecular biology questions do not correspond to any of the usual parameterizations used in statistics for factorial designs. Suppose for example that we proceed in the usual statistical way,

```
> Strain <- factor(targets$Strain, levels = c("WT",
+     "Mu"))
> Treatment <- factor(targets$Treatment, levels = c("U",
+     "S"))
> design <- model.matrix(~Strain * Treatment)
```

This creates a design matrix which defines four coefficients with the following interpretations:

Coefficient	Comparison
Intercept: baseline level of unstimulated wt	`WT.U`
StrainMu: unstimulated strain effect	`Mu.U-WT.U`
TreatmentS: stimulation effect for wt	`WT.S-WT.U`
StrainMu:TreatmentS: interaction	`(Mu.S-Mu.U)-(WT.S-WT.U)`

This is called the *treatment-contrast* parameterization. Note that one of our comparisons of interest, `Mu.S-Mu.U`, is not represented and instead the comparison `Mu.U-WT.U`, which might not be of direct interest, is included. We need to use contrasts to extract all the comparisons of interest:

```
> fit <- lmFit(eset, design)
> cont.matrix <- cbind(WT.SvsU = c(0, 0, 1, 0),
+     Mu.SvsU = c(0, 0, 1, 1), Diff = c(0, 0, 0,
+         1))
> fit2 <- contrasts.fit(fit, cont.matrix)
> fit2 <- eBayes(fit2)
```

This extracts the WT stimulation effect as the third coefficient and the interaction as the fourth coefficient. The mutant stimulation effect is extracted as the sum of the third and fourth coefficients of the original model. This analysis yields the same results as the previous analysis. It differs from the previous approach only in the parameterization chosen for the linear model, i.e., in the coefficients chosen to represent the four distinct RNA targets.

23.11   Time course experiments

Time course experiments are those in which RNA is extracted at several time points after the onset of some treatment or stimulation. Simple time course experiments are similar to experiments with several groups covered in Section 23.8. Here we consider a two-way experiment in which time course profiles are to be compared for two genotypes. Consider the targets frame

FileName	Target
File1	wt.0hr
File2	wt.0hr
File3	wt.6hr
File4	wt.24hr
File5	mu.0hr
File6	mu.0hr
File7	mu.6hr
File8	mu.24hr

The targets are RNA samples collected from wild-type and mutant animals at 0-, 6- and 24-hour time points. This can be viewed as a factorial experiment but a simpler approach is to use the group-mean parameterization.

```
> lev <- c("wt.0hr", "wt.6hr", "wt.24hr", "mu.0hr",
+     "mu.6hr", "mu.24hr")
> f <- factor(targets$Target, levels = lev)
> design <- model.matrix(~0 + f)
> colnames(design) <- lev
> fit <- lmFit(eset, design)
```

Which genes respond at either the 6-hour or 24-hour times in the wild-type? We can find these by extracting the contrasts between the wild-type times.

```
> cont.wt <- makeContrasts("wt.6hr-wt.0hr", "wt.24hr-wt.6hr",
+     levels = design)
> fit2 <- contrasts.fit(fit, cont.wt)
> fit2 <- eBayes(fit2)
```

Choose genes so that the expected false discovery rate is less than 5%.

```
> sel.wt <- p.adjust(fit2$F.p.value, method = "fdr") <
+     0.05
```

Any two contrasts between the three times would give the same result. The same gene list would be obtained had `"wt.24hr-wt.0hr"` been used in place of `"wt.24hr-wt.6hr"` for example.

Which genes respond in the mutant?

```
> cont.mu <- makeContrasts("mu.6hr-mu.0hr", "mu.24hr-mu.6hr",
+     levels = design)
> fit2 <- contrasts.fit(fit, cont.mu)
> fit2 <- eBayes(fit2)
> sel.mu <- p.adjust(fit2$F.p.value, method = "fdr") <
+     0.05
```

Which genes respond *differently* in the mutant relative to the wild-type?

```
> cont.dif <- makeContrasts(Dif6hr = (mu.6hr - mu.0hr) -
+     (wt.6hr - wt.0hr), Dif24hr = (mu.24hr - mu.6hr) -
+     (wt.24hr - wt.6hr), levels = design)
> fit2 <- contrasts.fit(fit, cont.dif)
> fit2 <- eBayes(fit2)
> sel.dif <- p.adjust(fit2$F.p.value, method = "fdr") <
+     0.05
```

23.12   Statistics for differential expression

limma provides functions `topTable` and `decideTests`, which summarize the results of the linear model, perform hypothesis tests and adjust the *p*-values for multiple testing. Results include (log) fold changes, standard errors, t-statistics, and p-values. The basic statistic used for significance analysis is the *moderated t-statistic*, which is computed for each probe and for each contrast. This has the same interpretation as an ordinary t-statistic except that the standard errors have been moderated across genes, i.e., shrunk toward a common value, using a simple Bayesian model. This has the effect of borrowing information from the ensemble of genes to aid with inference about each individual gene (Smyth, 2004). Moderated t-statistics lead to p-values in the same way that ordinary t-statistics do except that the degrees of freedom are increased, reflecting the greater reliability associated with the smoothed standard errors. Chapter 25 demonstrates the effectiveness of the moderated t approach on a test data set for which the differential expression status of each probe is known.

A number of summary statistics are presented by `topTable` for the top genes and the selected contrast. The M-value (`M`) is the value of the contrast. Usually this represents a $\log_2$-fold change between two or more experimental conditions although sometimes it represents a $\log_2$-expression level. The A-value (`A`) is the average $\log_2$-expression level for that gene across all the arrays and channels in the experiment. Column `t` is the moderated t-statistic. Column `p-value` is the associated p-value after adjustment for multiple testing. The most popular form of adjustment is `"fdr"`, which is Benjamini and Hochberg's method to control the false discovery rate (Benjamini and Hochberg, 1995). The meaning of `"fdr"` adjusted p-values is as follows. If all genes with p-value below a threshold, say 0.05, are selected as differentially expressed, then the expected proportion of false discoveries in the selected group is controlled to be less than the threshold value, in this case 5%.

The B-statistic (`lods` or `B`) is the log-odds that the gene is differentially expressed (Smyth, 2004, Section 5). Suppose for example that $B = 1.5$. The odds of differential expression is $\exp(1.5) = 4.48$, i.e, about four and a half to one. The probability that the gene is differentially expressed is $4.48/(1 + 4.48) = 0.82$, i.e., the probability is about 82% that this gene is differentially expressed. A B-statistic of zero corresponds to a 50-50 chance that the gene is differentially expressed. The B-statistic is automatically adjusted for multiple testing by assuming that 1% of the genes, or some other percentage specified by the user in the call to `eBayes`, are expected to be differentially expressed. The p-values and B-statistics will normally rank genes in the same order. In fact, if the data contains no missing values or quality weights, then the order will be precisely the same.

As with all model-based methods, the p-values depend on normality and other mathematical assumptions which are never exactly true for microarray data. It has been argued that the p-values are useful for ranking genes even in the presence of large deviations from the assumptions (Smyth et al., 2003, 2005). Benjamini and Hochberg's control of the false discovery rate assumes independence between genes, although Reiner et al. (2003) have argued that it works for many forms of dependence as well. The B-statistic probabilities depend on the same assumptions but require in addition a prior guess for the proportion of differentially expressed genes. The p-values may be preferred to the B-statistics because they do not require this prior knowledge.

The `eBayes` function computes one more useful statistic. The moderated F-statistic (F) combines the t-statistics for all the contrasts into an overall test of significance for that gene. The F-statistic tests whether any of the contrasts are non-zero for that gene, i.e., whether that gene is differentially expressed on any contrast. The denominator degrees of freedom is the same as that of the moderated-t. Its p-value is stored as `fit$F.p.value`. It is similar to the ordinary F-statistic from analysis of variance except that the denominator mean squares are moderated across genes.

23.13 Fitted model objects

The output from `lmFit` is an object of class *MArrayLM*. This section gives some mathematical details describing what is contained in such objects, following on from the Section 23.3. This section can be skipped by readers not interested in such details.

The linear model for gene j has residual variance σ_j^2 with sample value s_j^2 and degrees of freedom f_j. The output from `lmFit`, fit say, holds the s_j in component `fit$sigma` and the f_j in `fit$df.residual`. The covariance matrix of the estimated $\hat{\beta}_j$ is $\sigma_j^2 \mathbf{C}^T (\mathbf{X}^T \mathbf{V}_j \mathbf{X})^{-1} \mathbf{C}$ where $\mathbf{V}_j$ is a weight matrix determined by prior weights, any covariance terms introduced by correlation structure, and any iterative weights introduced by robust estimation. The square-roots of the diagonal elements of $\mathbf{C}^T (\mathbf{X}^T \mathbf{V}_j \mathbf{X})^{-1} \mathbf{C}$ are called unscaled standard deviations and are stored in `fit$stdev.unscaled`. The ordinary t-statistic for the kth contrast for gene j is $t_{jk} = \hat{\beta}_{jk} / (u_{jk} s_j)$ where u_{jk} is the unscaled standard deviation. The ordinary t-statistics can be recovered by

```
> tstat.ord <- fit$coef/fit$stdev.unscaled/fit$sigma
```

after fitting a linear model if desired.

The empirical Bayes method assumes an inverse Chi-square prior for the σ_j^2 with mean s_0^2 and degrees of freedom f_0. The posterior values for the residual variances are given by

$$\tilde{s}_j^2 = \frac{f_0 s_0^2 + f_j s_j^2}{f_0 + f_j}$$

where f_j is the residual degrees of freedom for the jth gene. The output from `eBayes` contains s_0^2 and f_0 as `fit$s2.prior` and `fit$df.prior` and the

$\tilde{s}_j^2$ as fit\$s2.post. The moderated t-statistic is

$$\tilde{t}_{jk} = \frac{\hat{\beta}_{jk}}{u_{jk}\tilde{s}_j}$$

This can be shown to follow a t-distribution on $f_0 + f_j$ degrees of freedom if $\beta_{jk} = 0$ (Smyth, 2004). The extra degrees of freedom f_0 represent the extra information that is borrowed from the ensemble of genes for inference about each individual gene. The output from eBayes contains the $\tilde{t}_{jk}$ as fit\$t with corresponding p-values in fit\$p-value.

23.14   Preprocessing considerations

This section discusses some aspects of preprocessing that are often neglected but that are important for linear modeling and assessing differential expression for two-color data. The construction of spot quality weights has already been briefly addressed in Section 23.4. Other important issues are the type of background correction used and the treatment of control spots on the arrays.

Background correction is more important than often appreciated because it impacts markedly on the variability of the log-ratios for low intensity spots. Chapter 4 shows an MA-plot for the $\beta 7$ data illustrating the fanning out of log-ratios at low intensities when ordinary background subtraction is used. Many more spots are not shown on the plot because the background corrected intensities are negative leading to NA log-ratios. Fanning out of the log-ratios is undesirable for two reasons. First, it is undesirable than any log-ratios should be very variable, because this might lead those genes being falsely judged to be differentially expressed. Second, the empirical Bayes analysis implemented in eBayes delivers most benefit when the variability of the log-ratios is as homogeneous as possible across genes. Chapter 4 shows that simply ignoring the background is a viable option. Another option is vsn normalization, a model-based method of stabilizing the variances that includes background correction (Huber et al., 2002, 2003). Here we illustrate a third option using the $\beta 7$ data, the model-based background correction method "normexp" implemented in the background-Correct function. This method uses the available background estimates but avoids negative corrected intensities and reduces variability in the log-ratios. Background correction is still an active research area and the optimal method has not yet been determined, but the adaptive methods "normexp" and "vsn" have been found to perform well in many cases.

For convenience, we read in the $\beta 7$ data again using the limma function read.maimages. A filter f is defined so that any spot that is flagged as "bad" or "absent" is given zero weight.

```
> beta7.dir <- system.file("beta7", package = "beta7")
> targets <- readTargets("TargetBeta7.txt", path = beta7.dir)
> f <- function(x) as.numeric(x$Flags > -75)
> RG <- read.maimages(targets$FileName, source = "genepix",
+     path = beta7.dir, wt.fun = f)
> RG$printer <- getLayout(RG$genes)
```

Here RG is an *RGList* data object. The data read by read.maimages differs slightly from read.GenePix because read.maimages reads mean foreground intensities for each spot, whereas read.GenePix reads median foreground intensities, although this difference should not be important here. The following code applies "normexp" background correction and then applies an offset of 25 to the intensities to further stabilize the log-ratios.

```
> RGne <- backgroundCorrect(RG, method = "normexp",
+     offset = 25)
```

Now normalize and prepare for a linear model fit as in Section 23.4.

```
> MA <- normalizeWithinArrays(RGne)
> design <- cbind(Dye = 1, Beta7 = c(1, -1, -1,
+     1, 1, -1))
```

It is usually wise to remove uninteresting control spots from the data before fitting the linear model. Control spots can be identified on arrays by setting the controlCode matrix in the marray package before using read.GenePix or by using controlStatus in the limma package. For the $\beta7$ data, control codes have already been set in the mraw object, so we can restrict the fit to interesting probes by

```
> isGene <- maControls(mraw) == "probes"
> fit <- lmFit(MA[isGene, ], design)
> fit <- eBayes(fit)
> tab <- topTable(fit, coef = "Beta7", adjust = "fdr")
> tab$Name <- substring(tab$Name, 1, 20)
> tab[, -(1:3)]
```

```
               ID               Name      M      A      t
5626   H200019655 KIAA0833 - KIAA0833  -1.481  8.63 -11.40
6029   H200017286 GPR2 - G protein-cou -1.977  8.18 -10.89
10115  H200018884 Homo sapiens cDNA FL -1.044  7.25 -10.79
10488  H200015303 CCR9 - Chemokine (C-  1.309 10.59  10.45
19217  H200001929 EPLIN - Epithelial p -0.864  8.86  -9.18
19346  H200008015 PTPRJ - Protein tyro -0.855 11.05  -8.84
20200  H200005842 GFI1 - Growth factor  0.762 11.66   8.57
10500  H200015731 SCYA5 - Small induci  1.540 11.12   8.33
3561   H200007572 Homo sapiens, clone   0.858  7.42   8.33
18292  H200000831 LRRN3 - Leucine-rich  0.803  9.55   8.01
       P.Value    B
5626    0.0109 5.30
6029    0.0109 5.03
```

```
10115   0.0109 4.98
10488   0.0109 4.79
19217   0.0261 3.99
19346   0.0300 3.74
20200   0.0327 3.55
10500   0.0327 3.36
3561    0.0327 3.36
18292   0.0390 3.10
```

Comparing this table to that in Section 23.4 shows more significant results overall, suggesting that the adaptive background correction has reduced variability and improved power. The vsn method could have been applied here by substituting

```
> MA <- normalizeBetweenArrays(RG, method = "vsn")
```

for the background correction and normalization steps above. This also gives good results for the $\beta 7$ data, with fewer significant results but with less attenuated fold change estimates compared to "normexp."

23.15   Conclusion

This chapter has demonstrated the ability of the linear modeling approach to handle a wide range of experimental designs. The method is applicable to both one- and two-channel microarray platforms. The method is flexible and extensible in principle to arbitrarily complex designs. Some ability has also been demonstrated to accommodate both technical and biological replication in the assessment of differential expression, although here only simple experimental structures can so far be accommodated. The survey of different designs given in this chapter complements the treatments of individual data sets given in other chapters of the book.

24

Classification with Gene Expression Data

M. Dettling

Abstract

A survey is given of tasks related to the construction and evaluation of classifiers applied to a renal cell cancer data set. Balanced sample splitting, non-specific filtering, linear discriminant analysis, nearest-neighbor prediction, and support vector machines are all concretely illustrated using the MLInterfaces package. Evaluations based on single and multiple random splits of data are compared. The entire presentation is given in a very generic programming format, to facilitate the adaptation and variation, by other investigators, of the techniques used here.

24.1 Introduction

The field of *class prediction* with microarray data has seen a lot of research activity in recent years. Owing to this effort, gene expression profiling is getting more and more established in clinical practice. The most prominent applications lie within cancer research: microarrays are often used to support exact phenotyping in early stages of the disease, which potentially allows for tailored treatment and better cure rates.

Class prediction is depicted as a simplified flowchart in Figure 24.1. It requires a training data set, consisting of both gene expression profiles x_i and phenotypic information y_i for a large enough number I of patients' samples. By statistical learning, we establish a class prediction rule $\mathcal{C}$ that reveals the connection between the outcome y_i and the gene profile x_i. The rule $\mathcal{C}$ can for example be based on a classifier like a *support vector machine* (SVM). The learning process then includes the choice of predictor variables, the tuning of parameters, and the fitting of the SVM. Once the rule C is determined, it can be used in a prospective manner to predict the unknown phenotypes of new, independent samples, e.g., in the clinic on a new cancer patient, whose yet unknown phenotype has to be predicted on the basis of his gene expressions x_ν.

This chapter contains an easy-to-follow generic recipe for class prediction with R. It starts by explaining how to retrieve data and terminates by showing how to summarize class prediction results. The main focus lies on learning classification rules, which, because it usually requires user interaction, is the most complex and laborious step in a class prediction analysis.

24.2   Reading and customizing the data

We here rely on the fully preprocessed renal cell cancer data set of Sült-mann et al. (2005). It is available as package kidpack and contains 74 gene expression profiles with 4224 genes each, obtained from patients that suffer from one of three renal cancer subtypes. We set the *random seed* (allowing for reproducibility of the analysis that follows) and load the data via

```
> set.seed(32)
> library("kidpack")
> data(eset)
```

We continue with an optional, non-mandatory step in the classification analysis. For illustrative purposes later in Section 24.5, we remove the first 10 samples from the data set.

```
> test <- (1:10)
> train <- (1:length(eset$type))[-test]
> trEset <- eset[, train]
```

We regard trEset, containing gene expression data and additional phenotypic information about 64 samples, as our training data set.

$$
\boxed{
\begin{array}{c}
\text{Training data } \mathcal{T} = \{(x_1, y_1), \ldots, (x_I, y_I)\} \\
\downarrow \\
\text{Learn prediction rule } \mathcal{C}(\cdot) \\
\downarrow \\
\text{Apply prediction rule to test data } x_\nu \\
\downarrow \\
\text{Yields prediction } \widehat{y}_\nu = \mathcal{C}(x_\nu)
\end{array}
}
$$

Figure 24.1. Class prediction as a simplified flowchart

24.3   Training and validating classifiers

In this chapter, we describe how to learn the prediction rule from the training data set. Above all, it is very important to note that the comparison
of classification methods cannot be done on the basis of in-sample errors.
This always favors the more complex methods and will lead to non-reliable
conclusions. Thus, we need to mimic a training/test-situation on the training data set. Here, we rely on *random divisions* (Dudoit et al., 2002), where
an arbitrary two thirds of the training data are used as learning set, while
the remaining third will serve as a validation set. Other popular choices for
defining learning and validation sets include k-fold cross validation (Ambroise and McLachlan, 2002) and out-of-bag estimation in conjunction with
bootstrapping (Efron and Tibshirani, 1997). For our random divisions, we
first define the sample size of the data chunks.

```
> t.size <- length(trEset$type)
> l.size <- round((2/3) * t.size)
> v.size <- t.size - l.size
```

Especially when only a small number of samples (or classes with very
few samples) are present, it is beneficial to rely on balanced sampling for
the random division into learning and validation set. This can be done as
follows:

```
> K <- nlevels(factor(trEset$type))
> l.samp <- NULL
> props <- round(l.size/t.size * table(trEset$type))
> props[1] <- l.size - sum(props[2:K])
> for (k in 1:K) {
+     y.num <- as.numeric(factor(trEset$type))
+     l.samp <- c(l.samp, sample(which(y.num ==
+         k))[1:props[k]])
+ }
> v.samp <- (1:t.size)[-l.samp]
```

The objects l.samp and v.samp now contain the sample indices assigned to
the learning and validation sets, respectively. A quick check shows that the
class distribution is indeed balanced

```
> table(trEset$type[l.samp])

ccRCC chRCC  pRCC
   31     5     7

> table(trEset$type[v.samp])

ccRCC chRCC  pRCC
   14     3     4
```

As the next step, we will perform gene selection. Due to the presence of
noisy genes, this generally has a positive effect on the predictive perfor-

mance. Another plus is that gene selection saves much computing time. We here rely on the simple F-statistic, also known as "between to within sums-of-squares ratio" (Dudoit et al., 2002). The package multtest contains an implementation of this procedure. Please note that this step can easily be replaced by a different *variable selection* procedure of choice, proposals in the literature are vast.

```
> library("multtest")
> yl.num <- as.numeric(factor(trEset$type[l.samp])) -
+     1
> xl.mtt <- exprs(trEset[, l.samp])
> f.stat <- mt.teststat(xl.mtt, yl.num, test = "f")
> best.genes <- rev(order(f.stat))[1:200]
> trselEset <- trEset[best.genes, ]
```

The function mt.teststat does not take instances of the exprSet-class as input. Because it only works with numerical input, we have to extract and transform our x- and y-variables. Moreover, it is very important to note that the variable selection is done on the learning set only. If it was done on the entire training data set, we would necessarily introduce a selection bias and inhibit a fair, reliable class prediction. We here preselect an arbitrary number of 200 genes, a number that was found to be reasonable according to several publications (Dudoit et al., 2002; Dettling and Bühlmann, 2003). Though strictly, the number of genes is a tuning parameter which would need to be optimized, i.e., it may be necessary to generate predictions with different gene sets.

We proceed to fitting several classifiers. We begin with *diagonal linear discriminant analysis* [DLDA, Dudoit et al. (2002)], a method derived from classical LDA. It relies on the assumption of a common, diagonal covariance matrix for the two classes. Both the assumptions of zero correlation between genes and equal correlation across classes may not reflect the truth, but due to the complexity of gene expression data and the usually very small sample size, DLDA has shown very promising empirical results (Dudoit et al., 2002; Dettling and Bühlmann, 2004; Dettling, 2004). An implementation of DLDA can be found in the package sma, but we will access this classifier through the interface that is provided by the package MLInterfaces.

```
> library("sma")
> library("MLInterfaces")
> l.samp <- as.integer(l.samp)
> dlda.predic <- stat.diag.daB(trselEset, "type",
+     l.samp)
> conf.matrix <- confuMat(dlda.predic)
> error.rate <- function(cm) 1 - sum(diag(cm))/sum(cm)
> dlda.error <- error.rate(conf.matrix)
```

We calculate the mean by comparing the predicted to the actual y-values and store them in dlda.error. For comparison, we will now evaluate the k

nearest neighbor classifier [*k*NN, Dudoit et al. (2002)], which is available in R after loading the package class. However, we will again rely on MLInterfaces for accessing this classifier. When presented with a sample, *k*NN works by identifying the *k* closest observations in the 200-dimensional input space and assigns the class label that is prevalent among the neighbors. The number of neighbors *k* is a tuning parameter. In the following, we will evaluate the performance of $k \in \{1, 3, 5\}$.

```
> library("class")
> knn.error <- numeric(3)
> for (k in c(1, 3, 5)) {
+       i <- ((k - 1)/2) + 1
+       knn.predic <- knnB(trselEset, "type", l.samp,
+           k = k, prob = FALSE)
+       knn.error[i] <- error.rate(confuMat(knn.predic))
+ }
```

Tuning the number of neighbors *k* requires a for-loop over the candidate set $\{1, 3, 5\}$. Again, we compare the predicted to the actual *y*-values and compute the error rate. Finally, we will consider a third, more modern and sophisticated classification procedure. This is a support vector machine [SVM, Burges (1998)]. It tries to separate the samples through a hyperplane, usually in a transformed, high-dimensional feature space. An implementation is available in the R package e1071, but again, we are using the interface which is provided by MLInterfaces. SVMs are flexible classifiers, allowing for a variety of tuning options. We rely here on a radial basis kernel and show how to optimize the cost parameter *c*, which determines the penalization of outlying learning samples that are on the 'wrong' side of the hyperplane, as well as the parameter γ that regulates the shape of the radial basis kernel. As shown below, coding an optimization over the two parameters *c* and γ already requires two nested loops.

```
> library("e1071")
> svm.error <- matrix(0, nrow = 3, ncol = 3)
> for (cost in 0:2) {
+       for (gamma in (-1):1) {
+           i <- cost + 1
+           j <- gamma + 2
+           svm.fit <- svmB(trselEset, "type", l.samp,
+               cost = 2^cost, gamma = 2^gamma/nrow(exprs(trselEset)),
+               type = "C-classification")
+           svm.error[i, j] <- error.rate(confuMat(svm.fit))
+       }
+ }
```

Again, the *y*-values on the validation sample are predicted, the error rate is computed and stored in the matrix svm.error.

Meanwhile, we have applied three different classification procedures and have generated 13 predictions of the y-values of the validation set. The time has come for an evaluation of error rates evaluation of the error rates.

```
> dlda.error
```

```
[1] 0
```

```
> knn.error
```

```
[1] 0.0476 0.0000 0.0476
```

```
> svm.error
```

```
        [,1]    [,2]    [,3]
[1,] 0.0476 0.0476 0.0952
[2,] 0.0476 0.0476 0.0952
[3,] 0.0476 0.0476 0.0952
```

We observe that DLDA yields the best result, with error-free prediction on the validation sample. The most accurate kNN classifier uses $k = 3$ neighbors and also results in perfect predictions. The best SVM, at $c = 2^2$ and $\gamma = \frac{2^1}{200}$ performs with an estimated error rate of 4.76% on the validation set. According to these results, we conjecture that either DLDA or kNN are most suitable for the training data set.

However, a somewhat bitter taste remains. Our conclusion depends on a single, random division of the training data set into learning and validation sets. The error rates and our decision might have been different for another random partition. A way to reduce the variability and haphazardness of our decision is to consider multiple random divisions. The next chapter discusses their implementation and evaluation.

24.4   Multiple random divisions

In fact, we do nothing else than repeating the code presented in Section 24.3. Due to random sampling, we will create different learning and validation sets in every run, and so can we expect the error rates to be. This section deals with how to optimally implement these repetitions in R. A special focus will be laid on storing, displaying, and analyzing the error rates. A convenient way to take is wrapping all the code from Section 24.3 in a function that is called randiv. Its input consists of the training data stored in trEset, whereas the output is a list containing the error rate objects dlda.error, knn.error and svm.error.

```
> randiv <- function(trEset) {
+      "#body suppressed; repeat all the code from Section 24.3"
+      list(dlda = dlda.error, knn = knn.error, svm = svm.error)
+ }
```

We will now define arrays for storing the errors rates and then create a loop for running the `randiv` function 50 times. Please note that running this chunk of code may take a few dozens of seconds.

```
> runs <- 50
> dlda.errors <- numeric(runs)
> knn.errors <- matrix(0, nrow = 3, ncol = runs)
> svm.errors <- array(0, dim = c(3, 3, runs))
> for (r in 1:runs) {
+       results <- randiv(trEset, r)
+       dlda.errors[r] <- results$dlda
+       knn.errors[, r] <- results$knn
+       svm.errors[, , r] <- results$svm
+       cat("This was run", r, "of", runs, "\n")
+ }
```

For a first and quick overview of the predictive performance, we average the error estimates over the 50 runs, which, as before in Section 24.3, leaves us with a single number, the estimated error rate. Note that computing means in arrays is best done via the `apply`-command.

```
> mean(dlda.errors)

[1] 0.0467

> apply(knn.errors, 1, mean)

[1] 0.0590 0.0571 0.0600

> apply(svm.errors, c(1, 2), mean)

        [,1]   [,2]   [,3]
[1,] 0.0657 0.0810 0.139
[2,] 0.0610 0.0705 0.129
[3,] 0.0562 0.0695 0.129
```

Indeed, the error-free classification of DLDA and kNN in Section 24.3 was a more or less lucky punch, based on a fortunate random partition into learning and validation set. When averaging over 50 such divisions, we observe that still DLDA yields the lowest mean error rate, but at 4.67%. Again, $k = 3$ is the best choice for the NN classifier, with an error rate of 5.71%. The best SVM is again the one with parameters $c = 2^2$ and $\gamma = 2^{-1}/200$, resulting in 5.62% false predictions. However, the analysis of mean error rates alone only superficially reflects the huge amount of computing we have done. For a better evaluation of variability and performance of the classifiers, a graphical presentation of the results is very beneficial. We do so by following the suggestions of Hothorn et al. (2004b) in their paper about how to perform benchmark studies. Figure 24.2 presents the results for DLDA, 3NN, and the two best SVMs. The left panel displays a *boxplot*, with the median error rate highlighted in red. The right panel shows density curves, where the vertical red lines show the mean error rate.

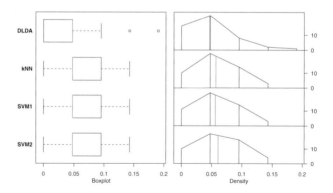

Figure 24.2. Analysis of the error rates on the renal cancer data set for Diagonal Linear Discriminant Analysis (DLDA), the three nearest neighbor method (kNN) and two Support Vector Machines (SVM1 with $c = 2^2$ and $\gamma = 2^1/200$; SVM2 with $c = 2^2$ and $\gamma = 2^0/200$). The left panel contains boxplots, where the median error rate is highlighted in red. The right panel shows barplots, where the mean error rate is displayed in red.

Our search for the most suitable classifier on the renal cancer training data set trEset can now be regarded as complete. By looking at the distribution of the error measures in Figure 24.2, we notice that the actual differences between the classifiers are not as big as the differences in the mean error rates might suggest. Even without performing a formal analysis, it becomes obvious that the differences among the classifiers do not reach statistical significance. Moreover, DLDA has not only the smallest mean error rate, but also the largest variability, bearing a small but potentially serious risk of performing clearly worse than kNN and SVM.

Finally, I would like to put emphasis on the fact that an analysis as presented in this chapter can never serve to choose the generally *correct* or *best* classifier for test data prediction. Rather, we collected empirical evidence that DLDA may be a suitable choice for the renal cancer data set. Additionally, our analysis has been designed for illustrative purposes and is not overly complete. Parameters which may clearly affect the performance, such as the variable selection procedure and the number of genes, have not been broadly evaluated. There may be other classification methods that outperform all the methods that have been considered here.

24.5   Classification of test data

At the beginning of our analysis in Section 24.2, we put aside the first ten samples from the renal cancer data set. Here and now, they are considered as additional test samples whose (unknown) phenotype needs to be predicted. According to the conclusion from Section 24.4, we are using DLDA with 200 genes, selected by the F-statistic. Because we now predict independent test samples, gene selection and classifier fitting are done on the entire training data set.

```
> yt.num <- as.numeric(factor(trEset$type)) - 1
> xt.mtt <- exprs(trEset)
> f.stat <- mt.teststat(xt.mtt, yt.num, test = "f")
> best.genes <- rev(order(f.stat))[1:200]
> selEset <- eset[best.genes, ]
> dlda.predic <- stat.diag.daB(selEset, "type",
+       train)
> dlda.predic@RObject$pred - 1

 [1] 0 0 0 0 0 2 0 2 1 0
```

In a clinical setting, these predictions may be used to decide about treatment. Regarding the importance of the decisions, it would be very useful to provide a confidence measure for the predictions, i.e., a conditional probability estimate. Unfortunately, neither of the standard implementations of DLDA, kNN and SVM do easily or reliably allow to do so.

If in the later course of the disease, the phenotype of the patients became apparent, or could easily be determined by additional testing, it would be possible to verify the microarray-based predictions. Because we are in a cartoon situation with given test set labels, we can check the predictive accuracy rightaway. An excellent means of visualization are confusion matrices.

```
> confuMat(dlda.predic)

        predicted
given   1 2 3
  ccRCC 7 0 0
  chRCC 0 1 0
  pRCC  0 0 2
```

We are happy to observe that DLDA made correct assignments to all our test patients.

24.6 Conclusion

In this chapter, we have shown an exemplary class prediction case study with microarray data. Besides the methods that have been used here, many other variable selection tools, classification procedures, and error rate estimation schemes exist, and may be used instead. The presented code should be generic enough so that it is obvious how to implement such alternative methods. The main reason for only presenting a limited amount of classifiers, variable selection tools, and tuning parameters were to keep this case study simple, to provide easily understandable code and to fulfill restrictions in code running time. For real applications however, an extensive search is desirable.

I will conclude this chapter by emphasizing again that the key to a reliable class prediction lies in avoiding selection bias, i.e., by properly splitting the training data into learning and validation sets, and choosing all the parameters *before* the optimal classifier is applied to test data.

25

From CEL Files to Annotated Lists of Interesting Genes

R. A. Irizarry

Abstract

One of the most popular applications of microarray technology is the identification of genes that are differentially expressed in two populations. With Affymetrix GeneChip technology, there are several steps between hybridization and the selection of interesting genes. The steps of preprocessing to improve signal to noise ratios, choosing a summary statistic for appropriate ranking of genes, and deciding on a final filter for candidate genes are largely statistical in nature. In this chapter, we demonstrate Bioconductor tools useful for creating such lists. We start from the raw probe level data (CEL files) and conclude with the creation of annotated reports.

25.1   Introduction

The identification of genes that are differentially expressed in two populations is a popular application of Affymetrix GeneChip technology. Due to the cost of this technology, experiments using a small number of arrays are common. A situation we often see is the case where three arrays are used for each population. In this chapter, we give an example of how to quickly create lists of genes that are interesting in the sense that appear to be differentially expressed, starting from the raw probe level data (CEL files). See Chapters 14 and 23 for more on the analysis of differential expression. In Section 25.2, we briefly describe the functions necessary to import the data into Bioconductor. In Section 25.3 we talk about preprocessing. In Section 25.4, we describe ways in which to rank genes and decide on a cutoff. Finally, in Section 25.5 we describe how to make annotated reports and examine the PubMed literature related to the genes in our list.

25.2   Reading CEL files

The affy package is needed to import the data into Bioconductor.

```
> library("affy")
```

As described in Chapter 2, we believe better results can be obtained by starting your analysis from the probe level data as opposed to the expression level data, provided by the Affymetrix software suite MAS 5.0. These data are contained in what we call the CEL files. These usually have extension .CEL. The function ReadAffy can be used to import these data into instances of the AffyBatch described in Chapter 2.

In this chapter, we do not include an example of how to read in the data. Instead we use an already created AffyBatch object, named spikein95, available through the package SpikeInSubset.

```
> library("SpikeInSubset")
> data(spikein95)
```

These data are a six array subset (2 sets of triplicates) from a calibration experiment performed by AffymetrixFor the purpose of this experiment, we replace the current phenoData, describing the calibration experiment, with information needed for our example dealing with two populations.

```
> pd <- data.frame(population = c(1, 1, 1, 2, 2,
+     2), replicate = c(1, 2, 3, 1, 2, 3))
> rownames(pd) <- sampleNames(spikein95)
> vl <- list(population = "1 is control, 2 is treatment",
+     replicate = "arbitrary numbering")
> phenoData(spikein95) <- new("phenoData", pData = pd,
+     varLabels = vl)
```

The assignment to phenoData above, can also be done using the function read.phenoData. For more details on this object, please refer to Chapter 7.

25.3   Preprocessing

The next step is to preprocess the data in spikein95 to obtain expression measures for each gene on each of the six arrays. In Chapter 2, we described various options for doing this. Here we simply use RMA (Irizarry et al., 2003b).

```
> eset <- rma(spikein95)
```

The rma function will background correct, normalize, and summarize the probe level data, as described by Irizarry et al. (2003b), into expression level data. The expression values are in log base 2 scale. The information is saved in an instance of the exprSet class, which contains similar information to the `AffyBatch` class. Specifically, the `exprSet` contains expression level as opposed to probe level data. Details about the `exprSet` and `AffyBatch` classes can be found in Chapter 1.

A matrix with the expression information is readily available. The following code extracts the expression data and demonstrates the dimensions of the matrix storing the data:

```
> e <- exprs(eset)
> dim(e)
```

```
[1] 12626     6
```

We can see that in this matrix, rows represent genes and columns represent arrays. To know which columns go with what population, we can rely on the `phenoData` information inherited from `spikein95`.

```
> pData(eset)
```

```
                 population replicate
1521a99hpp_av06       1         1
1532a99hpp_av04       1         2
2353a99hpp_av08       1         3
1521b99hpp_av06       2         1
1532b99hpp_av04       2         2
2353b99hpp_av08r      2         3
```

We can conveniently use `$` to access each column and create indexes denoting which columns represent each population:

```
> Index1 <- which(eset$population == 1)
> Index2 <- which(eset$population == 2)
```

We will use this information in the next section.

Notice we are only demonstrating the use of RMA. Other options are available through the functions `mas5` and `expresso`. If you find `expresso` too slow, the package affyPLM provides an alternative, `threestep`, that is faster owing to use of C code.

25.4   Ranking and filtering genes

Now we have, in e, a measurement x_{ijk} of log (base 2) expression from each gene j on each array i for both populations $k = 1, 2$. For ranking purposes, it is convenient to quantify the average level of differential expression for each gene. A naive first choice is to simply consider the average log fold-change:

$$d_j = \bar{x}_{j2} - \bar{x}_{j1}$$

with

$$\bar{x}_{j2} = (x_{1j2} + x_{2j2} + x_{3j2})/3 \text{ and } \bar{x}_{j1} = (x_{1j1} + x_{2j1} + x_{3j1})/3, \text{ for } j = 1, \ldots, J.$$

To obtain d, we can use the `rowMeans` function, which provides a much faster alternative to the commonly used function `apply`:

```
> d <- rowMeans(e[, Index2]) - rowMeans(e[, Index1])
```

Various authors have noticed that the variability of fold-change measurements usually depends on over-all expression of the gene in question. This means that the evidence that large observed values of d provide, should be judged by conditioning on some value representing over-all expression. An example is average log expression:

```
> a <- rowMeans(e)
```

The *MA-plot* 25.1A shows d plotted against a. We restrict the y-axis to log fold changes of less than 1 in absolute value (fold change smaller than 2). We do this because in this analysis only one gene reached a fold change larger than 2. To see this we can use the following line of code:

```
> sum(abs(d) > 1)
```

```
[1] 1
```

25.4.1   Summary statistics and tests for ranking

In Figure 25.1A we can see how variance decreases with the value of `a`. Also, different genes may have different levels of variation. Should we then consider a ranking procedure that takes variance into account? A popular choice is the t-statistic. The t-statistic is a ratio comparing the effect size estimate d to a sample based within group estimate of the standard error:

$$s_j^2 = \frac{1}{2}\sum_{i=1}^{3}(x_{ij2}-\bar{x}_{j2})^2/3 + \frac{1}{2}\sum_{i=1}^{3}(x_{ij1}-\bar{x}_{j1})^2/3, \text{ for } j=1,\ldots,J.$$

To calculate the the t-statistic, d_j/s_j, we can use the function `rowttests` from the package **genefilter**, as demonstrated in the following code:

```
> library("genefilter")
> tt <- rowttests(e, factor(eset$population))
```

The first argument is the data matrix, the second indicates the two groups being compared.

Do our rankings change much if we use the t-statistic instead of average log fold change? The volcano plot is a useful way to see both these quantities simultaneously. This figure plots p-values (more specifically $-\log_{10}$ the p-value) versus effect size. For simplicity we assume the t-statistic follows a t-distribution. To create a volcano plot, seen in Figure 25.1B, we can use the following simple code:

```
> lod <- -log10(tt$p.value)
> plot(d, lod, cex = 0.25, main = "B) Volcano plot for $t$-test")
> abline(h = 2)
```

This Figure demonstrates that the *t*-test and the average log fold change give us different answers.

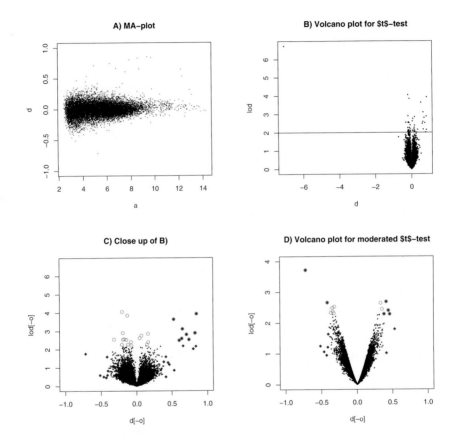

Figure 25.1. A) MA-plot. B) Volcano plot for the *t*-test. C) As B) but with restricted *x*-axis, blue diamonds denoting the top 25 genes ranked by average fold change and red circles to denoting the top 25 genes ranked by smallest *p*-value. D) As C) but for the moderated *t*-test.

Figure 25.1C is like Figure 25.1B but we restrict the *x*-axis to $[-1, 1]$ to get a better view of most of the data. We also add blue diamonds to denote the top 25 genes ranked by average fold change and red circles to denote the top 25 genes ranked by smallest *p*-value. The following lines of code permit us to create this figure:

```
> o1 <- order(abs(d), decreasing = TRUE)[1:25]
> o2 <- order(abs(tt$statistic), decreasing = TRUE)[1:25]
> o <- union(o1, o2)
> plot(d[-o], lod[-o], cex = 0.25, xlim = c(-1,
+      1), ylim = range(lod), main = "C) Close up of B)")
> points(d[o1], lod[o1], pch = 18, col = "blue")
> points(d[o2], lod[o2], pch = 1, col = "red")
```

There is a relatively large disparity. Possible explanations are 1) Some genes have larger variance than others. Large variance genes that are not differentially expressed have a higher chance of having large log fold changes. Because the t-statistics take variance into account, these do not have small p-values. 2) With only three measurements per group the estimate of the standard error of the effect size is not stable and some genes have small p-values only because, by chance, the denominator of the t-statistic was very small. Figure 25.1C demonstrates both of these explanations are possible.

Do we use average fold change or the t-statistic to rank? They both appear to have strengths and weaknesses. As mentioned, a problem with the t-statistic is that there is not enough data to estimate variances. Many researchers have proposed alternative statistics that borrow strength across all genes to obtain a more stable estimate of gene-specific variance. The resulting statistics are referred to as *modified* t-statistics. Because they reduce the possibility of large values they are also referred to as *penalized, attenuated,* or *regularized* t-statistics. The are many versions of modified t-statistics, for example the statistic used by SAM (Tusher et al., 2001), and it is impossible to describe them all in this chapter.

An example of such a modified t-statistic is given by Smyth (2004) and is available in the limma package. It is denoted the *moderated* t-statistic. This statistic is based on an empirical Bayes approach, described in detail by the above mentioned references, can be implemented with the following code:

```
> library("limma")
> design <- model.matrix(~factor(eset$population))
> fit <- lmFit(eset, design)
> ebayes <- eBayes(fit)
```

For more details on the above code please see Chapter 23. One detail we will point out is that, as part of its output, the lmFit function produces the values we have stored in d and tstat previously computed.

Figure 25.1D demonstrates the empirical Bayes approach improves on the t-statistic in terms of not giving high rank to genes only because they have small sample variances. To see this notice that we now have fewer genes with small p-values (large in the y-axis) that have very small log fold changes (values close to 0 in the x-axis).

25.4.2   Selecting cutoffs

We have presented three statistics that can be used to rank genes. Now we turn our attention to deciding on a cutoff. A naive approach is to consider genes attaining p-values less than 0.01. However, because we are testing multiple hypotheses, the p-values no longer have the typical meaning. Notice that if all the 12626 (number of genes on array) null hypotheses are true (no genes are differentially expressed) we expect $0.01 \times 12626 = 126.26$ false positives. In our data set we have

```
> sum(tt$p.value <= 0.01)
```

```
[1] 46
```

Various approaches have been suggested for dealing with the multiple hypothesis problem. . The suggested procedures usually provide adjusted p-values that can be used to decide on appropriate cutoffs. However, different procedures result in different interpretations of the resulting lists. For a detailed discussion, please see Chapter 15. In Section 25.5, we demonstrate simple code that can be used, along with the results of the eBayes function described above, to construct lists of genes based on adjusted p-values.

An alternative *ad hoc* procedure is to use the MA and volcano plots to look for groups of genes that appear to be departing from the majority. For example, in Figure 25.1C we can restrict our attention to the small cluster of genes in the upper right corner having small p-value and large fold changes.

25.4.3   Comparison

We deviate slightly from our analysis to demonstrate that the moderated t-statistic appears to perform better than average log fold change and the t-test.

Because we are using a spike-in experiment we can assess the three competing statistics. First we obtain the names of the genes that were spiked-in. This is available in the original phenoData.

```
> data(spikein95)
> spikedin <- colnames(pData(spikein95))
> spikedIndex <- match(spikedin, geneNames(eset))
```

Because in this experiment replicate RNA, except for the spiked-in material, was hybridized to all arrays, only the 16 spiked-in genes should be differentially expressed. This means that a perfect ranking procedure would assign ranks from 1 to 16 to the 16 spiked-in genes. The following code shows that the moderated t-statistic is performing better than its three competitors.

```
> d.rank <- sort(rank(-abs(d))[spikedIndex])
> t.rank <- sort(rank(-abs(tt$statistic))[spikedIndex])
```

```
> mt.rank <- sort(rank(-abs(mtstat))[spikedIndex])
> ranks <- cbind(mt.rank, d.rank, t.rank)
> rownames(ranks) <- NULL
> ranks
```

```
       mt.rank d.rank t.rank
 [1,]        1      1      1
 [2,]        2      2      3
 [3,]        3      3      5
 [4,]        6      4      8
 [5,]        7      5     13
 [6,]        8      6     17
 [7,]        9      9     19
 [8,]       10     11     27
 [9,]       11     12     28
[10,]       12     14     29
[11,]       16     16     45
[12,]       25     53     70
[13,]       28     86     71
[14,]       48    226     93
[15,]       77    331    390
[16,]      465    689    900
```

The results above permit us to see how many true positives we would have in a list of any size ranked by each of the three statistics. To do this, we simply count the number of spiked-in genes with ranks smaller or equal to the size of the list. As an example notice that in a list of the top 25 genes, we would obtain 12, 11, and 7 true positives for the moderated t-statistic. This demonstrates that the choice of ranking statistic can have an effect on your final results. Also notice that this is just one analysis, and one would expect different results with a new data set. We encourage users to run this analysis on other data sets and assess the performance of different alternative procedures.

25.5   Annotation

Let us now consider a list of interesting genes and create a report. As an example we will consider the top 10 genes as ranked by the moderated t-statistic. To do this we can use the `topTable` function from the limma package in the following way:

```
> tab <- topTable(ebayes, coef = 2, adjust = "fdr",
+      n = 10)
```

This function creates a table showing various interesting statistics. This line of code shows us the first 5 entries:

```
> tab[1:5, ]
```

	ID	M	A	t	P.Value	B
779	1708_at	-7.061	7.95	-73.53	9.87e-13	8.65
6252	36202_at	0.853	9.37	9.98	3.12e-03	4.59
6362	36311_at	0.832	8.56	8.36	1.27e-02	3.57
3284	33264_at	0.712	4.92	7.43	2.71e-02	2.84
2674	32660_at	0.655	8.68	7.36	2.71e-02	2.77

By specifying coef as 2 we define d as our parameter of interest. The argument n defines how many top genes to include in the table. Finally, through adjust we are instructing topTable to calculate the false discovery rate adjustment for the p-values obtained from eBayes using the procedure described by Benjamini and Hochberg (1995). See Chapter 23 for details.

In the remainder of this Chapter, we will use the Affymetrix identifiers to obtain biological meta-data information about these 10 genes. We can obtain these using the ID column of tab:

```
> genenames <- as.character(tab$ID)
```

25.5.1 PubMed abstracts

Chapters 8 and 9 describe powerful tools for combining important biological information about the genes we are studying with the results obtained with simple statistical analysis as the one presented here. In this section, we present a simple example.

We start by identifying the appropriate annotation packages we need:

```
> annotation(eset)
```

```
[1] "hgu95a"
```

Bioconductor contains the annotation for both HG-U95A and HG-U95Av2 chips in the annotation package hgu95av2.

```
> library("hgu95av2")
```

The annotate package contains many functions that can be used to query the PubMed database and format the results. Here we show how some of these can be used to obtain abstracts with references to a list of genes and report findings in the form of a Web page. Specifically, to get the abstract information we can use the pm.getabst in the annotate package (which requires the XML package):

```
> library("XML")
> library("annotate")
> absts <- pm.getabst(genenames, "hgu95av2")
```

The return value absts is a list of the same length as the genenames argument. Each of the list's components corresponds to a gene and is itself a list. Each component of this list is an abstract for that gene. To see the fourth abstract for the first gene, we type the following:

```
> absts[[1]][[4]]
```

```
An object of class 'pubMedAbst':
Title: Cyclin-dependent kinase 5 prevents neuronal
      apoptosis by negative regulation of c-Jun
      N-terminal kinase 3.
PMID: 11823425
Authors: BS Li, L Zhang, S Takahashi, W Ma, H Jaffe,
      AB Kulkarni, HC Pant
Journal: EMBO J
Date: Feb 2002
```

Please be careful using the function pm.getabst because it queries PubMed directly. Too many queries of PubMed can get you banned.

To only see the titles for, say, the second probeset, we can type the following.

```
> titl <- sapply(absts[[2]], articleTitle)
> strwrap(titl, simplify = FALSE)
```

```
[[1]]
[1] "Generation and initial analysis of more than 15,000"
[2] "full-length human and mouse cDNA sequences."
```

```
[[2]]
[1] "Primary-structure requirements for inhibition by the"
[2] "heat-stable inhibitor of the cAMP-dependent protein"
[3] "kinase."
```

```
[[3]]
[1] "Identification of an inhibitory region of the"
[2] "heat-stable protein inhibitor of the cAMP-dependent"
[3] "protein kinase."
```

```
[[4]]
[1] "Structure of a peptide inhibitor bound to the"
[2] "catalytic subunit of cyclic adenosine"
[3] "monophosphate-dependent protein kinase."
```

```
[[5]]
[1] "Inhibition of protein kinase-A by overexpression of"
[2] "the cloned human protein kinase inhibitor."
```

```
[[6]]
[1] "Isolation and characterization of cDNA clones for an"
[2] "inhibitor protein of cAMP-dependent protein kinase."
```

Here we used the strwrap function to format the text to fit the page width.

Having these abstracts as R objects is useful because, for example, we can search for words in the abstract automatically. The code below searches for the word "protein" in all the abstracts and returns a logical value:

```
> pro.res <- sapply(absts, function(x) pm.abstGrep("[Pp]rotein",
+       x))
> pro.res[[2]]
```

```
[1] FALSE TRUE  TRUE  TRUE  TRUE  TRUE
```

A convenient way to view the abstracts is to create a Web page:

```
> pmAbst2HTML(absts[[2]], filename = "pm.html")
```

The above code creates an HTML file with abstracts for the second probeset in genenames.

25.5.2 Generating reports

The object top gives us valuable information about the top 10 genes. However, it is convenient to create a more informative table. We can easily add annotation to this table and create an HTML document with hyperlinks to sites with more information about the genes. The following code obtains the locus link and symbol for each of our selected genes.

```
> ll <- getLL(genenames, "hgu95av2")
> sym <- getSYMBOL(genenames, "hgu95av2")
```

With these values available, we can use the following code to creates an HTML page useful for sharing results with collaborators.

```
> tab <- data.frame(sym, tab[, -1])
> htmlpage(ll, filename = "report.html", title = "HTML report",
+       othernames = tab, table.head = c("Locus ID",
+             colnames(tab)), table.center = TRUE)
```

The above HTML report only connects with locus link. To create a report with more annotation information, we can use the annaffy package. Below are a few lines of code that create a useful HTML report with links to various annotation sites.

```
> library("KEGG")
> library("GO")
> library("annaffy")
> atab <- aafTableAnn(genenames, "hgu95av2", aaf.handler())
> saveHTML(atab, file = "report2.html")
```

To see the type of Web page that is created by the functions used above, please see Chapters 9 and 18.

25.6   Conclusion

We have demonstrated how one can use Bioconductor tools to go from the raw data in CEL files to annotated reports of a select group of interesting genes. We presented only one way of doing this. There are many alternatives also available via Bioconductor. We hope that this chapter serves as an example of a useful analysis and also of the flexibility of Bioconductor.

Appendix A

Details on selected resources

A.1 Data sets

Here we describe three data sets that are used in many different examples throughout the book.

A.1.1 ALL

The ALL data were reported by Chiaretti et al. (2004) The data were normalized using quantile normalization, and expression estimates were computed using RMA (Irizarry et al., 2003b). We consider the comparison of the 37 samples from patients with the BCR/ABL fusion gene resulting from a chromosomal translocation (9;22) with the 42 samples from the NEG group.

They are available in the R package ALL.

A.1.2 Renal cell cancer

The kidpack package contains gene expression data from 74 renal cell carcinoma (RCC) patient biopsy samples (Sültmann et al., 2005). The samples were hybridized to two-color cDNA arrrays with PCR products of 4224 clones spotted in duplicate. These clones had been selected by their presumed relevance to cancer and on the basis of a previous genome-wide array study of kidney cancer. A pool of various RCC samples was used as a reference sample, which was always labeled with the red dye. The data were normalized with the vsn package, and we base our analysis on the generalized log-ratio values quantifying the difference between the tumor samples and the reference sample. The *exprSet* object esetSpot contains the normalized expression values and patient data. The RCC samples belong to three different histological types, clear cell (ccRCC), papillary (pRCC), and chromophobe (chRCC).

A.1.3 Estrogen receptor stimulation

The data are from eight samples from an estrogen receptor positive breast cancer cell line. After serum starvation, four samples were exposed to estrogen and then harvested for analysis with Affymetrix human genome U-95Av2 genechips after 10 hours for two samples and 48 hours for the other two. The remaining four samples were left untreated and harvested at corresponding times. The data are explained in more detail by Scholtens et al. (2004) and are available in the estrogen package.

A.2   URLs for projects mentioned

Below, we list URLs for different projects and computational resources that were discussed in the text.

AmiGO http://www.godatabase.org/

BioCarta www.biocarta.com

Bioconductor www.bioconductor.org

BioPAX www.biopax.org

BOOST www.boost.org

cMAP http://cmap.nci.nih.gov

CRAN http://cran.r-project.org

Entrez http://www.ncbi.nlm.nih.gov/entrez

GEO www.ncbi.nlm.nih.gov/geo

GO www.geneontology.org
The evidence codes: www.geneontology.org/GO.evidence.html

Graphviz www.graphviz.org

KEGG KEGG: www.kegg.org
The dbget archive: www.genome.jp/dbget/dbget.links.html
The SOAP API: http://www.genome.ad.jp/kegg/soap

libcurl http://curl.haxx.se

NAR Nucleic Acids Research on-line category listing http://www3.oup.co.uk/nar/database/c

Omegahat http://www.omegahat.org

OMIM http://www.ncbi.nlm.nih.gov/Omim

PubMed www.ncbi.nlm.nih.gov/entrez

R http://www.r-project.org

Reactome www.reactome.org

Resourcerer http://pga.tigr.org/tigr-scripts/magic/r1.pl

W3C http://w3c.org

References

Affymetrix. *Affymetrix Microarray Suite Users Guide.* Affymetrix, Santa Clara, CA, version 4.0 edition, 1999a. 19, 20

Affymetrix. *Affymetrix Microarray Suite Users Guide.* Affymetrix, Santa Clara, CA, version 5.0 edition, 2001b. 14, 19, 20, 26, 37

Affymetrix. Statistical algorithms reference guide. Technical report, Affymetrix, Santa Clara, CA, 2001. 19

Affymetrix. Statistical algorithms description document. Technical report, Affymetrix, Santa Clara, CA, 2002. 19

Affymetrix. *GeneChip® Expression Analysis Technical Manual.* Affymetrix, Santa Clara, CA, rev 4.0 edition, 2003c. 14

D. Amaratunga and J. Cabrera. Analysis of data from viral DNA microchips. *Journal of the American Statistical Association,* 96:1161–1170, 2001. 22

C. Ambroise and G. McLachlan. Selection bias in gene extraction on the basis of microarray gene-expression data. *Proc. Natl. Acad. Sci. of the U.S.A.,* 99:6562–6566, 2002. 423

M. Astrand. Contrast normalization of oligonucleotide arrays. *Journal of Computational Biology,* 10:95–102, 2003. 24

K. A. Baggerly, J. S. Morris, and K. R. Coombes. Reproducibility of SELDI-TOF protein patterns in serum: comparing datasets from different experiments. *Bioinformatics,* 20:777–85, 2004. 93, 100

R. Balasubramanian, T. LaFramboise, D. Scholtens, et al. A graph theoretic approach to testing associations between disparate sources of functional genomics data. *Bioinformatics,* 20:3353–3362, 2004. 346, 369, 370

P. Baldi and A. Long. A Bayesian framework for the analysis of microarray expression data: regularized t-test and statistical inferences of gene changes. *Bioinformatics,* 17:509–519, June 2001. 232

D. M. Bates and D. G. Watts. *Nonlinear regression analysis and its applications*. Wiley, New York, 1988. 277

G. D. Battista, P. Eades, R. Tamassia, et al. *Graph Drawing: Algorithms for the visualization of graphs*. Prentice Hall, Upper Saddle River, 1999. 360

Y. Benjamini and Y. Hochberg. Controlling the false discovery rate: a practical and powerful approach to multiple testing. *Journal of the Royal Statistical Society, Series B*, 57:289–300, 1995. 234, 238, 254, 266, 416, 439

Y. Benjamini and D. Yekutieli. The control of the false discovery rate in multiple hypothesis testing under dependency. *Annals of Statistics*, 29:1165–88, 2001. 266

C. Berge. *Graphs and Hypergraphs*. North-Holland, Amsterdam, 1973. 329, 337, 342

J. Bertin. *Semiologie Graphique*. Walter de Gruyter, Inc., Berlin, 2 edition, 1973. 162

M. Bittner, P. Meltzer, Y. Chen, et al. Molecular classification of cutaneous malignant melanoma by gene expression profiling. *Nature*, 406:536–540, 2000. 217

B. M. Bolstand, R. A. Irizarry, M. Astrand, et al. A comparison of normalization methods for high density oligonucleotide array data based on variance and bias. *Bioinformatics*, 19:185–193, 2003. 20, 21, 23, 64

I. Borg and P. Groenen. *Modern Multidimensional Scaling*. Springer-Verlag, New York, 1997. 173

S. P. Borgatti, M. G. Everett, and P. R. Shirey. LS sets, lambda sets and other cohesive subsets. *Social Networks*, 12:337–357, 1990. 345

M. Boutros, A. A. Kiger, S. Armknecht, et al. Genome-wide RNAi analysis of growth and viability in Drosophila cells. *Science*, 303: 832–835, 2004. 72, 74

J. M. Bower and H. Bolouri, editors. *Computational Modeling of Genetic and Biochemical Networks*. MIT Press, Cambridge, 2001. 125

A. Brazma, P. Hingamp, J. Quackenbush, et al. Minimum information about a microarray experiment (MIAME): toward standards for microarray data. *Nature Genetics*, 29:365–371, 2001. 8

L. Breiman. Bagging predictors. *Machine Learning*, 24:123–140, 1996a. 295

L. Breiman. Out-of-bag estimation. Technical report, Statistics Department, University of California Berkeley, 1996b. URL `ftp://ftp.stat.berkeley.edu/pub/users/breiman/`. 299

L. Breiman. Random forests. *Machine Learning*, 45:5–32, 2001. 279, 296

L. Breiman, J. H. Friedman, R. A. Olshen, et al. *Classification and Regression Trees*. Wadsworth, 1984. 279, 294

C. A. Brewer. Color use guidelines for mapping and visualization. In A. MacEachren and D. Taylor, editors, *Visualization in Modern Cartography*. Elsevier Science, Tarrytown, NY, 1994a. 162

C. A. Brewer. Guidelines for use of the perceptual dimensions of color for mapping and visualization. In J. Bares, editor, *Color Hard Copy and Graphic Arts III*, volume 2171, pages 54–63. Proceedings of the International Society for Optical Engineering (SPIE), Bellingham, 1994b. 162

P. Bühlmann and B. Yu. Analyzing bagging. *Annals of Statistics*, 30:927–961, 2002. 295

C. Burges. A tutorial on support vector machines for pattern recognition. *Knowledge Discovery and Data Mining*, 2:121–167, 1998. 425

A. Butte and I. Kohane. Mutual information relevance networks: Functional genomic clustering using pairwise entropy measurements. *Pacific Symposium on Biocomputing*, 5:415–426, 2000. 195

E. Camon, M. Magrane, D. Barrell, et al. The Gene Ontology annotation (GOA) database: sharing knowledge in Uniprot with Gene Ontology. *Nucleic Acids Res*, 32:D262–D266, 2004. 123

A. E. Carpenter and D. M. Sabatini. Systematic genome-wide screens of gene function. *Nature Reviews in Genetics*, 5:11–22, 2004. 71

D. B. Carr, R. Littlefield, W. Nicholson, et al. Scatterplot matrix techniques for large N. *Journal of the American Statistical Association*, 83:424–436, 1987. 163

J. M. Chambers. *Programming with Data: A Guide to the S Language*. Springer-Verlag, New York, 1998. 150

D. Chessel, A. B. Dufour, and J. Thioulouse. The ade4 package — I: One-table methods. *R News*, 4(1):5–10, 2004. URL `http://CRAN.R-project.org/doc/Rnews`. 360

S. Chiaretti, X. Li, R. Gentleman, et al. Gene expression profile of adult T-cell acute lymphocytic leukemia identifies distinct subsets of patients with different response to therapy and survival. *Blood*, 103: 2771–2778, 2004. 201, 232, 249, 252, 262, 300, 374, 443

R. Cho, M. Campbell, E. Winzeler, et al. A genome-wide transcriptional analysis of the mitotic cell cycle. *Molecular Cell*, 2:65–73, 1998. 370

D. Cilloni, A. Guerrasio, E. Giugliano, et al. From genes to therapy: the case of Philadelphia chromosome-positive leukemias. *Annal of the New York Academy of Sciences*, 963:306–312, 2002. 201

Ciphergen. *ProteinChip System Users Guide*. Ciphergen Biosystems, 2000. 92

W. S. Cleveland. *Visualizing Data*. Hobart Press, Summit, New Jersey, 1993. 162

W. S. Cleveland. *The Elements of Graphing Data (Revised)*. Hobart Press, Summit, New Jersey, 1994. 162

W. Conover. *Practical Nonparametric Statistics*. John Wiley& Sons, New York, 1971. 193

R. D. Cook and S. Weisberg. *Residuals and Influence in Regression*. Chapman & Hall, New York, 1982. 196

L. Cope, R. Irizarry, H. Jaffee, et al. A benchmark for Affymetrix GeneChip expression measures. *Bioinformatics*, 20:323–331, 2004. 31, 32

T. Cormen, C. Leiserson, and R. Rivest. *Introduction to Algorithms*. McGraw-Hill, New York, 1990. 348, 357

A. Cornish-Bowden. Nomenclature for incompletely specified bases in nucleic acid sequences: recommendations 1984. *Nucleic Acids Res*, 13:3021–3030, 1985. 145

T. Cox and M. Cox. *Multidimensional Scaling*. Chapman & Hall CRC, 2001. 173

F. Dannegger. Tree stability diagnostics and some remedies for instability. *Statistics in Medicine*, 19:475–491, 2000. 307

A. David and B. Mathieu. *Tulip: a system dedicated to the visualization of huge graphs*, 2004. http://www.tulip-software.org. 349

M. Dettling. Bagboosting for tumor classification with gene expression data. *Bioinformatics*, 24:3583–93, 2004. 424

M. Dettling and P. Bühlmann. Boosting for tumor classification with gene expression data. *Bioinformatics*, 19:1061–1069, 2003. 297, 298, 424

M. Dettling and P. Bühlmann. Finding predictive gene groups from microarray data. *Journal of Multivariate Analysis*, 90:106–131, 2004. 297, 424

E. W. Dijkstra. A note on two problems in connexion with graphs. *Numerische Mathematik*, 1:269–271, 1959. 357

B. Ding and R. Gentleman. Testing gene associations using co-citation. Technical report, The Bioconductor Project, Boston, 2004. 379, 380, 381, 383

P. Domingos. The role of Occam's razor in knowledge discovery. *Data Mining and Knowledge Discovery*, 3:409–425, 1999. 283

S. W. Doniger, N. Salomonis, K. D. Dahlquist, et al. Mappfinder: using Gene Ontology and GenMAPP to create a global gene-expression profile from microarray data. *Genome Biology*, 4:R7, 2003. 127, 331, 387, 390

R. O. Duda, P. E. Hart, and D. G. Stork. *Pattern Classification*. Wiley, New York, 2001. 191, 274, 277, 283

S. Dudoit, J. Fridlyand, and T. P. Speed. Comparison of discrimination methods for the classification of tumors using gene expression data. *Journal of the American Statistical Association*, 97:77–87, 2002. 423, 424, 425

S. Dudoit, J. P. Shaffer, and J. C. Boldrick. Multiple hypothesis testing in microarray experiments. *Statistical Science*, 18:71–103, 2003. 231

S. Dudoit and M. J. van der Laan. *Multiple Testing Procedures and Applications to Genomics*. Springer, New York, 2004. (in preparation). 250, 253, 254, 255, 258, 271

S. Dudoit, M. J. van der Laan, and M. D. Birkner. Multiple testing procedures for controlling tail probability error rates. Technical Report 166, Division of Biostatistics, University of California, Berkeley, 2004a. URL www.bepress.com/ucbbiostat/paper166. 250, 258, 271

S. Dudoit, M. J. van der Laan, and K. S. Pollard. Multiple testing. Part I. Single-step procedures for control of general Type I error rates. *Statistical Applications in Genetics and Molecular Biology*, 3 (1):Article 13, 2004b. URL www.bepress.com/sagmb/vol3/iss1/art13. 250, 253, 255, 256

S. Dudoit and Y. H. Yang. Bioconductor R packages for exploratory analysis and normalization of cDNA microarray data. In

G. Parmigiani, E. S. Garrett, R. A. Irizarry, and S. L. Zeger, editors, *The Analysis of Gene Expression Data: Methods and Software*. Springer-Verlag, New York, 2003. 51, 56

B. P. Durbin, J. S. Hardin, D. M. Hawkins, et al. A variance-stabilizing transformation for gene-expression microarray data. *Bioinformatics*, 18 Suppl. 1:S105–S110, 2002. 11, 194

D. Edwards. *Introduction to Graphical Modelling*. Springer-Verlag, New York, 2000. 343

B. Efron. The estimation of prediction error: Covariance penalties and cross-validation. *Journal of the American Statistical Association*, 99: 619–632, 2004. 282

B. Efron and R. Tibshirani. Improvements on cross-validation: The .632+ bootstrap method. *Journal of the American Statistical Association*, 92:348–360, 1997. 423

M. B. Eisen, P. T. Spellman, P. O. Brown, et al. Cluster analysis and display of genome-wide expression patterns. *Proc. Natl. Acad. Sci. of the U.S.A.*, 95:14863–14868, 1998. 166, 192, 193, 209, 212, 214

J. D. Emerson and D. C. Hoaglin. Analysis of two-way tables by medians. In D. C. Hoaglin, F. Mosteller, and J. W. Tukey, editors, *Understanding robust and exploratory data analysis*, pages 166–206. John Wiley & Sons, Inc., New York, 1983. 27, 42

P. Erdös and A. Rényi. On random graphs. *Publicationes Mathematicae*, 6:290–297, 1959. 343, 371

C. Fraley and A. Raftery. How many clusters? Which clustering method? Answers via model-based cluster analysis. *Computer Journal*, 41(8):578–588, 1998. 210

C. Fraley and A. Raftery. Model-based clustering, discriminant analysis, and density estimation. Technical Report 380, Univ. of Washington, Dept. of Statistics, October 2000. 210

Y. Freund and R. E. Schapire. Experiments with a new boosting algorithm. In L. Saitta, editor, *Machine Learning: Proceedings of the Thirteenth International Conference*, pages 148–156, San Francisco, 1996. Morgan Kaufmann. 296

J. Fridlyand and S. Dudoit. Applications of resampling methods to estimate the number of clusters and to improve the accuracy of a clustering method. Technical Report 600, Statistics Department, University of California, 2001. 217

J. E. F. Friedl. *Mastering Regular Expressions*. O'Reilly, Sebastopol, 2 edition, 2002. 123, 160

J. Friedman. Greedy function approximation: a gradient boosting machine. *Annals of Statistics*, 29:1189–1232, 2000. 289

J. Friedman, T. Hastie, and R. Tibshirani. Additive logistic regression: A statistical view of boosting. *Annals of Statistics*, 28:337–407, 2000. with Discussion. 296

E. T. Fung and C. Enderwick. ProteinChip clinical proteomics: Computational challenges and solutions. *Computational Proteomics Supplement*, 32:34–41, 2002. 93, 100

G. Gallo, G. Longo, S. Nguyen, et al. Directed hypergraphs and applications. *Discrete Applied Mathematics*, 42:177–201, 1993. 329, 342

E. R. Gansner and S. C. North. An open graph visualization system and its applications to software engineering. *Software Practice and Experience*, 30:1203–1233, 1999. 348, 349, 360

H. Ge, Z. Liu, G. Church, et al. Correlation between transcriptome and interactome mapping data from Saccharomyces cerevisiae. *Nature Genetics*, 29:482–486, 2001. 331, 369, 370, 371, 372

R. Gentleman. Using GO for statistical analyses. In J. Antoch, editor, *Compstat 2004 – Proceedings in Computational Statistics*, pages 171–180, Heidelberg, 2004. Physica Verlag, Heidelberg, Germany. 374, 376

R. Gentleman and V. J. Carey. Visualization and annotation of genomic experiments. In G. Parmigiani, E. S. Garrett, R. A. Irizarry, and S. L. Zeger, editors, *The Analysis of Gene Expression Data: Methods and Software*. Springer-Verlag, New York, 2003. 195

R. Gentleman and J. Gentry. Querying PubMed. *R News*, 2(2):28–31, 2002. URL http://CRAN.R-project.org/doc/Rnews. 138

R. Gentleman and A. C. Vandal. Computational algorithms for censored data problems using intersection graphs. *Journal of Computational and Graphical Statistics*, 10:403–421, 2001. 103

J. Gentry, V. Carey, E. Gansner, and R. Gentleman. Laying out pathways with Rgraphviz. *R News*, 4(2):14–18, 2004. URL http://CRAN.R-project.org/doc/Rnews. 391

D. Gerlich and J. Ellenberg. 4D imaging to assay complex dynamics in live specimens. *Nature Cell Biology*, Suppl:14–19, 2003. 73

D. E. Getz G., Levine E. Coupled two-way clustering analysis of gene microarray data. *Proc. Natl. Acad. Sci. of the U.S.A.*, 97:12079–12084, 2000. 215

G. Giaever, A. M. Chu, L. Ni, et al. Functional profiling of the Saccharomyces cerevisiae genome. *Nature*, 418:387–391, 2002. 72

T. R. Golub, D. K. Slonim, P. Tamayo, et al. Molecular classification of cancer: Class discovery and class prediction by gene expression monitoring. *Science*, 286:531–537, 1999. 190, 291

P. Good. *Permutation Tests: A Practical Guide to Resampling Methods for Testing Hypotheses*. Springer-Verlag, 1994. 380

A. D. Gordon. *Classification*. Chapman & Hall CRC, 2nd edition, 1999. 191

E. Graf, C. Schmoor, W. Sauerbrei, et al. Assessment and comparison of prognostic classification schemes for survival data. *Statistics in Medicine*, 18:2529–2545, 1999. 308

J. Gross and J. Yellen. *Graph Theory and its Applications*. CRC Press, 1999. 337

R. G. Halgren, M. R. Fielden, C. J. Fong, et al. Assessment of clone identity and sequence fidelity for 1189 IMAGE cDNA clones. *Nucleic Acids Res*, 29:582–588, 2001. 12

D. Hampel, C. Sansome, M. Sha, et al. Toward proteomics in uroscopy: urinary protein profiles after radiocontrast medium administration. *Journal of the American Society for Nephrology*, 12: 1026–1035, 2001. 92

T. Hastie and R. Tibshirani. *Generalized Additive Models*. Chapman & Hall, Boca Raton, 1990. 278

T. Hastie, R. Tibshirani, and J. H. Friedman. *The Elements of Statistical Learning : Data Mining, Inference, and Prediction*. Springer-Verlag, New York, 2001. 274, 278, 299

T. Hothorn, B. Lausen, A. Benner, et al. Bagging survival trees. *Statistics in Medicine*, 23:77–91, 2004a. 307

T. Hothorn, F. Leisch, A. Zeileis, et al. The design and analysis of benchmark experiments. Technical report, SFB Adaptive Informations Systems and Management in Economics and Management Science, 2004b. URL http://www.wu-wien.ac.at/am/reports.htm. 299, 427

P. J. Huber. *Robust statistics*. John Wiley & Sons, Inc, New York, New York, 1981. 42

W. Huber, A. von Heydebreck, H. Sültmann, et al. Variance stabilization applied to microarray data calibration and to the quantification of differential expression. *Bioinformatics*, 18 Suppl. 1:S96–S104, 2002. 6, 11, 24, 56, 64, 65, 194, 231, 418

W. Huber, A. von Heydebreck, H. Sültmann, et al. Parameter estimation for the calibration and variance stabilization of microarray data. *Statistical Applications in Genetics and Molecular Biology*, 2 (1), 2003. 24, 418

W. Huber, A. von Heydebreck, and M. Vingron. *Encyclopedia of Genetics, Genomics, Proteomics and Bioinformatics*, chapter Error models for microarray intensities. John Wiley & Sons, 2004. 11

L. H. Hubert. *Assignment Methods in Combinatorial Data Analysis*. Dekker, 1987. 380

T. Hughes, M. Marton, A. Jones, et al. Functional discovery via a compendium of expression profiles. *Cell*, 102:109–126, 2000. 215, 217

J. Huisken, J. Swoger, F. Del Bene, et al. Optical sectioning deep inside live embryos by selective plane illumination microscopy. *Science*, 305:1007–1009, 2004. 73

T. Ideker, V. Thorsson, A. Siegel, et al. Testing for differentially expressed genes by maximum-likelihood analysis of microarray data. *Journal of Computational Biology*, 7:805–818, 2000. 9

R. A. Irizarry, B. M. Bolstad, F. Collin, et al. Summaries of Affymetrix GeneChip probe level data. *Nucleic Acids Res*, 31:e15, Feb 2003a. 27

R. A. Irizarry, B. Hobbs, F. Collin, et al. Exploration, normalization, and summaries of high density oligonucleotide array probe level data. *Biostatistics*, 4:249–64, 2003b. 11, 18, 27, 232, 241, 432, 443

P. Jaccard. The distribution of flora in the alpine zone. *The New Phytologist*, 11:37–50, 1912. 380

T.-K. Jenssen, A. Laegreid, J. Komorowski, et al. A literature network of human genes for high-throughput analysis of gene expression. *Nature Genetics*, pages 21–28, 2001. 204

H. Joe. Relative entropy measures of multivariate association. *Journal of the American Statistical Association*, 84:157–164, 1989. 197

M. E. Johnson and C. J. Nachtsheim. Some guidelines for constructing exact D-optimal designs on convex design spaces. *Technometrics*, 25:271–277, 1983. 282

N. L. Johnston-Wilson, C. M. Bouton, J. Pevsner, et al. Emerging technologies for large-scale screening of humdan tissues and fluids in the study of severe psychiatric disease. *International Journal of Neuropsychopharmacology*, 4:83–92, 2001. 92

G. Joshi-Tope, M. Gillespie, I. Vastrik, et al. Reactome: a knowledge-base of biological pathways. *Nucleic Acids Res*, 33 Database Issue: 428–432, 2005. 333

R. S. Kamath, A. G. Fraser, Y. Dong, et al. Systematic functional analysis of the *Caenorhabditis elegans* genome using RNAi. *Nature*, 421:231–237, 2003. 72

M. Kanehisa. A database for post-genome analysis. *Trends in Genetics*, 13:375–376, 1997. 114

M. Kanehisa and S. Goto. KEGG: Kyoto encyclopedia of genes and genomes. *Nucleic Acids Res*, 28:27–30, 2000. 136, 142, 332, 333

M. Kanehisa, S. Goto, S. Kawashima, et al. The KEGG resources for deciphering the genome. *Nucleic Acids Res*, 32:D277–D280, 2004. 114

L. Kaufman and P. J. Rousseeuw. *Finding Groups in Data*. Wiley, 1990. 190, 191, 212, 217

S. Keleş, M. J. van der Laan, S. Dudoit, et al. Multiple testing methods for ChIP-Chip high density oligonucleotide array data. Technical Report 147, Division of Biostatistics, University of California, Berkeley, 2004. URL www.bepress.com/ucbbiostat/paper147. 250

C. Kendziorski, M. Newton, H. Lan, et al. On parametric empirical Bayes methods for comparing multiple groups using replicated gene expression profiles. *Statistics in Medicine*, 22:3899–3914, 2003. 232

M. Kerr and G. A. Churchill. Bootstrapping cluster analysis: Assessing the reliability of conclusions from microarray experiments. *Proc. Natl. Acad. Sci. of the U.S.A.*, 98:8961–8965, 2001. 215

M. K. Kerr, M. Martin, and G. A. Churchill. Analysis of variance for gene expression microarray data. *Journal of Computational Biology*, 7:819–837, 2000. 6

S. Klamt and E. D. Gilles. Minimal cut sets in biochemical reaction networks. *Bioinformatics*, 20:226–234, 2004. 331, 360

J. Knight. When the chips are down. *Nature*, 410:860–861, 2001. 12

L. Krishnamurthy, J. Nadeau, G. Ozsoyoglu, et al. Pathways database system: an integrated system for biological pathways. *Bioinformatics*, 19:930–937, 2003. 125, 331, 360

Y. K.Y., H. D.R., and R. W.L. Validating clustering for gene expression data. *Bioinformatics*, 17:309–318, 2001. 215

S. L. Lauritzen. *Graphical Models*. Clarendon Press, 1996. 343

G. Lennon, C. Auffray, M. Polymeropoulos, et al. The IMAGE consortium: an integrated molecular analysis of genomes and their expression. *Genomics*, 33:151–152, 1996. 222

C. Li and W. H. Wong. Model-based analysis of oligonucleotide arrays: Expression index computation and outlier detection. *Proc. Natl. Acad. Sci. of the U.S.A.*, 98:31–36, 2001a. 26, 196

C. Li and W. H. Wong. Model-based analysis of oligonucleotide arrays: model validation, design issues and standard error application. *Genome Biology*, 2(8):RESEARCH0032, 2001b. 21

A. Liaw and M. Wiener. Classification and regression by randomForest. *R News*, 2(3):18–22, 2002. URL http://CRAN.R-project.org/doc/Rnews. 301

U. Liebel, V. Starkuviene, H. Erfle, et al. A microscope-based screening platform for large-scale functional protein analysis in intact cells. *FEBS Letters*, 554:394–398, 2003. 73

D. M. Lin, Y. H. Yang, J. A. Scolnick, et al. Spatial patterns of gene expression in the olfactory bulb. *Proc. Natl. Acad. Sci. of the U.S.A.*, 101:12718–12723, 2004. 199

R. J. Lipshutz, S. Fodor, T. Gingeras, et al. High density synthetic ologonucleotide arrays. *Nature Genetics*, Suppl. 21:20–24, 1999. 14

C. Loader. *Local Regression and Likelihood*. Springer, New York, 1999. 89

D. Lockhart and E. Winzeler. Genomics, gene expression and DNA arrays. *Nature*, 405:827–836, 2000. 215

I. Lönnstedt and T. Speed. Replicated microarray data. *Statistica Sinica*, 12:31–46, 2002. 232

D. R. Mani and M. Gillette. Proteomic data analysis: Pattern recognition for medical diagnosis and biomarker discovery. In M. M. Kantardzic and J. Zurada, editors, *New Generation of Data Mining Applications*, Piscataway, 2004. IEEE Press. 101

K. Mardia, J. Kent, and J. Bibby. *Multivariate Analysis*. Academic Press, 1979. 190, 191

D. R. Masys, J. B. Welsh, J. L. Fink, et al. Use of keyword hierarchies to interpret gene expression patterns. *Bioinformatics*, 17:319–326, 2001. 204

G. Meister and T. Tuschl. Mechanisms of gene silencing by double-stranded RNA. *Nature Genetics*, 431:343–9, 2004. 72

K. Melhorn and S. Näher. *LEDA*. Cambridge University Press, 1999. 331

G. W. Milligan and M. C. Cooper. An examination of procedures for determining the number of clusters in a data set. *Psychometrika*, 50: 159–179, 1985. 217

G. A. Milliken and D. E. Johnson. *Analysis of Messy Data Volume 1: Designed Experiments*. Chapman & Hall, New York, 1992. 403

F. Naef, D. A. Lim, N. Patil, and M. A. Magnasco. From features to expression: High-density oligonucleotide array analysis revisited. http://xxx.lanl.gov/abs/physics/0102010, 2001. 19

J. Nevins, E. Huang, H. Dressman, et al. Towards integrated clinico-genomic models for personalized medicine: combining gene expression signatures and clinical factors in breast cancer outcomes prediction. *Human Molecular Genetics*, 12:R153–R157, 2003. 216

M. E. J. Newman, S. H. Strogatz, and D. J. Watts. Random graphs with arbitrary degree distributions and their applications. Technical report, Cornell University, 2001. 343

M. E. J. Newman, D. J. Watts, and S. H. Strogatz. Random graph models of social networks. *Proc. Natl. Acad. Sci. of the U.S.A.*, 99: 2566–2572, 2002. 343

E. Paradis, J. Claude, and K. Strimmer. APE: Analyses of phylo-genetics and evolution in R language. *Bioinformatics*, 20:289–290, 2004. 360

F. Pesarin. *Multivariate Permutation Tests with Applications in Biostatistics*. Wiley, 2001. 240

A. Peters, T. Hothorn, and B. Lausen. ipred: Improved predictors. *R News*, 2(2):33–36, 2002. URL http://CRAN.R-project.org/doc/Rnews. ISSN 1609-3631. 308

E. F. Petricoin, A. M. Ardekani, B. A. Hitt, et al. Use of proteomic patterns in serum to identify ovarian cancer. *Lancet*, 359:572–577, 2002. 92

F. Piano, A. J. Schetter, D. G. Morton, et al. Gene clustering based on RNAi phenotypes of ovary-enriched genes in *C. elegans*. *Current Biology*, 12:1959–1964, 2002. 71

R. Picard and K. Berk. Data splitting. *American Statistician*, 44: 140–147, 1990. 282

K. Pollard and M. van der Laan. A method to identify significant clusters in gene expression data. In *SCI2002 Proceedings*, volume II, pages 318–325, Orlando, 2002a. International Institute of Informatics and Systemics. 217, 219

K. Pollard and M. van der Laan. Statistical inference for simultaneous clustering of gene expression data. *Mathematical Biosciences*, 176(1): 99–121, 2002b. 216

K. S. Pollard, S. Dudoit, and M. J. van der Laan. Multiple testing procedures and applications to genomics. Technical Report 164, Division of Biostatistics, University of California, Berkeley, 2004. URL www.bepress.com/ucbbiostat/paper164. 250

K. S. Pollard and M. J. van der Laan. Choice of a null distribution in resampling-based multiple testing. *Journal of Statistical Planning and Inference*, 125:85–100, 2004. 250, 252, 254, 255, 256

J. Pontius, L. Wagner, and G. Schuler. *UniGene: a unified view of the transcriptome. In: The NCBI Handbook.* National Center for Biotechnology Information, Bethesda, 2003. 7

H. C. Purchase. Effective information visualization: a study of graph drawing aesthetics and algorithms. *Interacting with Computers*, 13: 147–162, 2000. 362

A. Reiner, D. Yekutieli, and Y. Benjamini. Identifying differentially expressed genes using false discovery rate controlling procedures. *Bioinformatics*, 19:368–375, 2003. 416

B. Ripley. *Pattern Recognition and Neural Networks.* Cambridge University Press, Cambridge, 1996a. 173, 190, 274, 277, 278, 279, 281

B. D. Ripley. *Pattern Recognition and Neural Networks.* Cambridge University Press, Cambridge, 1996b. 282

D. M. Rocke and B. Durbin. A model for measurement error for gene expression arrays. *Journal of Computational Biology*, 8:557–569, 2001. 9, 186

D. M. Rocke and B. Durbin. Approximate variance-stabilizing transformations for gene-expression microarray data. *Bioinformatics*, 19: 966–972, 2003. 24

M. Rodriguez, A. C. Paquet, Y. Yang, et al. Differential gene expression by memory/effector T helper cells bearing the gut-homing receptor integrin $\alpha 4\beta 7$. *BMC Immunology*, 4:5–13, 2004. 50

K. Rose, E. Gurewitz, and G. Fox. Statistical mechanics and phase transitions in clustering. *Physical Review Letters*, 65:945–948, 1990. 217

P. Rosenbaum. *Observational studies.* Springer, 1995. 282

M. E. Ross, X. Zhou, G. Song, et al. Gene expression profiling of pediatric acute myelogenous leukemia. *Blood*, 102:2951–2959, 2004. 34

C. Rosty, L. Christa, S. Kuzdzal, et al. Identification of hepatocarcinoma-intestine-pancreas / pancreatitis-associated protein I as a biomarker for pancreatic ductal adenocarcinoma by protein biochip technology. *Cancer Research*, 62:1868–1875, 2002. 92

T. Ryan. *Modern Regression Methods*. Wiley, New York, 1997. 194

E. E. Schadt, C. Li, B. Ellis, et al. Feature extraction and normalization algorithms for high-density oligonucleotide gene expression array data. *Journal Cellular Biochemistry*, Suppl. 37:120–125, 2001. 5, 21, 24

B. Schölkopf and A. Smola. *Learning with Kernels*. MIT Press, Cambridge, 2001. 274, 281

D. Scholtens and R. Gentleman. Making sense of high-throughput protein-protein interaction data. *Statistical Applications in Genetics and Molecular Biology*, 3:Article 39, 2004. 331

D. Scholtens, A. Miron, F. Merchant, et al. Analyzing factorial designed microarray experiments. *Journal of Multivariate Analysis*, 90: 19–43, 2004. 241, 444

A. Scott and M. Simmons. Clustering methods based on likelihood ratio criteria. *Biometrics*, 27:387–397, 1971. 217

L. C. Seamer, C. B. Bagwell, L. Barden, et al. Proposed new data file standard for flow cytometry, version FCS 3.0. *Cytometry*, 28: 118–122, 1997. 77

R. Sedgewick. *Algorithms, 3rd Edition*. Addison Wesley, Boston, 2002. 337, 348

I. A. Sidorov, D. A. Hosack, D. Gee, et al. Oligonucleotide microarray data distribution and normalization. *Information Sciences*, 146:67–73, 2002. 22

J. G. Siek, L.-Q. Lee, and A. Lumsdaine. *The Boost Graph Library*. Addison Wesley, Boston, 2002. 331, 348, 352

J. C. Simpson, R. Wellenreuther, A. Poustka, et al. Systematic subcellular localization of novel proteins identified by large-scale cDNA sequencing. *EMBO Reports*, 1:287–292, 2000. 72

S. Singh-Gasson, R. D. Green, Y. Yue, et al. Maskless fabrication of light-directed oligonucleotide microarrays using a digital micromirror array. *Nature Biotechnology*, 17:974–978, 1999. 14

M. Sirava, T. Schaäfer, M. Kaufmann, et al. BioMiner – modeling, analyzing, and visualizing biochemical pathways and networks. *Bioinformatics*, 18, Suppl 2:S219–S230, 2002. 125

G. Smyth. Linear models and empirical Bayes methods for assessing differential expression in microarray experiments. *Statistical Applications in Genetics and Molecular Biology*, 3:Article 3, 2004. 68, 232, 397, 399, 416, 418, 436

G. K. Smyth, J. Michaud, and H. Scott. The use of within-array replicate spots for assessing differential expression in microarray experiments. *Bioinformatics*, 21:to appear, 2005. 237, 397, 403, 406, 416

G. K. Smyth, Y. H. Yang, and T. Speed. Statistical issues in cDNA microarray data analysis. *Methods in Molecular Biology*, 224:111–36, 2003. 416

P. H. A. Sneath and R. R. Sokal. *Numerical Taxonomy: The Principles and Practice of Numerical Classification.* Freeman, 1973. 170

A. M. Snijders, N. Nowak, R. Segraves, et al. Assembly of microarrays for genome-wide measurement of DNA copy number. *Nature Genetics*, 29:263–264, 2001. 179

S. Solinas-Toldo, S. Lampel, S. Stilgenbauer, et al. Matrix-based comparative genomic hybridization: Biochips to screen for genomic imbalances. *Genes Chromosomes Cancer*, 20:399–407, 1997. 179

L. Stein. Creating a bioinformatics nation. *Nature*, 417:119–120, 2002. 135

C. J. Stone, M. H. Hansen, C. Kooperberg, et al. Polynomial splines and their tensor products in extended linear modeling (disc: P1425-1470). *The Annals of Statistics*, 25:1371–1425, 1997. 278

K. F. Storch, O. Lipan, I. Leykin, et al. Extensive and divergent circadian gene expression in liver and heart. *Nature*, 417:78–83, 2002. 331

H. Strasser and C. Weber. On the asymptotic theory of permutation statistics. *Mathematical Methods of Statistics*, 8:220–250, 1999. 301

R. Strausberg, E. Feingold, L. Grouse, et al. Generation and initial analysis of more than 15,000 full-length human and mouse cDNA sequences. *Proc Natl Acad Sci U S A.*, 99:16899–903, 2002. 379, 382, 385

S. H. Strogatz. Exploring complex networks. *Nature*, 410:268–276, 2001. 329

A. I. Su, T. Wiltshire, S. Batalov, et al. A gene atlas of the mouse and human protein-encoding transcriptomes. *Proc. Natl. Acad. Sci. of the U.S.A.*, 101:6062–6067, 2004. 174

H. Sültmann, A. von Heydebreck, W. Huber, et al. Gene expression in kidney cancer is associated with novel tumor subtypes, cytogenetic abnormalities and metastasis formation. *Clinical Cancer Research*, 11:646–655, 2005. 236, 304, 422, 443

Supplement to *Nature Genetics* (1999). *The Chipping Forecast*, volume 21, January 1999. 50

N. Suzuki, J. Nakayama, I. M. Shih, et al. Expression of trophinin, tastin, and bystin by trophoblast and endometrial cells in human placenta. *Biol Reprod.*, 60:621–7, 1999. 382

D. F. Swayne, D. T. Lang, A. Buja, et al. GGobi: Evolving from xgobi into an extensible framework for interactive data visualization. *Computational Statistics and Data Analysis*, 43:423–444, 2003. 84, 163

M. G. Tadesse, J. G. Ibrahim, and R. Gentleman. Bayesian error–in–variables survival model for the analysis of GeneChip arrays. *Biometrics*, in press, 2005. 194

R. E. Tarjan. Depth first search and linear graph algorithms. *SIAM Journal on Computing*, 1:146–160, 1975. 355

T. M. Therneau and E. J. Atkinson. An introduction to recursive partitioning using the rpart routine. Technical report, Section of Biostatistics, Mayo Clinic, Rochester, 1997. URL http://www.mayo.edu/hsr/techrpt/61.pdf. 294

R. Tibshirani, G. Walther, and T. Hastie. Estimating the number of clusters in a dataset via the gap statistic. Technical report, Department of Statistics, Stanford University, 2000. 217

D. M. Titterington. Bayesian methods for neural networks and related models. *Statistical Science*, 19:128–139, 2004. 278

A. H. Y. Tong, G. Lesage, G. D. Bader, et al. Global mapping of the yeast genetic interaction network. *Science*, 303:808–813, 2004. 71

P. Törönen, M. Kolehainen, G. Wong, et al. Analysis of gene expression data using self-organizing maps. *FEBS Letters*, 451:142–146, 1999. 212

E. Tufte. *Envisioning Information (2e)*. Graphics Press, Cheshire, 1990. 162

E. Tufte. *The Visual Display of Quantitative Information (2e)*. Graphics Press, Cheshire, 2001. 162

V. Tusher, R. Tibshirani, and G. Chu. Significance analysis of microarrays applied to the ionizing radiation response. *Proc. Natl. Acad. Sci. of the U.S.A.*, 98:5116–5121, 2001. 232, 436

M. van der Laan and J. Bryan. Gene expression analysis with the parametric bootstrap. *Biostatistics*, 2:445–461, 2001. 215, 216

M. van der Laan and K. Pollard. Hybrid clustering of gene expression data with visualization and the bootstrap. *Journal of Statistical Planning and Inference*, 117:275–303, 2003. 209, 212, 214, 222

M. J. van der Laan, S. Dudoit, and K. S. Pollard. Augmentation procedures for control of the generalized family-wise error rate and tail probabilities for the proportion of false positives. *Statistical Applications in Genetics and Molecular Biology*, 3(1):Article 15, 2004a. URL www.bepress.com/sagmb/vol3/iss1/art15. 250, 258, 259, 266

M. J. van der Laan, S. Dudoit, and K. S. Pollard. Multiple testing. Part II. Step-down procedures for control of the family-wise error rate. *Statistical Applications in Genetics and Molecular Biology*, 3(1): Article 14, 2004b. URL www.bepress.com/sagmb/vol3/iss1/art14. 250, 255, 257

V. Vapnik. *Statistical Learning Theory*. Wiley, 1998. 274

W. N. Venables and B. D. Ripley. *Modern Applied Statistics with S (4e)*. Springer, New York, 2002. 42

A. Vlahou, C. Laronga, L. Wilson, et al. A novel approach toward development of a rapid blood test for breast cancer. *Clinical Breast Cancer*, 4:203–209, 2003. 92

A. Vlahou, P. F. Schellhammer, S. Mendrinos, et al. Development of a novel proteomic approach for the detection of transitional cell carcinoma of the bladder in urine. *American Journal of Pathology*, 154:1491–1502, 2001. 92

A. von Heydebreck, W. Huber, and R. Gentleman. Differential expression with the Bioconductor project. In *Encyclopedia of Genetics, Genomics, Proteomics and Bioinformatics*. Wiley, New York, 2004. 375

Scalable Vector Graphics (SVG) 1.1 Specification W3C Recommendation. W3C, 14 January 2003. http://www.w3.org/TR/SVG11/. 158

G. Wahba. *Spline models for observational data*. SIAM Press, Philadelphia, 1990. 278

S. Wang, D. L. Diamond, G. M. Hass, et al. Identification of prostate specific membrane antigen (PSMA) as the target of monoclonal antibody 107-1a4 by ProteinChip array surface-enhanced laser desorption/ionization (SELDI) technology. *International Journal of Cancer*, 92:871–876, 2001. 92

C. Ware, H. Purchase, L. Colpoys, et al. Cognitive measurements of graph aesthetics. *Information Visualization*, 1:103–110, 2002. 362

J. A. Warrington, S. Dee, and M. Trulson. Large-scale genomic analysis using Affymetrix GeneChip. In M. Schena, editor, *Microarray Biochip Technology*, chapter 6, pages 119–148. BioTechniques Books, Natick, 2000. 14

S. Wasserman and K. Faust. *Social Network Analysis, Methods and Applications*. Cambridge University Press, Cambridge, 1994. 345, 379, 381

P. Westfall and S. Young. *Resampling-based multiple testing: examples and methods for p-value adjustment*. Wiley, New York, 1993. 233, 255

S. Wiemann, D. Arlt, W. Huber, et al. From ORFeome to Biology: A Functional Genomics Pipeline. *Genome Research*, 14:2136–2144, 2004. 72, 77

G. Wills. Nicheworks - interactive visualization of very large graphs. In *Graph Drawing '97 Conference Proceedings*. Springer-Verlag Lecture Notes in Computer Science, New York, 1997. 362

A. Winter. Exchanging Graphs with GXL. Technical Report 9/2001, Universität Koblenz-Landau, Institut für Informatik, 2001. URL http://www.uni-koblenz.de/fb4. 349

C. Workman, L. J. Jensen, H. Jarmer, et al. A new non-linear normalization method for reducing variability in DNA microarray experiments. *Genome Biology*, 3(9):research0048, 2002. 22

Z. Wu, R. Irizarry, R. Gentleman, F. Martinez Murillo, and F. Spencer. A model based background adjustement for oligonucleotide expression arrays. *Journal of the American Statistical Association*, in press, 2005. 27, 28

Y. Xiao, M. R. Segal, and Y. H. Yang. Stepwise normalization of two-channel spotted microarrays. *Statistical Applications in Genetics and Molecular Biology*, 4(1), 2005. 62

Y. H. Yang, M. J. Buckley, S. Dudoit, et al. Comparison of methods for image analysis on cDNA microarray data. *Journal of Computational and Graphical Statistics*, 11:108–136, 2002a. 5

Y. H. Yang, S. Dudoit, P. Luu, et al. Normalization for cDNA microarray data: a robust composite method addressing single and multiple slide systematic variation. *Nucleic Acids Res*, 30:e15, 2002b. 23, 24

Y. H. Yang and T. P. Speed. Design and analysis of comparative microarray experiments. In T. P. Speed, editor, *Statistical Analysis of Gene Expression Microarray Data*, pages 35–91. Chapman & Hall/CRC Press, Boca Raton, 2003. 399

Y. H. Yang and N. Thorne. Normalization for two-color cDNA microarray data. In *Science and Statistics: A Festschrift for Terry Speed*, volume 40, pages 403–418. IMS Lecture Notes, Monograph Series, 2003. 64

Y. Yasui, M. Pepe, M. L. Thompson, et al. A data-analytic strategy for protein biomarker discovery: profiling of high-dimensional proteomic data for cancer detection. *Bioinformatics*, 20:777–785, 2004. 93, 103

J. Zhang, V. Carey, and R. Gentleman. An extensible application for assembling annotation for genomic data. *Bioinformatics*, 19(1): 155–56, 2003. 115

L. Zhang, M. F. Miles, and K. D. Aldape. A model of molecular interactions on short oligonucleotide mic roarrays: implications for probe design and data analysis. *Nature Biotechnology*, 21(7):818–821, 2004. 28, 29

X. Zhou, M. C. Kao, and W. H. Wong. Transitive functional annotation by shortest-path analysis of gene expression data. *Proc. Natl. Acad. Sci. of the U.S.A.*, 99:12783–8, 2002. 331

Index

Statistical Methods in Bioinformatics:
An Introduction
Second Edition

W. Ewens and G. Grant

Advances in computers and biotechnology have had a profound impact on biomedical research, and as a result complex data sets can now be generated to address extremely complex biological questions. This book provides an introduction to some of these new methods. The main biological topics treated include sequence analysis, BLAST, microarray analysis, gene finding, and the analysis of evolutionary processes. The main statistical techniques covered include hypothesis testing and estimation, Poisson processes, Markov models and Hidden Markov models, and multiple testing methods. The second edition features new chapters on microarray analysis and on statistical inference, including a discussion of ANOVA, and discussions of the statistical theory of motifs and methods based on the hypergeometric distribution. Much material has been clarified and reorganized.

2004. 588 p. (Statistics for Biology and Health) Hardcover ISBN 0-387-40082-6

Computational Genome Analysis:
An Introduction

R. Deonier, S. Tavaré, and M. Waterman

Computational Genome Analysis: An Introduction presents the foundations of key problems in computational molecular biology and bioinformatics. It focuses on computational and statistical principles applied to genomes, and introduces the mathematics and statistics that are crucial for understanding these applications. The book is appropriate for a one-semester course for advanced undergraduate or beginning graduate students, and it can also introduce computational biology to computer scientists, mathematicians, or biologists who are extending their interests into this exciting field.

2005. 512 p. (Statistics for Biology and Health) Hardcover ISBN 0-387-98785-1